Principles and Procedures of Plant Protection

Third Edition

SB Chattopadhyay

Former Vice-Chancellor
Bidhan
Chandra Krishi Viswa
West Bengal

Oxford & IBH Publishing Co. Pvt. Ltd.

New Delhi

(A Unit of CBS Publishers & Distributors Pvt Ltd)

CBS

CBS Publishers & Distributors Pvt Ltd

New Delhi • Bengaluru • Chennai • Kochi • Kolkata • Mumbai
Bhubaneswar • Hyderabad • Jharkhand • Nagpur • Patna • Pune • Uttarakhand

Principles and Procedures of Plant Protection

Third Edition

ISBN-13: 978-81-204-0202-7
ISBN-10: 81-204-0202-2

© 1980, 1985, 1991 SB Chattopadhyay
Reprint: 1993, 1997, 2000, 2018

OXFORD & IBH

New Delhi
(A Unit of CBS Publishers & Distributors Pvt Ltd)

Published by Satish Kumar Jain and Produced by Varun Jain for
CBS Publishers & Distributors Pvt Ltd
4819/XI Prahlad Street, 24 Ansari Road, Daryaganj, New Delhi 110 002, India.
Ph: 23289259, 23266861, 23266867   Website: www.cbspd.com
Fax: 011-23243014   e-mail: delhi@cbspd.com;
                                                                          cbspubs@airtelmail.in.
Corporate Office: 204 FIE, Industrial Area, Patparganj, Delhi 110 092, India
Ph: 4934 4934   e-mail: publishing@cbspd.com;
Fax: 4934 4935   publicity@cbspd.com

Branches

- **Bengaluru:** Seema House 2975, 17th Cross, K.R. Road, Banasankari 2nd Stage, Bengaluru 560 070, Karnataka
  Ph: +91-80-26771678/79   Fax: +91-80-26771680   e-mail: bangalore@cbspd.com
- **Chennai:** 7, Subbaraya Street, Shenoy Nagar, Chennai 600 030, Tamil Nadu
  Ph: +91-44-26680620, 26681266   Fax: +91-44-42032115   e-mail: chennai@cbspd.com
- **Kochi:** Ashana House, 39/1904, AM Thomas Road, Valanjambalam, Ernakulam 682 016, Kochi, Kerala
  Ph: +91-484-4059061-65,67   Fax: +91-484-4059065   e-mail: kochi@cbspd.com
- **Kolkata:** 6/B, Ground Floor, Rameswar Shaw Road, Kolkata-700014 (West Bengal), India
  Ph: +91-33-2289-1126, 2289-1127, 2289-1128   e-mail: kolkata@cbspd.com
- **Mumbai:** 83-C, Dr E Moses Road, Worli, Mumbai-400018, Maharashtra
  Ph: +91-22-24902340/41   Fax: +91-22-24902342   e-mail: mumbai@cbspd.com

Representatives

• **Bhubaneswar**	0-9911037372	• **Hyderabad**	0-9885175004	• **Jharkhand**	0-9811541605
• **Nagpur**	0-9021734563	• **Patna**	0-9334159340	• **Pune**	0-9623451994
• **Uttarakhand**	0-9716462459				

Printed at Chaman Enterprises, Daryaganj, New Delhi, India

PREFACE TO THE THIRD EDITION

It is gratifying for the author to note that the previous edition and its subsequent reprints have received recognition from those for whom the book has been intended.

While plant protection is still regarded as one of the essential inputs in augmenting and stabilizing the yield, still doubts have arisen in the mind of scientists, technologists regarding methodology adopted by users. This situation has arisen mainly due to consciousness among the people regarding toxic hazards of pesticides, role of pesticides in environmental pollution on one hand and the danger of over dependence on toxic chemicals and indiscriminate use of the same. But use of plant protection chemicals cannot be discontinued, as this will have a serious effect on productivity and production, as these are indispensible in tackling any epidemics of plants due to various pests. Hence a rational approach has to be made and strategies are being evolved combining different methods and a new methodology based on the concept of need based use of pesticides is being worked out. This needs concept of assessment of pests, losses and various factors involved. Greater attention is now being placed on safe use of pesticides including storage. Farmers have now become more interested than before in use of herbicides.

Keeping above changes in perspective and changed requirements, new chapters have been added, existing chapters have been revised and brought upto date wherever found necessary.

It is hoped that the revised edition will be found useful to the readers and serve the purpose for which it is intended.

In the preparation of this revised edition, I gratefully acknowledge the help and advice received from Adhyapaka Mrinal Kanti Dasgupta and Professor Sisir Kumar Mukhopadhyay, Palli Siksha Bhavana, Visva Bharati, Professor Narayan Aditya Choudhury and his research Scholars Sarbasree Hemanta Banerjee and Atri Singh Choudhury,

Bidhan Chandra Krishi Viswa Vidyalaya, West Bengal and Dr. D. Konar, Deputy Director of Agriculture (Plant Protection), Govt. of West Bengal. 1 am thankful to Brachmachari Sreemanta, Ram Krishna Mission Asram, Narendrapur, West Bengal for pointing out a discrepancy which has been duly taken care of.

S.B. CHATTOPADHYAY

PREFACE TO THE SECOND EDITION

The first edition of the book was well received by the users and was shortly out of print. It was particularly well-received by students, both at the undergraduate and postgraduate level, and by field extension workers.

The importance of plant protection in stabilising yield by prevention of losses due to diseases and pests is gaining momentum in recent years. Newer chemicals are appearing in the market and technology is also undergoing change. It has been felt that use of pesticides cannot be eliminated in present-day agriculture production programme. In recent years there has been greater awareness among all concerned about the toxic hazards of pesticides and their possible role in environmental pollution. While the use of chemicals for control of pests is on the increase, there is a greater demand to curb on the use of pesticides. In view of these current trends, in the second edition, information has been provided on the chemical nature of the pesticides, their toxic hazards, problem of residues, and care in handling of pesticides. It is important that students, extension workers and subject-matter specialists should have a clear understanding of the problem to be able to find a solution.

It is hoped that the present edition will be helpful to the readers and serve the purpose for which it is intended.

S.B. CHATTOPADHYAY

PREFACE TO THE FIRST EDITION

As a teacher and an organiser of field work in plant protection, I have felt the necessity of a book which will give a fairly comprehensive account of the basis of the principles and procedures of the subject so that students, extension works and subject matter specialists develop a clearer understanding and proper perspective. In recent years, plant protection has become one of the essential inputs in crop production. In the context of multiple cropping, introduction of exotic varieties, application of higher doses of fertilisers, diseases and pests have assumed a special significance. The technology of plant protection is also undergoing change in concept and methodology. Informations are not readily available in a concise form to many who are interested in the subject. This situation has prompted me to write this book which is intended to meet the requirements of students of agriculture at the undergraduate and post-graduate level, extension workers and subject matter specialists. The aim is to present the basic principles, procedures and techniques in a simple and analytical manner.

An attempt has been made to present the requistite background information on diverse plant pests—insects, nematodes, mites, fungi, bacteria, viruses and damages they cause, conditions under which they become pests and magnitude of loss due to pests. Different methods of control, namely, cultural, chemical, biological, mechanical, use of resistant varieties, genetical, autocidal, legislative have been described. The scope of different methods have been brought out. In respect of chemical methods, the range of chemicals with their uses and limitations have been dealt with. For obvious reasons chemical methods have received much greater attention.

The method of application of the pesticides is an equally important factor for the efficacy of pesticides. Hence an account has been given of different methods and equipments along with advantages, disadvantages and possible limitations. Damages caused by vertebrate pests and losses sustained in storage are often enormous particularly in tropical countries, hence the principles and techniques of their control have received special emphasis.

Though pesticides are indispensable for food production, their toxicity may prove to be detrimental to the applicator, user, consumer and the biological environment in which they are applied. Hence toxicity hazards accompanied with the use of different categories of pesticides and safety methods that need to be imposed through advisory and statutory measures have been dealt with so that students and extension workers become conversant with the inherent dangers and the safeguards against the same.

Developing concepts and practices in the management of pests as well as the future trends have been briefly discussed so that the students may have an idea of possible developments in near future. Pests are not limited by geographical or political boundaries and need an international approach for successful tackling. Hence a concise account of the cooperation existing at international levels has been given.

Developmental activities to a large extent are channelised through Government agencies and regulatory activities of the Government constitute an important factor in the gamut of plant protection. So an account of plant protection organisation at the Government level has been given.

The most satisfactory way of encouraging students to take interest and develop an insight in the subject is to have discussions on relevant topics either in the classroom or through home assignments. Hence at the end of each chapter, topics for discussion have been suggested. It is hoped that they will prove useful.

Individual diseases and pests have not been dealt, as they are beyond the scope of this book. Such information can easily be found in many other textbooks.

There is a very wide spectrum of literature on the subject that has been published in India and abroad. It is impossible to detail all references. Only pertinent and important works are included in the references.

Probably the book may not be as complete as some may have wished it to be. Suggestions for the same are welcome for future editions. I shall consider my efforts successful and gratifying if this book serves the purpose for which it is intended.

I deeply appreciate help I received from my former students Dr. Anil Kumar Chaudhuri, lecturer in Plant Pathology, Bidhan Chandra Krishi Viswa Vidyalaya, Dr. Ashoke Kumar Bera, lecturer in Botany, Bidhan Chandra Krishi Viswa Vidyalaya and Dr. Sunil Kumar Santra, lecturer in Agricultural Extension, Bidhan Chandra Krishi Viswa Vidyalaya in preparation of manuscript. My special thanks and appreciation are due to my former student Shri Samir Kumar Samanta, M.Sc. (Ag.), Research Scholar Department of Genetics and Plant Breeding for the illustrations which he has prepared so carefully and intelligently for this textbook. I acknowledge with thanks help rendered by Dr. B. D. Sharma, Plant Pathologist. All India Coordinated Rice Improvement Project, a former student of mine, Dr. S. K. Ghosh, Reader in Agricultural Entomology, Bidhan Chandra Krishi Viswa Vidyalaya for critically going through some chapters, Shri M. C. Muthaiyan, Officer-in-charge, Plant Quarantine and Fumigation Station, Calcutta, Directorate of Plant Protection and Quarantine, Ministry of Agriculture for providing some pertinent informations. I also thank Messrs American Spring and Pressing Works (Private) Ltd., Bombay and Messers Shaw Wallace & Co., Ltd,, Calcutta for allowing to use some illustrations of plant protection machineries. I express my gratitude to National Book Trust of India, Delhi for subsidising publication of this book. I thank Messrs Oxford & IBH Publishing Company for the interest and trouble taken in connection with publication of this book. Lastly, but not in the least I expres my thanks to my wife Mrs. Pratima Chattopadhyay for encouragement given to me in writing this book.

CALCUTTA
15th June, 1980   S. B. CHATTOPADHYAY

CONTENTS

LIST OF ILLUSTRATIONS

CHAPTER 1

INTRODUCTION

Importance of Plants

In the neolithic period sometimes between 12,000 and 10,000 B.C. men took to growing of plants for their subsistence from hunting and food gathering habits. Probably one of the most significant land marks in the history of civilisation that prompted men to cultivate plants was the discovery that the seeds dropped into the earth would grow into plants and give food.

The genesis of agriculture also lay on observations that seeds put into holes or dribbles or furrows specially made for the purpose by stone implements showed better germination and gave greater yield. The above observations required men to remain in one place long enough to harvest their crops and thus ancient civilisation originated in the fertile riverine tracts. Since then men have grown plants for their basic necessities of life such as food, fibre, wood and wood derivatives, edible oils, "essential" oils and related substances, beverages, narcotics, dyes, gums, latex products, drugs, masticatories etc.

The search for food producing and other useful plants led to exploration and discovery of new lands and introduction of plants from one country to the other. Most superficial study or casual survey would show increasing dependence of men on plants with increasing complexity of civilisation, in spite of the fact that rapid advances have been made in other disciplines of science. Plants are economically important in other ways, such as, checking soil erosion, improvement of soil fertility, provision of food and shelter to many animals. Apart from these utilitarian aspects, plants are grown for beautification and aesthetic pleasures.

Whenever plants are grown whether for subsistence or meeting other necessities or even for aesthetic purposes, growers are keenly interested in having an assured yield or return. Accordingly, constant attention is paid to ensure proper growth and production as successful cultivation of plants necessitates apart from suitable varieties, supply of inputs and efficient management of the same, constant care and protection from the dangers that may deter the cultivator from deriving full benefits from the economically important plants. Herein comes the question of protection from pests.

Importance of Pests

Pests have been known to cause damage to plants from ancient times, though the exact causes might not then be known. References of locusts, rusts, mildews are found in the Bible and similar other ancient scriptures or literatures. They were considered to appear due to the wrath of the gods. Locusts were reported to cause ravages periodically which caused immense human misery and often migration of people from one country to another.

Late blight of potato caused by *Phytophthora infestans* in Ireland in the 1840's resulted in a potato famine and the consequent migration of half a million people from Ireland to the U.S.A. One of the major causes of the Bengal famine of 1943, which was responsible for the death of a very large number of people was due to the failure of the rice crop succumbing to the brown spot disease incited by *Helminthosporium oryzae*. Severity of rust (*Hemileia vastatrix*) disease, led to the abandonment of coffee cultivation in Sri Lanka towards the end of the nineteenth century and the adoption of tea culture. Red rot of sugar cane caused by *Glomerella tucumanensis* appeared in such a virulent form in North Bihar and Eastern Uttar Pradesh between 1939 and 1942 that the very existence of the sugar industry was threatened. Panama disease incited by *Fusarium oxysporum* f. *cubense* and the bunchy top (virus) disease often have led to the abandonment of banana cultivation and limiting areas of cultivation under this crop where they appear in a virulent form. *Tristeza* (quick decline, a virus disease) of citrus plants has brought decline in orange plantations in many areas to such a extent that areas which once flourished in oranges are now depleted of these valuable plants. Similarly in West Africa, swollen shoot of cacao (a virus disease) has threatened the economic cultivation of this valuable crop. Instances of failure of wheat crop due to attack of rust diseases is too well known.

As far insect pests are concerned, apart from extensive damages caused by locusts, there are other instances of serious depredations. Termites or white ants, as they are popularly called are serious enemies of several agricultural crops, namely, sugar cane, wheat, maize, sorghum, groundnut, beans chillies as well as many forest trees and timbers of farmhouses, poles, supporting roofs and other woodwork. In many parts of India, a large number of hairy caterpillars appear in certain seasons and destroy all crops that come in their way as they move from field to field. They cross roads, drywater channels on their way. They cause damage to cereals, oil seeds, pulses, and fodders to such an extent that the crop may have to be resown. In recent years, gall midge (*Orseolia oryzae*) and brown plant hopper (*Nilaparvata lugens*) are reported to cause such infestation so that to ensure proper yield definite measures are to be taken. Stem borer (*Scirpophaga incertulas*) causes much damage in major rice growing areas necessitating positive control methods to be adopted. Pink bollworm *(Pectinophora gossypiella)* is one of the most destructive insect pests of cotton not only in India, but in all cotton growing countries of the world. In sugar cane, top borer (*Scirpophaga nivella*) and Pyrilla often cause epidemics resulting in extensive reduction in yield. Rhinoceros beetle (*Oryctes rhinoceros*) in coconut palm ; cottony cushion scale (*Icerya purchasi*) in citrus ; hoppers (*Amritodus atkinsoni* and *Idioscopus clypealis*) in mango ; fruifly (*Dacus dorsalis*) in mango and many other fruit plants ; American bollworm (*Heliothis armigera*) on cotton, pulses ; etc. are some examples how destructive insect pests can be to crops.

Amongst nematodes, root knot nematodes (*Meliodogyne* spp.) and burrowing nematodes (*Radopholus similis*) are known to cause serious damage to many plants.

Damages caused by rodents are also too well known to merit special mention. Numerous such instances can be cited to show that pests are potential threats to the successful cultivation of crops, and they can pose serious problems, if adequate steps are not taken to protect the crops. Even without taking into account extreme cases, it is estimated that under "normal" or "average" conditions, substantial loss is incurred due to attack of pests.

Definition of pest

Insect and disease producing organisms are normally recognized as pests. Pest has been defined as any organism detrimental to man or his property in causing damage significant of economic importance. This organism may be insects, arachnids, nematodes, disease—producing pathogens including fungi, bacteria, viruses, mycoplasma, weeds, angiospermic parasites, rodents, birds and other animals. Pests however convey the idea or concept that they are controllable by suitable methods.

Pest control may be defined as any method or procedure employed to reduce the pest population and prevent damages caused by them. Pesticides are substances used to control pests and they include insecticides, fungicides, weedicides, nematicides, rodenticides, etc. The control of pests signifies any action taken by men to mitigate or prevent losses caused by them. Regulation of population in nature without any manipulation by man normally does not come under the purview of pest control measures. Control measures adopted for pest control may not always aim at the destruction of pests, but also amelioration of damages caused by them.

How do pest problems arise

It may be recognised that out of two to three million organisms in nature, only a limited number have become pests. The situation has arisen primarily due to the efforts of men in improving and intensifying activities in agriculture to raise more crops and thus their interference with the ecosystem in nature. Cultivation of plants for deriving economic benefit is the focal point around which the interests of men and pests have clashed. One of the major causes of pest problems is the unlimited supply of food for the pests due to the intensified growing of crops. With the progress in agriculture in the selection of suitable varieties of crops, plants with higher yield potentials are often associated with very succulent growth that the cultivated plants have now become more suitable or agreeable to pests for their food. This situation has been further aggravated by multiple cropping which has provided for an unlimited food chain. On the other hand clean cultivation aiming to increase production per unit area has eliminated or considerably reduced natural fauna or flora which act as checks and balances or barriers in the maintainence of biological equilibrium and has resulted in an unstable simple ecosystem which is often very vulnerable to fluctuations in the environment. It is well known that a complicated ecosystem existent under natural conditions is very stable.

Nature tends to scatter the plant species making it difficult for pests to spread and survive, whereas agriculture tends to concentrate millions of plants of a particular species in a limited defined area year after year. Homogenisation with one particular variety over a wide cultivated area makes the situation still more favourable for pests. Under such conditions, easy survival and multiplication of pests take place.

Many obscure pest organisms at one time restricted in their area to wild plant hosts have now become adapted to economically important plants wherever they are cultivated. Colorado beetle of potato (*Leptinotarsa decemlineata*) once a pest of wild solanaceous plants has now become pests in major potato growing areas in Europe and North America. Similarly organisms often brought inadvertently into a country along with host plants have become pests in the introduced country particularly in the absence of any suppressing organism. Examples may be given of European corn borer (*Ostrinia nubilalis*), Dutch elm disease (*Ceratocystis ulni*) introduced into the U.S.A. from Europe to mention a few cases. Wart (*Synchytrium endobioticum*) golden nemaode (*Heterodera rostochiensis*) of potato was introduced into India from Europe.

Sometimes due to the introduction of new exotic varieties minor pests often assume importance. For example gall midge (*Orseolia oryzae*), brown plant hopper (*Nilaparvata lugens*), and bacterial blight (*Xanthomonas oryzae*) of rice which have become serious pests after the introduction of high yielding dwarf indica varieties of rice.

Due to mutation in nature, or due to the development of resistance against pesticides, often new races or strains of pests appear which are more virulent and destructive and cause extensive damage. Example may be cited of black stem rust of wheat (*Puccinia graminis* f. *tritici*) where new virulent races have been reported to occur.

These are some of the relevant examples to illustrate how pest problems often arise.

Damages caused by pests

Damages caused by pests may be quantitative or qualitative or both. Quantitative loss is recorded when there is an overall reduction in yield or outturn. In case of qualitative damages, the gross yield may not be affected but the presence of markings, blemishes, wart or offensive odour may fetch much less price and the net income from unit area is reduced. In case of scab infection (*Streptomyces scabies* or *Spongospora subterranea*) the yield is not adversely affected, but the presence of such markings or blemishes makes them less rapidly acceptable to consumers as a result less price is fetched, besides keeping or storage quality may also be low. Similarly banana fruits with markings caused by beetle or citrus fruits with surface damage caused by miners (*Phyllocnistis citrella*) fetch low prices. Bunt (*Tilletia* spp.) infection or infestation causes both reduction in yield and depreciation in quality. Similarly ergot (*Claviceps purpurea*) infection not only brings down the yield, but also makes infested grains unacceptable to consumers. Red rot of sugar cane (*Glomerella tucumanensis*) or damage by borer (*Chilotraea infuscatellus*) causes not only reduction in yield, but affects the quality of juice of sugar cane. In case of damage by some pests in jute, not only the out-turn, but the quality of fibres also deteriorates.

Damage caused to the plants due to attack of pests may again be either direct or indirect depending on the parts affected and the resultant effect. In the case of infestation of stem borer of rice, the entire panicle becomes sterile and white and the damage is said to be direct, so also in the case of ear cutting caterpillars (*Cirphis unipuncta*) when the entire earhead is cut off and destroyed. Whereas in the case or rice hispa, (*Hispa armigera*) leaf tissues are destroyed and the damage is indirect reflected through loss in photosynthetic area which may adversely affect total carbohydrate synthesis and grain formation. Similarly in the case of vascular wilt caused by a pathogen the entire plant is dead and the damage may be termed direct whereas in leaf spot, the effect is indirect again through loss of photosynthetic area.

The pest is said to be direct when the host is damaged by its direct action in a direct or indirect manner. If infection does not cause any damage to the host but it is affected in some other way, pest may be termed as indirect. Late blight organism (*Phytophthora infestans*) is a direct pest of potato as it affects leaves, young twigs and tubers, whereas aphid *Myzus persicae* may be called an indirect pest of potato in as much as it does not itself cause any damage to the plant, but it transmits leaf roll and other viruses from a diseased to healthy plant and thus is responsible for loss in an indirect manner. Similarly barberry plants (*Berberis vulgaris*) does not cause any direct damage to wheat plants, but they are alternate hosts of black stem rust (*Puccinia graminis* f. *tritici*) and thus may affect wheat plants by harbouring pathogen in the off season. Similarly weeds are not normally parasitic on cultivated plants, but they compete for space, light, water, and rob the cultivated plants of nutrients and thus are responsible for causing loss in an indirect manner.

A few instances may be cited in which infection apart from causing direct damage to the plant or plant parts may be responsible for further damage to many other organisms in a different manner. Grains or groundnut kernels infected with *Aspergillus flavus* show signs of positive direct damage and thus responsible for spoilage. Men or livestock consuming such spoiled grains or kernels show signs of poisoning because of production of a toxin known as "aflatoxin" in the damaged kernels or grains.

Usefulness of plant protection

Some degree of protection of crops was always necessary in the past, but in recent years it is an admitted fact that with the intensification in agriculture, a grower has to keep constant vigil against attack of pests to protect the crop and have a good harvest. Even in the recent past, a farmer could raise good crops without taking recourse to plant protection measures, when pest problems were comparatively few. But intensive agriculture is necessary to feed the increasing population and to supply raw materials for the expanding agro-business. If measures for controlling pests, diseases, rodents, weeds, etc. are not taken for the crops that are grown, they will not reach the desired level of production, as a consequence famine would likely engulf a huge chunk of population. To avoid famine, huge quantities of food would have to be imported and as a result the drain on the economy would be tremendous.

The National Council of Applied Economic Reasearch (1967) carried out benefit cost analysis of pesticide treatment. According to the report benefit cost was, in case of insecticidal treatment 4.6 in rice, 3.2 in wheat, 2.9 in jowar 2.8 in cotton, 3.6 in sugar cane and 9.5 in potato. In the case of fungicidal treatment, it was 2.9 in paddy, 3.1 in wheat, 3.1 in cotton and 6.8 in potato. According to various research findings attempted to find out the benifit cost ratio of pesticide treatment, ratios in case of food croops were from 3 to 7.2, vegetables 2.8 to 4.7, cotton 2 to 13.5, sugar cane 3.4 to 14, oil seeds 1.4 to 44, jute 4, coconut 2. In general, savings due to plant protection measures have been estimated to vary from 5 to 15 per cent. One of the most commonly repeated 'guess' estimates of losses due to pests and weeds amount to twenty per cent, monetary value being over 1000 crores. Probably the greatest benefit of crop protection is the additional power it gives the growers to control and stabilise production without leaving the same to vagaries of nature.

Methods of pest control

There have been attempts to classify different control measures that are employed for protection of crops against pests. But a rigid classification is difficult in view of the interdependence of methods upon one another. Broadly there may be two approaches, (1) method directed against the parasite to reduce population or inoculum to a very low level by chemical, biological, autocidal and cultural methods so that damage at the economic level does not take place ; or (2) measures may be taken to the host 'or' in the environment in which the host is growing so that host plant is in a position to ward off or counteract attack of different parasites in an effective manner. Besides steps may also be taken to exclude the pest in such a manner that the pest does not have any opportunity to come in contact with the host, consequently no damage can take place. Measures may also be directed to cure the host plants of infection after it has become established. Methods may be presented as follows :

(A) PROTECTION OF THE HOST (PREVENTIVE CONTROL MEASURES)

(1) Exclusion or prevention of the parasite to come into contact with the host or reduce such contact to the minimum.

(a) quarantine and legislative measures ;

(b) inspection and certification ;

(c) physical barriers (fences, fly screens ; insect-proof-packing) ;

(d) use of repellants, attractants, antifeedants ; and

(e) circumvention of attack by adjustment of sowing, selection of sites, crop rotation and such cultural methods.

(2) Reduction of the effects of contact between pest and host.

(a) use of resistant varieties ; and

(b) use of protective chemicals.

(B) DESTRUCTION OF THE PEST

(1) Cultural control.

(a) Cultivation and other methods to destroy the organism and to reduce biotic or inoculum potential, and

(b) sanitary measures like destruction of infected materials, etc.

(2) Ecological control—modification of the existing environment to curb the pest or to encourage natural enemies of the pest.

(3) Biological control.

(a) introduction of suitable parasites, predators, and pathogens of the pest into the environment to reduce the pest population ; and

(b) autocidal method—use of pest species itself or some closely related species which will mate with pest species to destroy the pest population or the use of some characteristics of the pest species to destroy the population.

(4) Destruction.

(a) chemical pesticides ; and

(b) physical agencies, e.g., electromagnetic or mechanical energy.

Chemical methods of pest control by application of pesticides.

These methods either aim to keep the population of pests down to low level so that economic injury does not take place or eliminate pests alotogether by the toxic action of chemicals. They may also be preventive in action, being applied as a prophylactic measure. There are a few instances, where chemicals act as therapeutants to get rid of an established infection in plants. Since World War II, there has been tremendous development in the field of chemical pesticides and a vast array of pesticides are now available in the market. At present chemical pesticides are used for control of insects, arachnids, molluscs, nematodes, rodents, fungi, bacteria, etc. The use of pesticides constitutes probably the most important method in spite of their limitations and adverse effect on the environment.

Cultural methods of pest control

Cultural methods of pest control aim in reduction of population of pests by suitable adjutment of farming practices. They also include measures which are eradicative in nature, namely destruction of crop residues, stubbles, weeds and infected plants or plant parts. These methods also include circumvention of pest attack by adjustment of sowing or planting dates and cultivation in such areas which do not have suitable climatic conditions for multiplication and dissemination of pests. They also aim at having proper and healthy growth of plants so that they can afford greater resistance to attack of pests.

Use of resistant varieties

Use of hosts which are not amenable to attack of pests or can resist the same is a very practical method of control. This method has been found to be the only one which can be profitably employed in combating infections of certain categories of pathogens.

Biological methods

Under this category, comes the mutual antagonism of one organism against another present in nature and is responsible for the maintainence of equilibrium in an ecosystem which is taken advantage of in the supression of pests. In this process, specific parasites or predators either naturally occuring or exotic ones introduced into the area of infections are used. In control of pathogenic diseases, a somewhat non-specific microbial antagonism is induced in the soil for control of soil-borne pathogens inciting diseases in plants.

Use of hormones, which are of natural occurrence or their synthetic analogues also come under purview of biological methods.

Autocidal methods

The control of insects by genetic manipulation which has been experimentally introduced in a few specific cases offer a promising new approach.

Physical and mechanical methods

These include use of traps, barriers, physical agents like heat, light, electricity etc.

Legislative measures

Quarantine regulations aim to exclude such pests which do not occur in a country from entering into the country, or preventing their spread within the country in case they have been introduced by restricting area of incidence by suitable legislative enactments. These may include certification where sale of infected material for planting purposes are prohibited or standards have been laid down regarding seeds and planting materials and same are enforced under provisions of law.

Integrated pest management

Intergrated pest management signifies the combination of all pertinent methods chemical, cultural, biological, use of resistant varieties, etc.

QUESTIONS FOR DISCUSSION

1. The statement that famine would engulf large chunk of population and economy will be adversely affected if pest controls are not adopted, may appear unrealistic. Is there any evidence to substantiate the statement ?

2. Why have more pest problems arisen in recent years as compared to the situation in the past ?

3. Why is protection of plants necessary ? How much loss is likely to be incurred due to attack of pests ?

4. Can you cite instance or instances where a minor pest has become a major one ?

5. Why is pest trouble expected when new planting materials or varieties are introduced ?

6. Why are certain measures called protective ? To whom do they give protection ?

7. What do you understand by biological control ? How can you aim at biological control ?

8. What is a direct damage ? How does it differ from indirect damage ?

9. What are quantitative and qualitative damages ?

10. Can you justify the statement that "with increasing complexity of civilisation, there has been greater dependence on plants"?

11. Is it correct to state that populations of organisms in nature are in a state of balance with one other ? How does agriculture affect such a balance ?

12. For increasing production in agriculture, attention must be paid to crop protection. Justify the statement.

13. Are crop protection measures always economically justified ?

14. Under what conditions would it not be worthwhile to control crop pests ?

CHAPTER 2

PESTS

For the intelligent and successful tackling of any plant pest, it is necessary to have some knowledge of the organisms which normally cause damage, the nature of damage, etc. ; and the factors contributing to the development of pest in a serious form. In this chapter, a brief account is given of different kinds of pests that cause damage to plants.

ANIMALS

In the animal kingdom, animals ranging from an elephant or a monkey to nematodes may be pests. In Table 2.1 an account is given of the various animals in different phyla in the animal kingdom which are injurious to plants.

TABLE 2.1

Classification of the Agriculturally Important Group of the Animal Kingdom (Injurious to Plants)

Phylum	Class	Characteristics	Examples
		Vertebrates	
Chordata	Mammalia	Hairy, fourfooted, milk secreting	monkey, cow, elephant mice, rabbit, etc.
	Aves	Wings and feathers	birds
		Invertebrates	
Arthropoda	Hexapoda	Body differentiated into head, thorax and abdomen, three pairs of legs, wings	insects, bugs, beetles

TABLE 2.1 (*Conted.*)

Phylum	Class	Characteristics	Examples
	Diplopoda	Head and abdomen many segmented. Two pairs of legs per segment	millipedes
	Arachnida	Head and thorax fused (cephalothorax), four pairs of walking legs, abdomen distinct, unsegmented, no antennae	mites
	Crustacea	Hard, limey chitinous exoskeleton, two pairs antennae, cephalothorax with biramous appendages	crabs
Mollusca	Gastropoda	Soft-bodied non-segmented, no jointed appendages, body enclosed in calcareous shell	snail, slugs
Nemata (Nemahelminthes)	Nematoda	Tiny, cylindrical, elongated unsegmented with tough cuticle, some forms visible under microscope	nematodes

Insects are tracheate arthropods, with distinct head, thorax, abodomen, a single pair of antennae, three pairs of working legs confined to the thorax and one or two pairs of wings. The integument is hardened to an exoskeleton covering the body muscles, with striated fibres and arranged segmentally.

Some insects are smaller than larger protozoa, whereas there are others which are larger than small mammals. Among the smallest insects are beetles Trichopterygidae (0.5 mm long) and Mymar (0.2 mm long) parasitic on eggs of insects. Some beetles like Dynastidae, Cerambycidae exceed the size of small mammals. Shapes of insects vary very much and they are often clothed with hairs, scales, spines, horns or other appendages.

The head is usually composed of six segments, fused immovably together ; thorax of three segments ; and abdomen is usually eleven segments. The alimentary canal is differntiated into foregut, midgut, hindgut. The heart is a dorsal tube in pericardial space with paired segmental ostia, but without arteries and veins. The circulatory system is open. Respiration is by the tracheal system, excretion is through malphigian tubules. The nervous sysem is an elaboration of the type found in annelids with a ventral chord three-paired ganglia in thorax and one pair of ganglia in each segment of abdomen except the last. The sensory organs include simple and compound eyes, olfactory, gustatory, tactile and auditory organs. Light and sound producing organs are not uncommon.

The sexes are separate. External genital aperture of the female is in the eighth and of the male in the ninth abdominal segment. Fertilisation is always internal. Reproduction is sexual, Parthenogenesis, vivipary and paedogenesis are also known. The development of insect is often attended with pronounced metamorphosis.

Metamorphosis refers to any change in form, structure and appearance of an animal between birth and maturity. Insects may be divided into three groups with reference to metamorphosis.

(a) Insects without metamorphosis

These insects on emerging from the egg are essentially like the adults except for size. They undergo a series of moults at intervals, each time increasing in size till attainment of sexually mature stage. These insects are wingless and are considered to be primitive types from the evolutionary standpoint. They have chewing type of mouth parts. In this group, is included Colemboila, which includes springtails, one species of which has been found to be a pest of garden plants in temperate countries.

(b) Insects with gradual or incomplete metamorphosis

All winged insects are to undergo metamorphosis, as no insect has visible wings when it emerges from the egg. Insects with gradual or incomplete metamorphosis on hatching out from the eggs show a resemblance to adults but do not possess any wings. Except for aquatic forms, these immature insects of same types

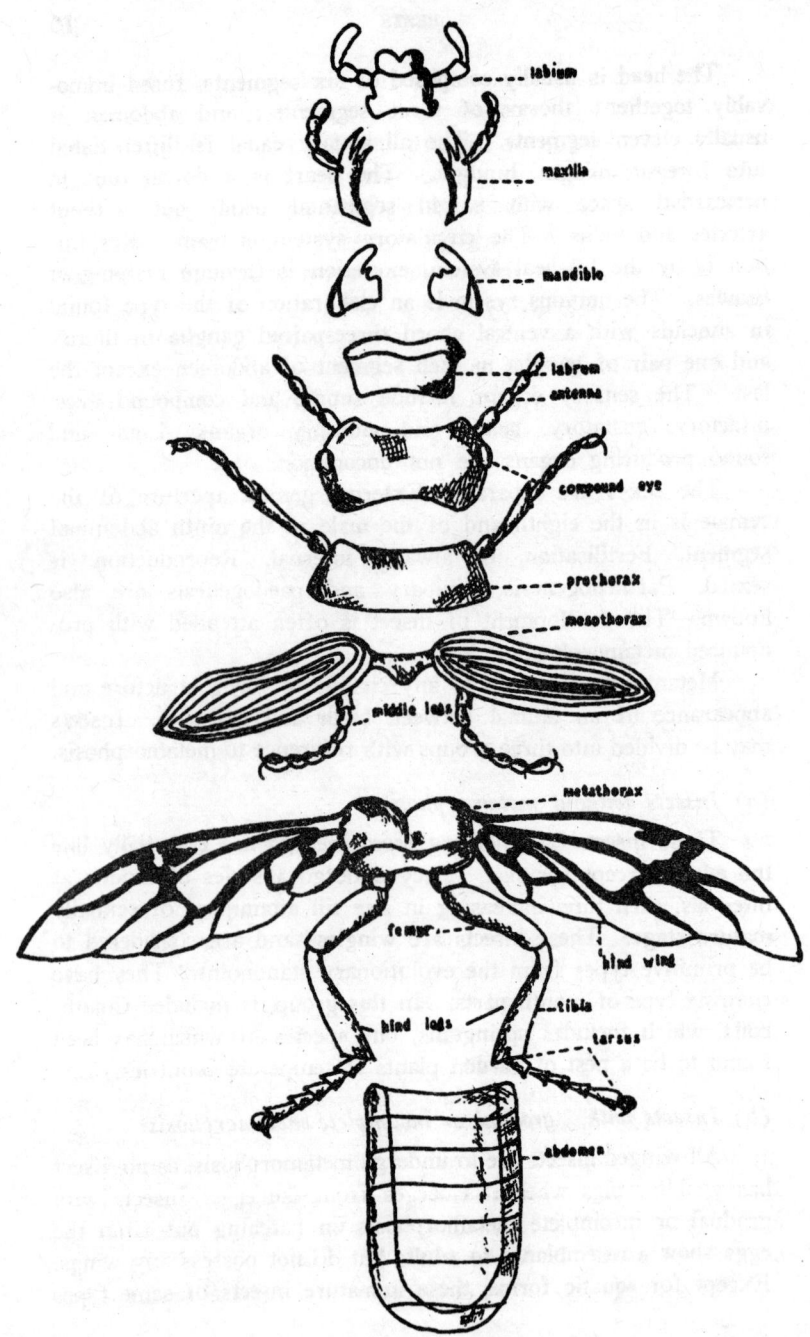

FIG. 1. Parts of an insect

of appendages, have same habits and food as the parents. The young or immature ones of the terrestrial forms are known as *nymphs*. Nymphs grow through a series of moults, each time becoming larger and the periods between moults are known as *instars*. Wing buds which are scarcely noticeable on the very young nymphs, become larger progressively with each moult till they expand to fully developed wings in the adult stage after final moulting. The number of moults and instars remains constant for the species, but varies in different species. The length of period of instars depends on temperature and nutrients.

Most insects of this group, except Orthoptera have piercing sucking mouth parts. These insects, which include many bad enemies of plant, e.g., grasshoppers, locusts, crickets, aphids, scale insects, leafhoppers, thrips, plant bugs are most easily controlled in the nymphal stages. Adults are likely to be more resistant or they may fly away on application of pesticides.

(c) Insects with complete or complex metamorphosis

The young stage of insects of this group do not show resemblance to adults, but look "wormlike" and are known as *larvae*. Larvae do not show any traces of wings externally during any period of growth and differ greatly from the adults in habit. Larvae also undergo moults and pass through instars. They normally have elongate bodies, accessory legs or prolegs and simple eyes. These insects cause maximum damage at the larval stage when feeding and growth take place. When a larva attains its full growth, it sheds its skin and the body parts of adult insect including wingbuds appear folded up against the body. This is known as *pupal stage* and the insect at this stage is referred to as *pupa*. The pupal stage of growth is concealed in the soil or in some protected place or bundled up in a silken case known as *cocoon*. During this stage changeover into adult organs takes place. When the changeover is complete, insects come out of the pupal stage expand their wing and are ready to fly after a short period of drying. The pupal stage is the most resistant to insecticides and the insect is least exposed at this stage. Hence control is to be directed usually at larval stages of growth.

Some types of larvae are known by definite names e.g.,

2

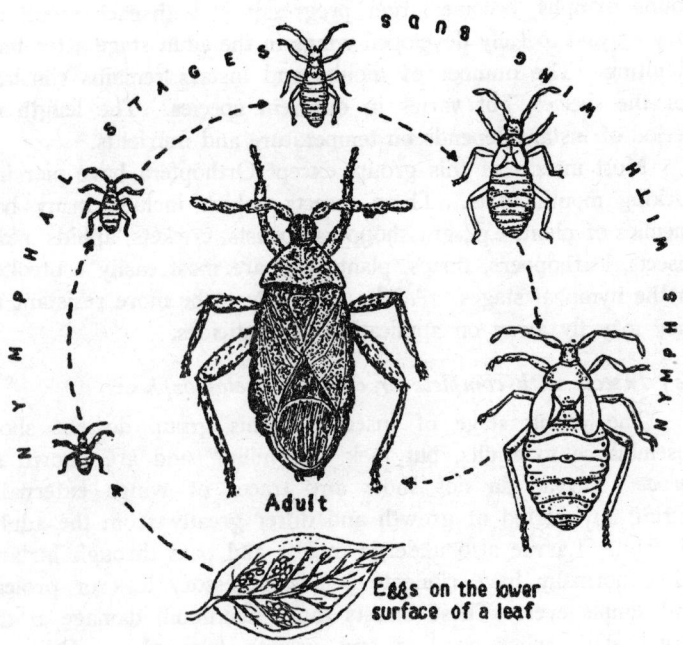

Adult

Eggs on the lower
surface of a leaf

Incomplete or Gradual Metamorphosis As Shown By Squash Bug
(Anasa   tristis)

Fig. 2

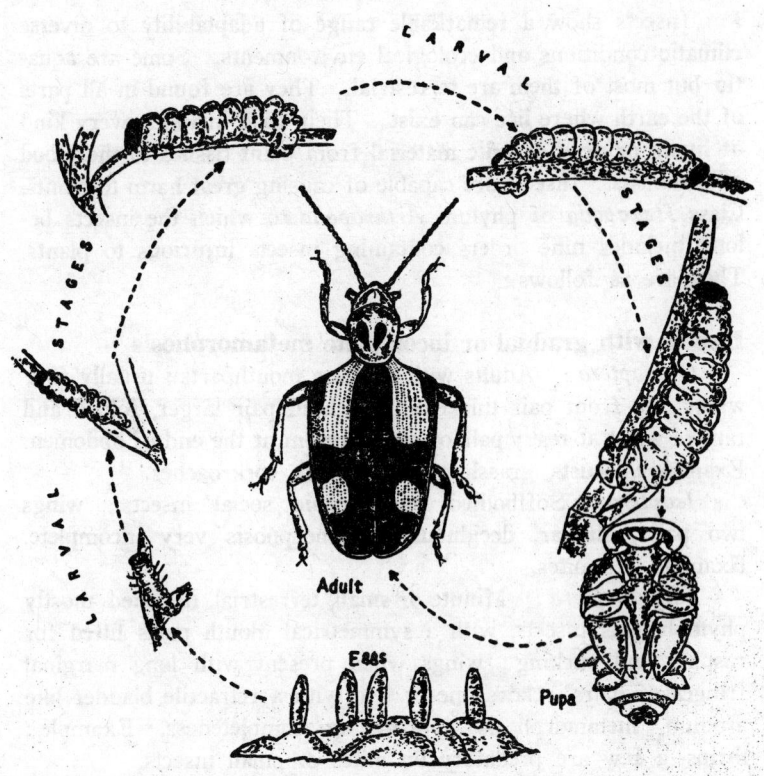

Complete or Complex Metamorphosis As Shown By
Common Asparagus Beetle
(*Crioceris asparagi*)

FIG. 3

caterpillars for larvae of moths and butterflies ; *maggots* larvae of flies ; *grubs* larvae of beetles ; *wireworms* larvae of click beetles ; *cutworms* larvae of certain moths, etc.

Insects show a remarkable range of adaptability to diverse climatic conditions and ecological environments. Some are aquatic, but most of them are terrestrial. They are found in all parts of the earth where life can exist. Their food includes every kind of living or dead organic material from plant tissues to the blood of mammals.. Insects are capable of causing great harm to plants. Class *Hexapoda* of phylum *Arthropoda* to which the insects belong includes nine orders containing insects injurious to plants. They are as follows :

Insects with gradual or incomplete metamorphosis

Orthoptera : Adults with chewing mouthparts ; usually four wings, the front pair thickened, the hind pair larger, folded and fanlike when at rest ; pair of cerci present at the end of abdomen. Example : locusts, grasshoppers, crickets, cockroaches.

Isoptera : Softbodied polymorphic social insects ; wings two pairs, similar, deciduous, metamorphosis very incomplete. Example : termites.

Thysanoptera : Minute or small, terrestrial, flattened mostly phytophagous insects, with a symmetrical mouth parts fitted for rasping and sucking ; wings when present with long marginal fringes of hairs ; claws one or two with a retractile bladder like arolium, metamorphosis approaching completeness. Example : thrips, a few are predatory on mites or small insects.

Heteroptera : Minute to large, terrestrial or secondarily aquatic, phytophagous or predacious insects ; with piercing— sucking mouth parts ; four wings held flatly over abdomen, forewings modified into hemelytra, hind wings membraneous, wing often reduced or secondarily absent ; cerci absent ; metamorphosis incomplete. Example : true bugs.

Homoptera : A vast assemblage of diversified and greatly specialised insects ; mouthparts suctorial, active forms with four wings in both the sexes ; sedentary form wings—one pair of wings in the male and female apterous ; forewings never modified into hemelytra. Example : aphids, scale insects, leaf hoppers, psyllids.

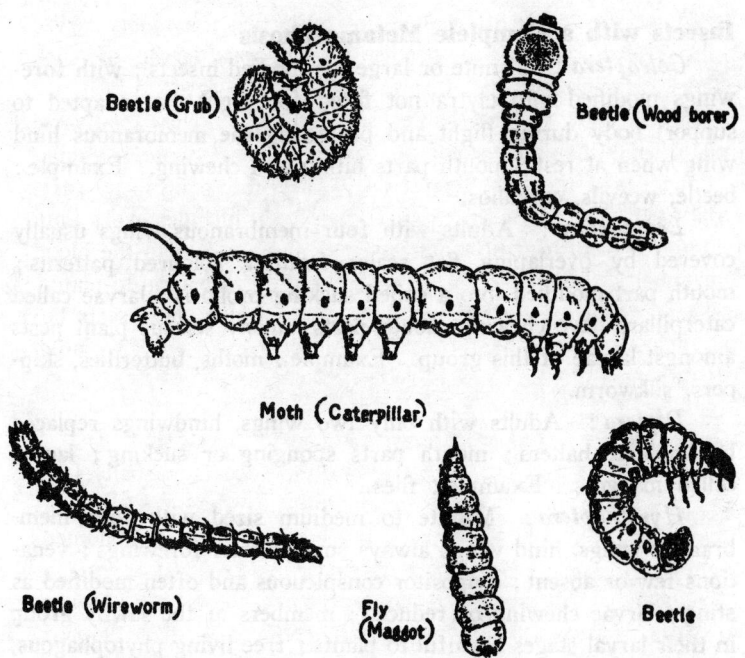

Beetle (Grub)

Beetle (Wood borer)

Moth (Caterpillar)

Beetle (Wireworm)

Fly (Maggot)

Beetle

Different Larvae Of Insects
FIG. 4

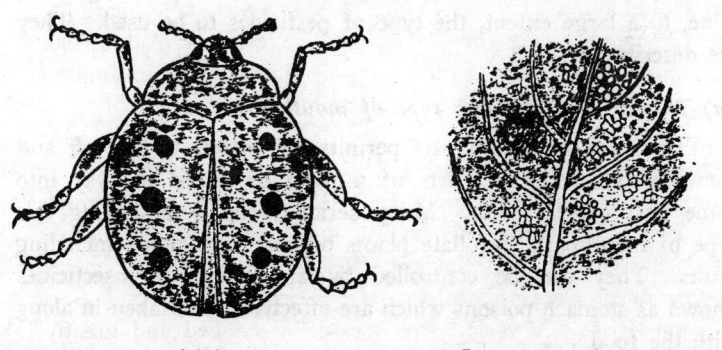

Adult

Eggs, larva and leaf injury

Mexican Bean Beetle – Epilachna varivestis
FIG. 5

Insects with a Complete Metamorphosis

Coleoptera :   Minute or large hardbodied insects ; with fore-wings modified into elytra not fitted for flight, but adapted to support body during flight and protecting the membranous hind wing when at rest ; mouth parts biting and chewing. Example : beetle, weevils, curculios.

Lepidoptera :   Adults with four membranous wings usually covered by overlaping flat scales forming coloured patterns ; mouth part modified into a coiled sucking probocis ; larvae called caterpillars have chewing mouth parts ; many serious plant pests amongst larvae of this group. Example : moths, butterflies, skippers, silkworm.

Diptera :   Adults with only two wings, hindwings replaced by knoblike halters ; mouth parts sponging or sucking ; larvae called maggots. Example : flies.

Hymenoptera :   Minute to medium sized with four membranous wings, hind wings always smaller than forewings ; venations few or absent ; ovipositor conspicuous and often modified as sting ; larvae chewing or reduced ; members of the sawfly group in their larval stages harmful to plants ; free living phytophagous, predatory, entomophagous parasitic or social insects with polymorphism. Example : sawflies, bees, ants, hornets.

Peculiarities of mouth parts and mechanism of feeding determine, to a large extent, the type of pesticides to be used. They are described below :

(a) *Insects with chewing type of mouthparts*

This type of mouth part permits the insects to bite off and chew on into external parts of a plant or tunnel its way into some part of the plant. Many serious crop pests having this type of mouthparts defoliate plants or bore into plants including fruits. They can be controlled by application of insecticides known as stomach poisons which are effective when taken in along with the food.

(b) *Insects with piercing sucking type of mouth parts*

This type of mouthpart which is considered to be evolved from chewing types is found in aphids, leafhoppers, scale insects, bugs and cicadas. These insects cause discolouration, curling of leaves and their eventual weakening and death of plant parts. They may also attack young twigs and other parts of plants and cause them to dry up. As these insects take their food from inside the plant, stomach poisons are not effective unless the insecticide is a systemic toxicant. Contact poisons are more effective.

(c) Insects with rasping-sucking type of mouth parts

Thrips are characterised by this type of mouthpart. Due to the peculiary of mouth parts and their mechanism of action in rasping the tissues, exudation of juice from inside the plant takes place which is sucked up. Parts damaged present a whitish mottled appearance. Such insects can be controlled both by stomach and contact poisons.

(d) Insects with sucking type of mouth parts

They are represented by moths and butterflies. The mouth part is nothing more than a long coiled tube which acts as simple sucking or siphoning. They are not harmful in their adult stages, but in the larval stage they cause extensive damage. Control is effective in the larval stage by stomach poisons.

(e) Insects with sponging type of mouth parts

They are represented by certain flies including houseflies. The mouth part consists of a hinged fleshy probocis partly concealed in a cavity in the head with a sponge-like organ at the end for sucking liquid. These insects first let out saliva for pre-digesting or dissolving the food and then suck the same up. They may be controlled by poisons taken internally or those that come into contact with the body.

(f) Insects with chewing-lapping type of mouth parts

e.g., honeybees. In these cases mandibles retain their original function of chewing, but maxillae and labium are modified into slim lapping organ for taking up liquids. As these insects are beneficial, insecticides are not applied on them.

Just like any other organism, insects are very much influenced by ecological factors which may be broadly divided into three categories : (a) biotic or bicoenetic factors, (b) abiotic

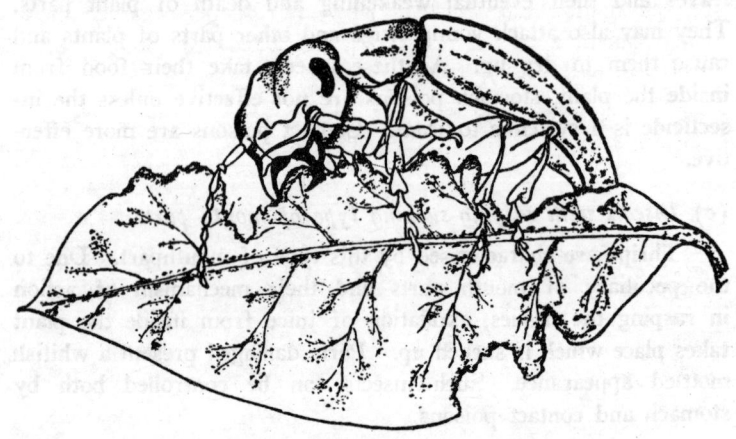

A Chewing Type Of Beetle Feeding On A Leaf

FIG. 6

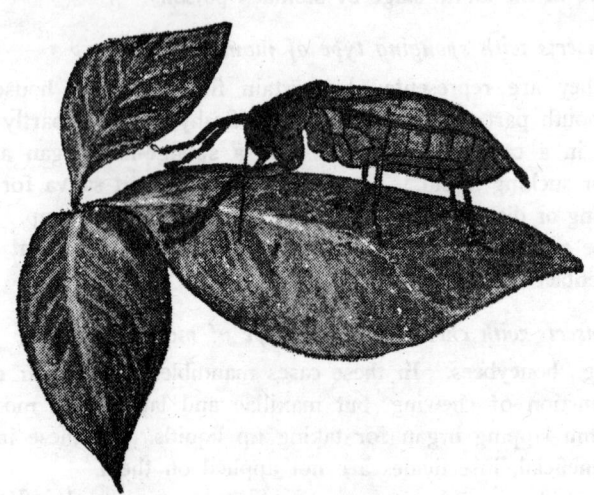

A Bug Sucking On A Leaf

FIG. 7

or physiographical factors, and (c) super-organic factors. Environment need not necessarily be outside the body, the body itself can be an inner environment.

TABLE 2.2

Some Important Pests that Chew off or Bite External Parts

Insect	Family	Scientific Name	Host plants
Cutworm	Noctudiae	*Agrotis ipsilon*	potato, wheat, pulses, etc.
Epilachna beetle	Coccinellidae	*Epilachna dodecastigama Henosepilachna vigintioctopunctata*	potato, eggplant
Red cucumber betle	Galerucidae	*Raphidopalpa foveicolis*	Cucurbit family
Cabbage white butterfly	Pieridae	*Pieris brassicae*	Cabbage
Bihar hairy caterpillar	Arctidae	*Dicrisia. obliqua*	jute, *Phaseolus spp.*, eggplant sunnhemp, cotton sesame, linseed, bajra, maize, etc
Mustard sawfly	Tenthridinidae	*Athalia lugensproxima*	mustard, cabbage, cauliflower

Some Important Pests that suck Plant Parts

Insect	Family	Scientific Name	Host plants
Aphid	Aphididae	*Aphis craccivora*	leguminous plants
Mango mealy bug	Margarodidae	*Myzus persicae Drosicha mangiferae*	potato mango
Scale insect	Diapsididae	*Aspidiotus perniciosus*	apple, peach etc.
Thrip	Thripidae (Thysanoptera)	*Thrips tabaci*	onion, cotton, etc.
		Scirotothrips dorsalis	chillies, egg plant, tomato, etc.

TABLE 2.2 (*Contd.*)

Insect	Family	Scientific Name	Host plants
White fly	Aleyrodidae	*Bemisia tabaci*	polyphagous cotton, bhendi, eggplant
Hopper	Cicadellidae	*Amritodus atkinsoni Idiocerus clypealis, I. niveasparsus*	mango
Leaf hopper	„	*Nephotettix nigropictus, N. virescens*	rice
Plant hopper	Delphacidae	*Nilaparvata lugens*	rice

Some Important Borers

Fruit, seed and bud borers

Fruit borer	Pyraustidae	*Leucnodes orobonalis*	egg plant
Pink bollworm	Gelechidae	*Pectinophora gossypiella*	cotton
Mango fruitfly	Tephritidae	*Dacus dorsalis*	mango, guava, jackfruit, plum, apricot, citrus, chillies, egg plant
Cucurbit fruitfly		*Dacus cucurbitae*	cucurbits

Herbaceous plant borers

Sugar cane borer		*Chilotraea infuscatellus*	sugar cane
Rice stem borer		*Scirpophaga incertulas*	rice
		Chilo partellus	jowar, maize, Sudan grass

TABLE 2.2 (Contd.)

Leaf miners

Insect	Family	Scientific Name	Host plants
Groundnut leaf miner	Gelechiidae	*Stomopterux nertaria*	groundnut
Citrus leaf miner	Gracillaridae	*Phyllocnistis citrella*	citrus, jasmine, willow.
Pea leaf mining fly	Agromyzidae	*Phytomyza atricornis*	pea, lentil, egg-plant, carrot, ber-seem, crucifers

Some Important Subterranean Pests

White grub	Melolonthidae	*Holotruchia consanguinea*	sugar cane
Onion maggot	Anthomyiidea	*Hylemya antigua*	onion
Termite	Macrotermitidae	*Microtermes obesi, Odontotermes obesus*	sugar cane, wheat etc.
Root borer	Pyralidae	*Emmolocera depressella*	sugar cane

Some Important Gall-forming Insects

Rice gall midge	Cicidomyiidae	*Orseolia oryzae*	rice
Sesamum gall fly	„	*Asphondylia sesami*	Sesame
Linseed blossom midge	„	*Diasincura lini*	Linseed

Biotic or biocoenetic factors relate to nutrition, reproduction and interactions with the living environment which may be with insects of the same species or between other species including parasites, predators, foreign organisms, natural enemies, etc. Social life in colonies, parental care, etc., also come under biotic interrelationship. Insects may also exhibit relationships with plants which may act as attractants and may be preferentially used for food or as repellants.

Abiotic factors in contrast to biotic factors originate from non-living environment and comprise climatic, geographic or local influences. The climate in an area is largely determined by its geographical position, latitude, altitude, temperature—constant and fluctuating, relative humidity, precipitation—its amount and duration, light, and wind. Local influences have their origin from location within an area namely proximity to water area or water sheds, forest belts, cultivation—monoculture or multiple or relay cropping, presence of physical barriers like hill slopes, etc. Abiotic factors exert profound influence on the growth and multiplication of insects. Seasonal and geographical variations can be ascribed to the role of climatological factors which represent the sum total of physiographical and meterological factors and determine relative suitability of a region for a particular insect for its habitation and thriving. Super-organic factors relate to the interference of man with nature, namely, deforestation, afforestation, change of a plant species in a vegetation, creation or arrangement of artificial water sources and irrigation, environmental pollution, agricultural or silvicultural system, etc.

For each environmental factor there are usually three cardinal points ; minimum level below which an organism cannot exist ; optimum level which results in highest density of population, and maximum level above which population is lost due to death of the organism. The width of the interval between these limits within which existence of an organism is possible is known as the ecological valence. Complex interactions of ecological factors along with biotic factors influence to a great extent fluctuations in the insect population which are of basic considerations in pest control.

Chapman, in 1928, introduced the concept of biotic potential for quantitative expression of the capacity of the insect to be in an environment. It is represented by the number of viable eggs in an unit area and generation time. The effect of environment is determined by finding the ratio of biotic potential and actual population in an environment. When the environmental factors play a determining role or become limiting, the ratio is large.

The determination of biotic potential and its relationship with actual population in an environment is very important in scheduling operations in pest control. Both active and passive means of dispersal have a differential effect on insect species and are influenced to a large extent by environmental factors. The capacity of dispersal and its methods have a bearing on the distribution of insects including pests.

ARTHROPODS OTHER THAN INSECTS

Arachnids

Mites and red spiders are common and destructive pests of many plant species. They cause damage to plant tissues which they pierce through with their sharp mouth parts for sucking the sap and destroying chlorophyll. The injury is often evident as symptoms of grey and brown mottling or the formation of galls or growths. Many species are beneficial as predators or parasitic on insects.

Mites are distinguished from spiders to which they are closely related in that the abdomen of the former is joined to the cephalothorax with no sign of division between these two regions and the whole body presents a sac-like appearance. Mites have one or more pairs of simple eyes. They have four pairs of legs originating from the cephalothorax. Mites attacking plants are broadly divided into two groups : mites or red spiders (Tarsonemidae, Tetranychidae) and gall, rust or blister mites (Eriophyidae).

Red spiders or mites are actively crawling and their eyes are recognisable as red spots in front of the body. Mouth parts are needle like. They commonly cover the leaf surface with fine webbing. Their life stages consist of egg, larva (with three pairs of legs), and two nymphal stages that resemble the adult, and finally the adult. Each moult is preceded by a quiescent stage during which period the mites are resistant to the action of chemicals. Mites prefer lower surfaces of foliage, but may be found on both sides in cases of heavy infestation. They normally flourish during dry warm weather and are not generally effective pests in cool wet weather.

Gall, rust and blister mites. These are extremely minute, and are designated as gall, rust or blister mites depending on the symptoms they produce. They produce abnormalities on the leaf surfaces that they attack and these are often mistaken for fungal infection. Their life cycle consits of egg, two nymphal stages and adult.

TABLE 2.3
Some Important Mites

Name	Family	Scientfiic Name	Host Plants
Red spider mite	Tetranychidae	*Tetranychus cinnabarinus*	cotton, castor, citrus, bhendi, egg-plant, cucurbits, etc.
Rice mite	„	*Oligonychus oryzae*	rice
Tea red mite	„	*O. coffeae*	tea, jute, coffee
Litchi mite	Eriophyidae	*Aceria litchi*	litchi
Yellow mite of jute	Tarsonemidae	*Polyphagus-tarsonemus latus*	jute, tea, chillies, citrus

THE SNAILS AND SLUGS (MOLLUSCA)

Slugs and snails may be harmful to plants. Slugs are nocturnal creatures and come out only after dark. They do not have any protective calcareous shell as found in snails. In the day time they normally stay in damp, dark places along hedgerows or in soil cracks or underneath damp refuse. They usually cause damage in cool damp weather.

Relatively few species of land snails are harmful to plants. Some of them commonly feed on foliage.

NEMATODES (NEMATA)

Plant parasite nematodes are slender, cylindrical, eel-shaped, 0.2 to 1.00 mm in length, generally less than 2 mm. In some genera, e.g. *Meliodogyne,* female forms may be pear or lemon-shaped. They possess a protruding stylet for feeding. The body of a nematode is covered by an impermeable cuticle which may be smooth or may have sculpturings or markings. It tapers at both ends. At the anterior, there is a mouth covered by

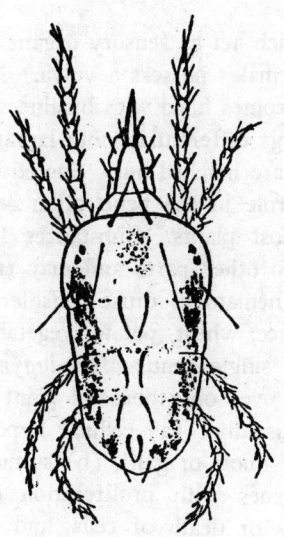

Two-Spotted Mite

FIG. 8

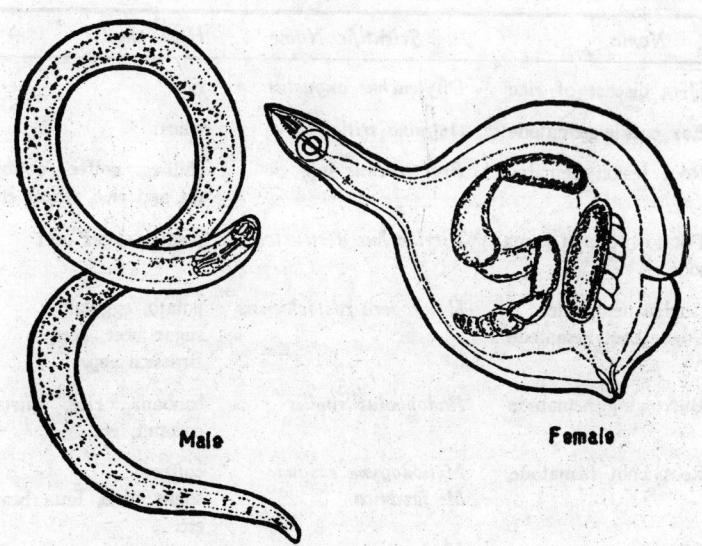

Male

Female

Root Knot Nematode - <u>Meloidogyne incognita</u>

FIG. 9

papillae or setae which act as sensory organs and at the opposite end is the anus ; Females possess a vulva. After copulation the body of a female becomes filled with hyaline and elliptical eggs in a jelly-like sac. Eggs differentiate into larvae and remain inside the mother till they are hatched out. Secretion from the salivary glands play a key role in the penetration and establishment of nematodes in the host plants. Substances from the saliva are easily translocated to other parts and may cause damage.

Plant parasitic nematodes cause considerable damage to important crops like rice, wheat, potato, vegetables, citrus, tea, etc. Different species of single genus *Meliodogyne* causing root-knot diseases may attack over one thousand plant species. Result of nematode attack is usually : (a) cellular hypertrophy and hyperplasia—formation of knots or galls, (b) stimulation of growth of a few selective tissues with proliferation of affected organs, (c) cellular necrosis or death of cells, and (d) suppression of mitosis, a fairly common phenomenon in root infection.

TABLE 2.4
List of Some Important Plant Nematodes

Name	Scientific Name	Host plants
Ufra disease of rice	*Ditylenchus angustus*	rice
Ear cockle nematode	*Anguina tritici*	wheat
Root lesion nematode	*Pratylenchus* spp.	chillies, coffee, cotton, tea and rice, wheat etc.
Potato rootknot nematode	*Ditylenchus destructor*	potato, sugar beet
Golden nematode Sugar beet nematode	*Heterodera rostochiensis*	potato, eggplant, sugar beet, beet, Brassica spp.
Burrowing nematode	*Radopholus similis*	bannana, rice, citrus, coconut, etc.
Root knot nematode	*Meliodogyne exigua*	coffee
	M. javanica	sugar cane, lima bean, etc.
	M. hapla	potato, tomato
	M. incognita var. acrita	jute, cotton

Plant

In the plant kingdom, the following groups of organisms are normally known to incite diseases ; fungi, bacteria, angiosperms and algae in a limited number of cases. Among them fungi play a very important part. Fungi are plants which lack chlorophyll. Vegetative body of most of them (*thallus*) is composed of thread-like filaments (*hyphae*) which are aggregated to form a much branched structure (*mycelium*). From the mycelium, special reproductive structures-asexual and sexual are formed. They bear or produce reproductive units normally known as spores. Reproductive structures and units show complexity as well as diversity of forms and they form the basis of classification. In a few cases, the vegetative body may be a plasmodium (*amoeboid*) and the entire protoplasmic content of the vegetative body may be involved in reproduction, in which case they are known as *holocarpic*. In other cases only a part of the thallus is used in reproduction and it is known as *eucarpic*.

As fungi lack chlorophyll, they cannot synthesise their essential food requirements. Consequently they have to depend on food already synthesised by other organisms. According to the mode of nutrition, they are broadly classified as *saprophytes* and *parasites*. Saprophytic fungi live on dead tissues of either plant or animal origin. They are present in animal or plant debris, soil or any other substratum where dead organic matter is found. Parasitic fungi obtain their food from the living tissues of plants or animals which they attack. They have special sucking organ known as *haustoria* for drawing nutrition from cells of host plants. Among parasites different grades of parasitism are noticed. They are some which have not been grown on dead or artificial material though axenic cultures of a few of them have been made, e.g., rusts, downy mildews, powdery mildews. They are known as *obligate parasites*. There are a number of fungi which are normally parasitic in their mode of life, but can pass a part of their life under saprophytic conditions, should there be any necessity for the same, e.g., smuts and leaf curl fungi. These are known as *facultative saprophytes*. Among the saprophytic fungi, there are some which can be parasites on plants under suitable conditions, and they are known as *facultative parasites*, *e.g.*, *Pythium, Phytophthora, Rhizopus*.

In some cases fungi live in symbiotic association with members of other groups of plants. In lichens, the fungus lives in close association with algae to the mutual benefit of each other—fungi providing for water and mineral elements in solution, and algae sugar and other products of photosynthesis. Association of certain fungi with roots of higher plants also occurs to a form what is known as *mycorrhiza (fungus roots)*.

On the basis of mycelial characteristics and methods of reproduction, fungi may be divided into four important classes : (1) *Phycomycetes,* (2) *Ascomycetes,* (3) *Basidiomycetes* and (4) *Deuteromycetes* or *Fungi Imperfecti.*

TABLE 2.5

Some of the Important Diseases Caused by Different Classes of Fungi

Common name	Casual organism	Host plants
Phycomycetes		
Club root	*Plasmodiophora brassicae*	Crucifers
Wart	*Synchytrium endobioticum*	Potato
Damping off	*Pythium debaryanum* and other species	Numerous species
Late blight	*Phytophthora infestans*	Potato
Foot rot	*Phytophthora nicotianae f. parasitica*	Betelvine and many other plants
Koleraga	*Phytophthora arecae*	Arecanut
Bud rot	*Phytophthora palmivora*	Coconut
Gummosis	*Phytophthora citrophora*	Citrus
Downy mildews	*Peronospora, Sclerospora Plasmopara viticola Pseudoperonospora cubense*	Grains, grasses, tobacco, etc. Grape Cucurbits

TABLE 2.5 (*Conted.*)

Common name	Casual organism	Host plants
White rust	*Albugo candida*	Crucifers
Soft rot	*Rhizopus nigricans*	Sweet potato, various vegetables, fruits specially in storage
Ascomycetes		
Leaf curl	*Taphrina deformans*	Peach
	T. maculans	Turmeric
Powdery mildews	*Erysiphe polygoni*	Pea, legumes,
	Erysiphe graminis	Cereals
	Uncinula necator	Grape
Stem rot of patwa	*Sclerotinia sclerotiorum*	Mesta (*patwa*)
Ergot	*Claviceps purpurea*	Rye
Scab	*Venturia inaequalis*	Apple,
	V. pyrina	Pear
Black spot	*Diplocarpon rosae*	Rose
Anthraenose	*Glomerella gosybi*	Cotton
	Glomerella cingulata	Chilli, fruits and a number of plants
Basidiomycetes		
Rust (Black stem)	*Puccinia graminis* f. trictici	Wheat
	P. graminis f. *avenae*	Oat
	P. graminis f. *secalis*	Rye
Stem and leaf rust	*Puccinia sorghii*	Jowar
Yellow rust	*Puccinia striformiis*	Wheat
Leaf rust	*Puccinia recondita*	Wheat
Bean rust	*Uromyces appendiculatus*	Bean
Pea rust	*Uromyces fabae*	Pea
Linseed rust	*Melampsora lini*	Linseed
Coffee rust	*Hemiliea vastatrix*	Coffee

TABLE 2.5 (*Conted.*)

Common name	Casual organism	Host plants
Smut		
Loose smut	*Ustilago nuda*	Wheat, barley
	Ustilago avenae	Oat
Covered smut	*Ustilago hordei*	Barley
grain smut	*Sphacelotheca sorghi*	Jowar
long smut	*Tolyposporium ehrenbergii*	Jowar
whip smut	*Ustilago scitaminea*	Sugar cane
flag smut	*Urocystis agropyri*	Wheat
root gall smut	*Urocystis brassicae*	Mustard
Bunt	*Tilletia foetida* T. caries	Wheat
Karnal bunt	*T. barclayana*	Wheat
Other Basidomycetes		
Blister blight	*Exobasidium vexans*	Tea
Pink disease	*Corticium salmonicolor*	Orange, rubber, tea, coffee, mango, cinchona
Fungi Imperfecti		
Brown spot	*Helminthosporium oryzae*	Rice
Blast	*Pyricularia oryzae*	Rice and some grasses
Early blight	*Alternaria solani*	Potato, tomato
Stem rot, root rot	*Macrophomina phaseolina*	Jute, guava, and may other plants
Wilt (Panama)	*Fusarium oxysporum* f. cubense	Banana
	F. oxysporum f. *ciceri* arietini	Gram
	F. oxysporum f. *pisi*	Pea
	F. solani	Guava
	F. oxysporum f. lycopersici	Tomato

TABLE 2.5 (*Conted.*)

Common name	Casual organism	Host plants
Tikka	*Cercospora personata*	Groundnut
	Cercospora arachidicola	Groundnut
Blight	*Ascochyta rabiei*	Gram
Stem canker	*Diplodia cajani*	Arhar
Red rot	*Colletotrichum falcatum*	Sugar cane
Grey blight	*Pestaliopsis theae*	Tea
Stem rot	*Sclerotium oryzae*	Rice

Symptoms produced in plants due to attack of pathogens and fungi in particular may be broadly classified into the following categories :

I. *Breakdown of tissues and utilisation of stored food* : Examples : soft rot, seedling blight, damping off, dry rot, etc. Pathogens secrete specific extracellular enzymes ; pectinolytic and celluloytic, which bring about dissolution of middle lamella and other components of the cell wall and bring about loss of coherence of cell walls and the collapse of cells. These conditions lead to leakage of nutrients from the host cells which are utilised by pathogens. These also include soft rot produced in stored plant products in the post-harvest stage.

II. *Effects on absorption of water and solutes* : Example : root rots. Pathogens invade mainly the epidermal layer containing root hairs and thin walled parenchyma forming cortical region around the central vascular core and destroy them so that the absorbing surface is greatly reduced. Consequently plants suffer from symptoms of shortage of water and mineral solutes.

III. *Effects on water conduction.* Example : vascular wilts. Pathogens enter into the vascular system, mainly through water conducting xylem vessels. They cause a blockade of the xylem vessels by their own growth, or the break down of cell and cell wall components, or induce formation of tyloses, etc., inside the vessels and affect the supply of water to the aerial plants thereby inducing symptom of wilting.

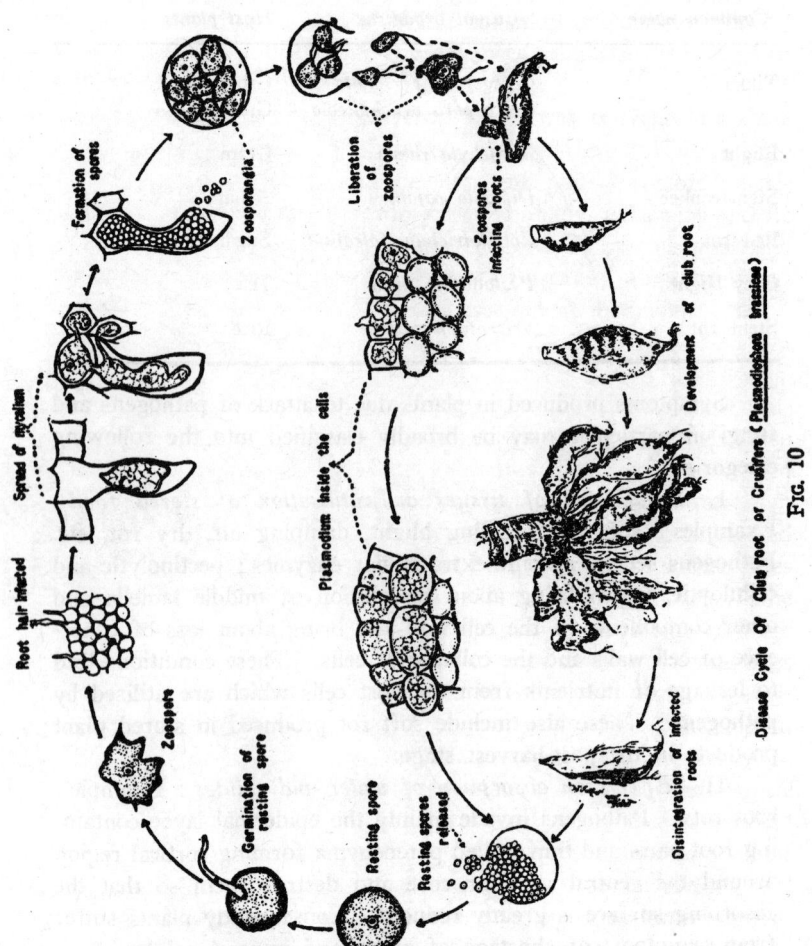

Formation of spore

Zoosporangia

Liberation of zoospores

Zoospores infecting roots

Development of club root

Spread of mycelium

Root hair infected

Zoospore

Plasmodium inside the cells

Disintegration of infected roots

Germination of resting spore

Resting spore released

Resting spore

Disease Cycle Of Clubroot or Crucifers (Plasmodiophora brassicae)

Fig. 10

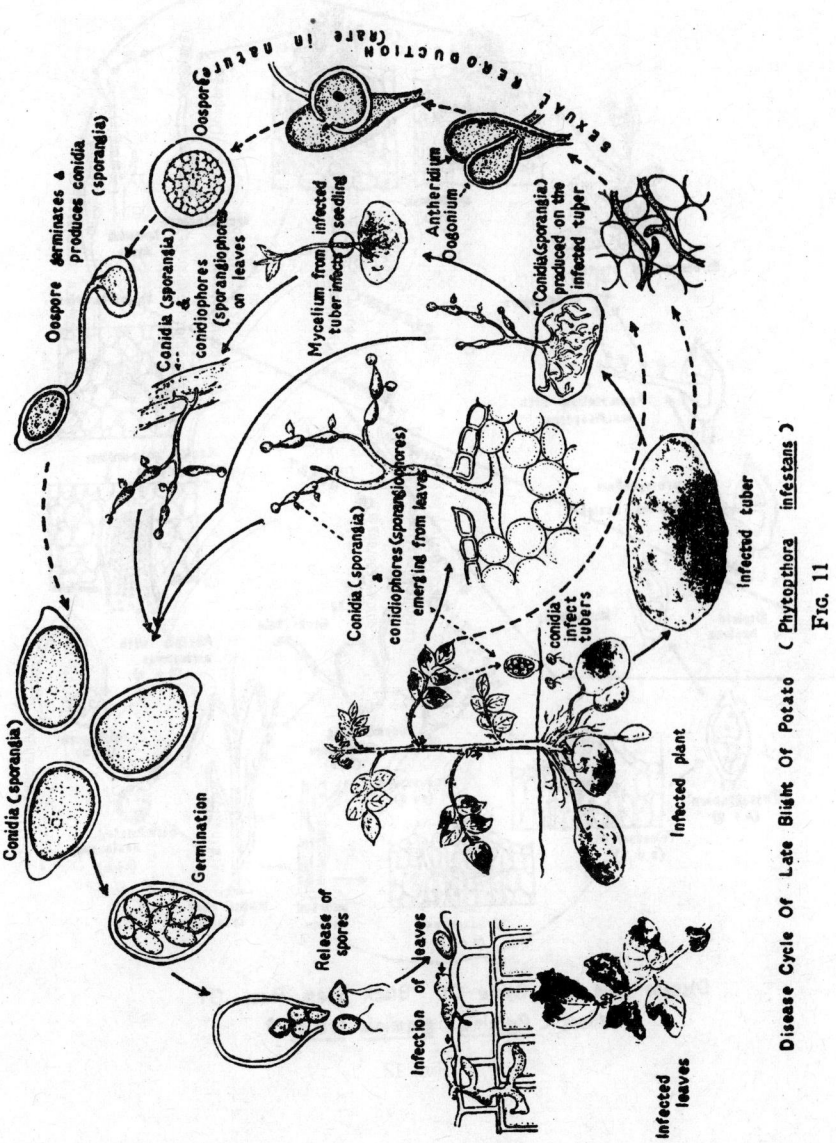

Disease Cycle Of - Late Blight Of Potato (*Phytopthora infestans*)

Fig. 11

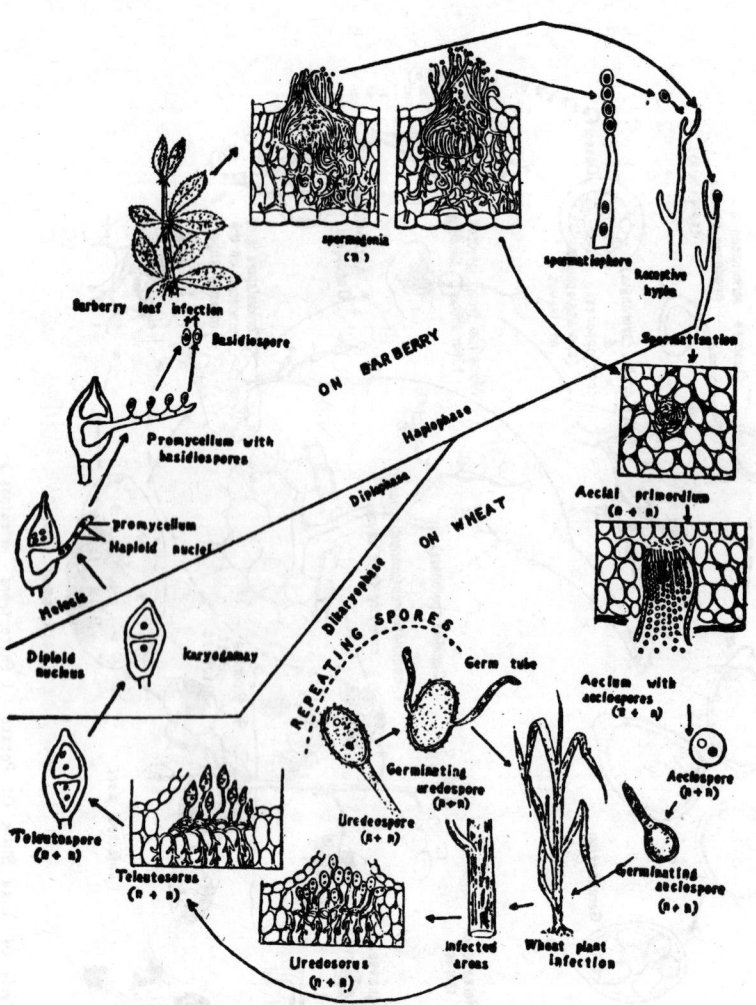

Disease cum Life Cycle Of Black Stem Rust Of
Wheat (*Puccinia graminis tritici*)

Fig. 12

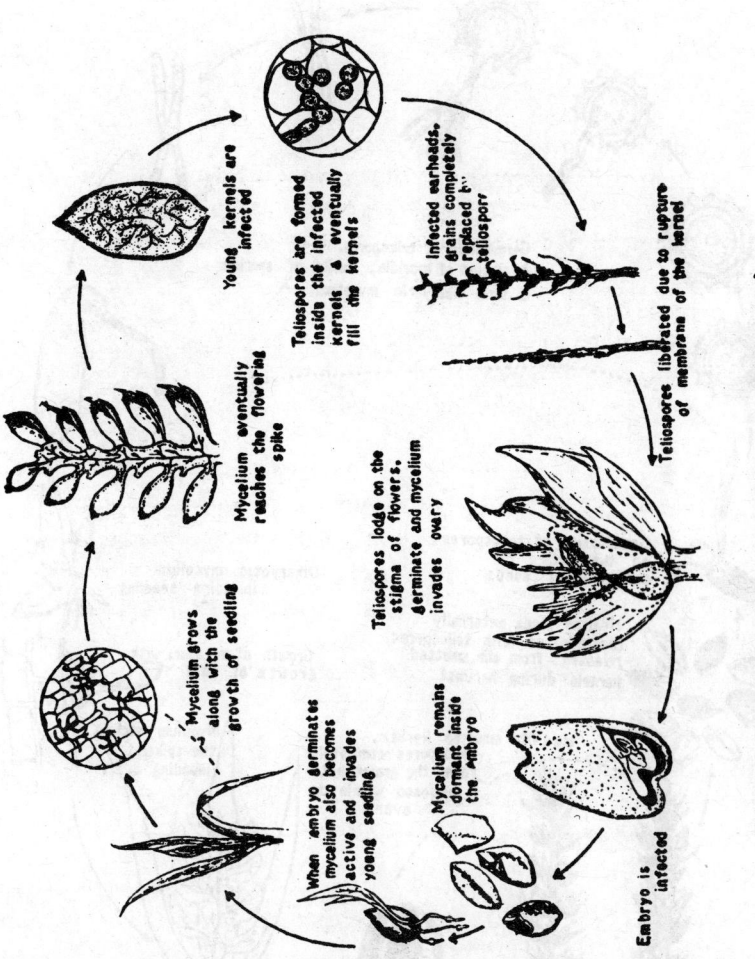

Young kernels are infected

Teliospores are formed inside the infected kernels & eventually fill the kernels

Infected earheads, grains completely replaced by teliospores

Teliospores liberated due to rupture of membrane of the kernel

Mycelium eventually reaches the flowering spike

Teliospores lodge on the stigma of flowers, germinate and mycelium invades ovary

Mycelium grows along with the growth of seedling

When embryo germinates mycelium also becomes active and invades young seedling

Mycelium remains dormant inside the embryo

Embryo is infected

Disease Cycle Of Loose Smut Of Wheat (Ustilago nuda)

Fig. 13

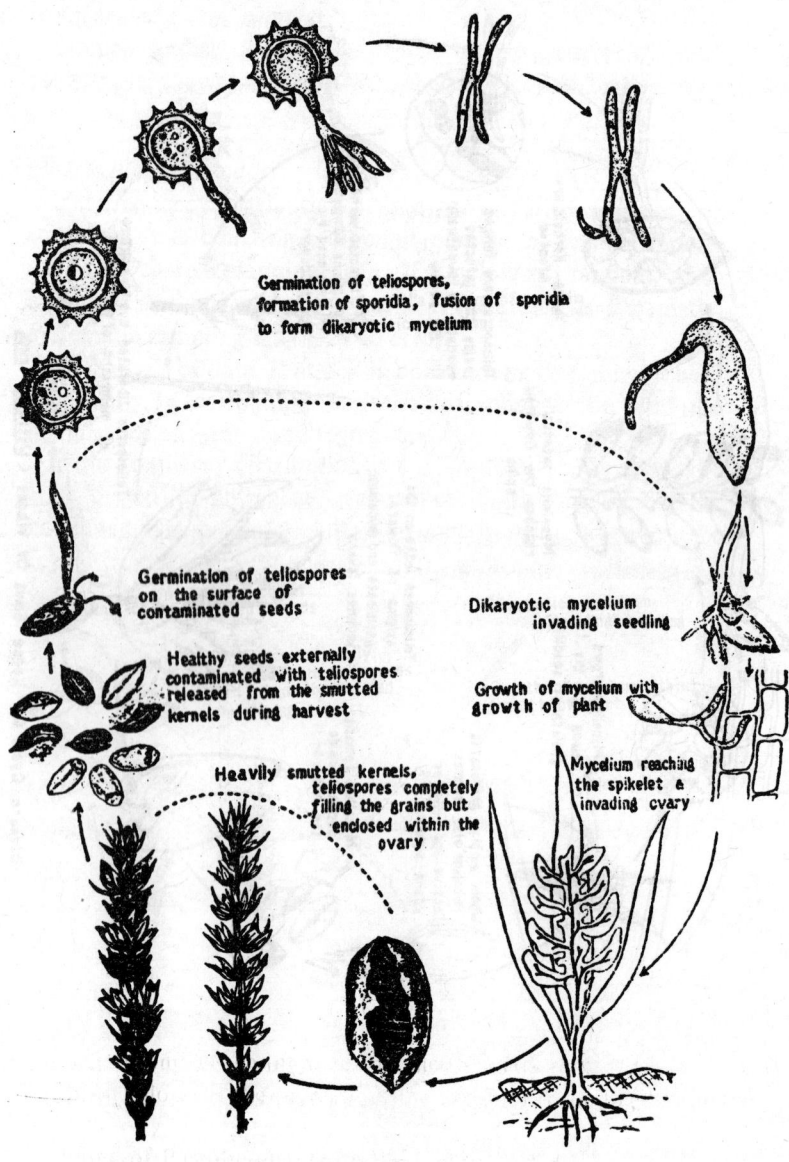

Germination of teliospores,
formation of sporidia, fusion of sporidia
to form dikaryotic mycelium

Germination of teliospores
on the surface of
contaminated seeds

Dikaryotic mycelium
invading seedling

Healthy seeds externally
contaminated with teliospores
released from the smutted
kernels during harvest

Growth of mycelium with
growth of plant

Heavily smutted kernels,
teliospores completely
filling the grains but
enclosed within the
ovary

Mycelium reaching
the spikelet &
invading ovary

Disease Cycle Of Bunt in Wheat (*Tilletia caries* — Seedling Infection)

FIG. 14

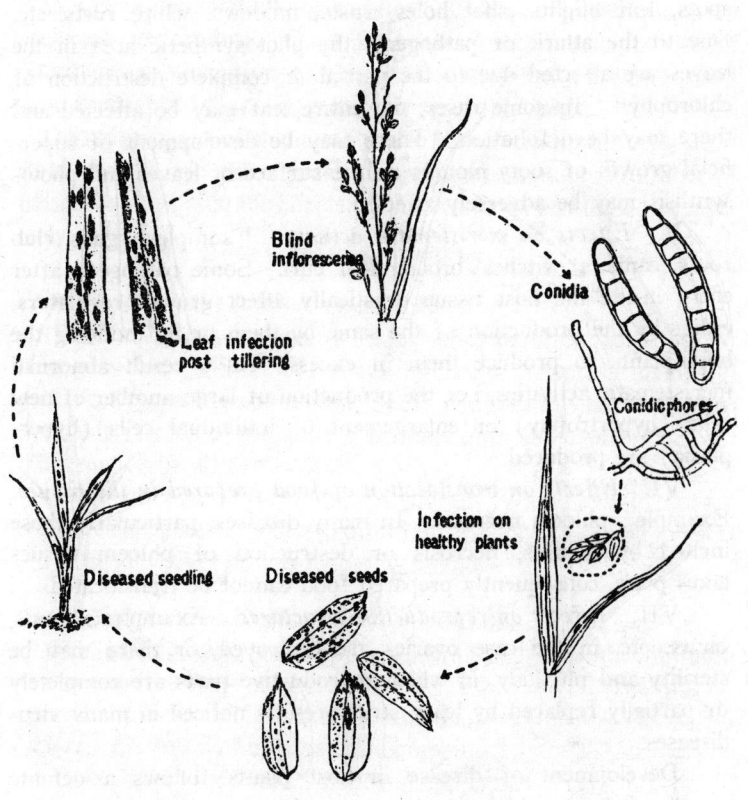

Brown Spot Disease Of Rice (Helminthosporium oryzae)

FIG. 15

IV. *Effects on photosynthetic activity.* Examples : leaf spots, leaf blights, shot holes, rusts, mildews, white rusts, etc. Due to the attack of pathogens, the photosynthetic area in the leaves are affected due to the partial or complete destruction of chlorophyll. In some cases, the entire leaf may be affected and there may be defoliation. There may be development of superficial growth of sooty moulds on the surface of leaves and photosynthsis may be adversely affected.

V. *Effects on meristematic activity.* Examples : galls, club roots, cankers, witches' broom, leaf curl. Some pathogens after entry inside the host tissue drastically affect growth regulators, either by the production of the same by them or by inducing the host plants to produce them in excess. As a result abnormal meristematic activities, i.e., the production of large number of new cells (hypertrophy) or enlargement of individual cells (hyperplasia) are produced.

VI. *Effects on translocation of food prepared, in the tissues.* Example : phloem necrosis. In many diseases, particularly those incited by viruses, necrosis or destruction of phloem tissues takes place, consequently prepared food cannot be translocated.

VII. *Effects on reproductive structures.* Examples : smuts, bunts, etc. in the case ovaries ,are destroyed, or there may be sterility and phyllody, in which reproductive parts are completely or partially replaced by leafy structures, as noticed in many virus diseases.

Development of disease in host plants follows a definite course of events which may be arranged in a sequence as follows : (a) *Infection* denoting the process from the entry of the organism and its initial invasion till it. has been able to establish itself inside the host tissue, (b) *Incubation*, the time period between the etablishment of infection and expression of symptoms of the disease, and (c) *Disease development*, the process involved after the establishment of infection till the disease is completely manifested.

In the development of disease in crops, the most important considerations are abundance of inoculum and its source. This is connected with the mode of survival of the pathogens between two successive crops. Pathogens may perennate in the seed, in the soil, infected plant parts in or above the soils or in plants itself in case of infections on perennial plant, or in other plants or weeds. Dissemination of pathogens may be effected by air, soil, water, seeds, affected plant parts, insects, etc.

Disease results from host-pathogen interaction causing interferences or disturbances in form or function or both. These are manifested in the form of definite symptoms. In the host-pathogen interaction, environmental factors play a very important part. Any consideration of a plant disease normally involves the following triangle

$$\text{Pathogen} \rightleftarrows \text{Host}$$
$$\Updownarrow \qquad \Updownarrow$$
$$\text{Environment}$$

Environmental factors may affect host pathogen and host-pathogen interaction. They may relate to temperature, precipitation, humidity, cloudiness, dew deposition, light intensity of the atmosphere ; or various soil factors namely soil pH, soil moisture, soil nutrients including soil organic matter, soil temperature, soil texture, soil biotic community.

Air-borne diseases are influenced by aerial factors, and the soil factors, in such cases, do not have much influence except through effect on host. Similarly in soil-borne disease, factors pertaining to soil play important parts, whereas ambient conditions do not appear to exert much influence. Hence these two groups of diseases should be considered separately in relation to environment. In soil apart from physical and chemical factors microbial population plays on important part. Condition in the small volume of soil surrounding the roots—rhizosphere, in many cases exert more influence. In insect-borne diseases, factors determining population and movement of insects, exert influence.

When there is a concentration of disease or pest infestation in a space and time, due to the presence of abundant inoculum or initial infestation, effective movement or dispersal of the same, and favourable environmental conditions for a long period to establish infection or infestation, an epidemic or epiphytotic or epizootic condition is created. In plant diseases, the appearance of epidemics is related with the progressive increase in the intensity of the disease which can be measured quantitatively by adoption of an appropriate scale, several of which have been developed and are in use.

Development of a disease can be mathematically expressed by an equation propounded by Van der Plank (1963) which is as follows :

$$X = X_{oe}rt$$

X = amount of disease on a plant in a given time
X_0 = amount of initial inoculum
r = average infection rate
t = time
e = base of logarithm

Hence the development of a disease in epidemic proportion is dependent on the initial inoculum and rate of progress, in which environmental factors play a very important part. Protection measures may be aimed at the reduction of inoculum and rate of progress by reducing effectively the population density of causal organism by interfering with production and dissemination of inoculum. Genetic basis of the host may also play an important part. In plant diseases, organisms in all cases excepting a few ones, enter inside the plants and cause damage, hence chemical control is to be applied before infection is established. In most cases of insect damage, insects remain outside and feed on plants. Control aims at keeping down insect population.

That bacteria can cause diseases in plants was first demonstrated in 1878 by T. J. Burrill of Illinois, U.S.A. who showed that the causative organism of fire blight of apples and pears was a bacterium. Since then a number of plant diseases have been demonstrated to be incited by pathogenic bacteria. Bacterial plant pathogens are normally facutative parasites. In physiology and nutrition, they have similarities to soil bacteria rather than to species pathogenic to animal.

Normally plant pathogenic bacteria incite three broad categories of symptoms in plants : (a) vascular diseases, (b) parenchyma diseases, and (c) hyperplastic diseases. In the first type, abundant growth of the bacterium in the lumen of xylem vessels as well as break down of cell components of xylem by enzymatic action of the bacterium causes the blocking of the vessels resulting in wilting. In the second category of diseases, the organism invades parenchymatous tissues and causes soft rot without inciting any abnormal growth of host tissue. In hyper-plastic diseases, there is increased meristematic activity of enlargement of cells of both due to interference with formation and accumulation of growth regulating substances.

Bacteria enter host plants through wounds, stomata, water-pores or lenticels. Entry may also take place through nectaries. They normally reside inside the cells within the host tissue. Nodule-producing bacteria (*Rhizobium* spp.) enter into symbiotic relationship with appropriate species of leguminous host plant.

TABLE 2.6
Some Important Diseases Incited by Bacteria

Common name	Organism	Host plants
Citrus canker	*Xanthomonas citri*	Citrus spp.
Black rot of crucifers	*Xanthomonas campestris*	Cabbage, cauliflower and other cruciferous vegetables
Bacterial blight of rice	*Xanthomonas oryzae*	Rice
Angular leaf spot of cotton	*Xanthomonas malvacearum*	Cotton
Leaf spot of chilli	*Xanthomonas vesicatoria*	Chillies
Bacterial wilt of solanaceous plants	*Pseudomonas solanacearum* and *P. solanacearum* var *asiatica*	Potato, tomato, tobacco, chilli, eggplant and others.
Crowngall	*Agrobacterium tumefaciens*	Fruit trees, shrubs
Fire blight	*Erwinia amylovora*	Apple, pear, etc.
Soft rot	*Erwinia carotovora*	Storage tissues of potato, carrot, etc.
Scab	*Streptomyces scabies*	Potato
Yellowing rot of wheat	*Corynebacterium rathayi*	Wheat

Mechanism of pathogenicity of plant pathogenic bacteria is similar to that in fungi. Pathogens are disseminated through seed, wind splashed rain, irrigation water, contaminated compost, insects, infected soil, etc. Important plant pathogenic genera of bacteria are *Xanthomonas, Pseudomonas, Erwinia, Streptomyces, Corynebacterium, Agrobacterium.*

In addition to fungi and bacteria, viruses may incite diseases in plants. Virus diseases in plants have been known for many years, e.g., breaking of tulips, now known to be a virus disease, was reported as early as 1576; degeneration of potatoes, now known to be due to infection of viruses was known in Europe in the eighteenth century. Existence of virus as an entity, distinct from microorganisms was shown by Iwanoswky, in 1882, in tobacco mosaic disease. It was later confirmed by Beijerinck in 1896. A number of destructive diseases are due to infection of viruses.

Viruses are nucleoproteins of various shapes. They may be of flexuous threads, or rigid rods, sphaeroids or polyhedra. Individual virus particle is too small to be seen except under electron microscope in purified preparations. Basically a virus particle (virion) consists of a central core of nucleic acid surrounded by a protein cost (capsid). Most plant viruses contain RNA (ribonucleic acid). At least one virus cauliflower mosaic virus, contains DNA (deoxyribnucleic acid). Multiplication of virus inside the host cell is dependent on the interaction of independent genomes of virus and the host. Host cells cannot manufacture virus, unless it is infected, virus can multiply only in appropriate host cells, when genome of the host fits for the purpose. On being introduced into the appropriate host plant, the virus interferes with metabolism of the host plant and produces characteristic symptoms. After the introduction of virus in the host tissue, viral nucleic acid is released from the viral protein coat. Virus nuclei acid contains necessary genetic information and initiates its own replication by manufacturing its own componests—nucleic acid of the core and protein of the coat—finally assembling the components into a virus particle. Virus nucleic acid is a genetic determinant and contains only the requisite structural information necessary for self replication as well as virus-specific proteins and expressing itself within genetic potentialities of the host plant. Virus nucleic acid is synthesised in the host nucleus and is deposited in the cytoplasm. Infection spreads from cell to cell through distribution of virus nucleic acid.

The presence of infection of viruses can be determined by the symptoms produced on the host plants, but symptoms should not be used as a basis for identifying the presence of specific viruses. In cases of infection of virus diseases, the whole plant may be systemically infected, but the developing or growing parts specially leaves show characteristic symptoms. The effect however is pronounced on all the organs of the plant. Infection of viruses causes general stunting, reduction in growth, vigour and yielding ability of the plant. Virus diseases in plants are normally not lethal but they cause severe damages.

Plant viruses are transmitted through grafts, (grafting between healthy stock and diseased scion), sap of infected plant and vegetative propagative stocks produced from infected plants. Most viruses are however dependent for transmission on the activity of the insects which are termed vectors.

TABLE 2.7
Some Important Plant Virus Diseases

Host plants	Disease	Vector
Wheat	Mosaic streak	*Rhopalosiphum maidis* *R. padi* *Brachycaudus heichrysi* *Sitobion avenae*
Maize	Mosaic	*Aphis maidis* *A. gossypii*
	Vein enation	*Cicadulina intoila*
Barley	Mosaic	*Rhopalosiphum maidis*
	Yellow dwarf	*R. maidis* *Macrosiphum aranarium*
Bajra	Mosaic	*Rhopalosiphum maidis*
Jowar	Yellowing	*Peregrinus maidis*
Rice	Tungro	*Nephotettix nigropictus* *N. virescens*
	Grassy stunt	*Nilaparvata lugens*

Table 2.7 (contd.)

Host plants	Disease	Vector
Potato	Potato necrosis virus (PVY),	*Aphis rhamni*
	Potato severe mosaic	*Myzus persicae*
	Leaf roll	*Myzus persicae*
	Super mild mosaic	*M. persicae*
		M. circumflexus
Cucurbits	Bottle gourd mosaic	*Aulacophora foveicollis*
	Vegetable marrow mosaic	*Aphis gossypii*
		Myzus persicae
	Filiform leaf mosaic	*M. persicae*
	Pumpkin mosaic	*M. persicae*
		Sitobion rosaeformis
Tomato	Mosaic	*Aphis gossypii*
		A. craccivora
		Myzus persicae
Tomato	Leaf curl	*Bemisia tabaci*
	Black ring spot	*Aphis craccivora*
		A. gossypii
		Myzus persicae
		Bemisia tabaci
Chilli	Mosaic	*Myzus persicae*
		Aphis gossypii
	Leaf curl	*Bemisia tabaci*
Bhendi	Yellow vein mosaic	*Bemisia tabaci*
Eggplant	Mosaic	*Epitrix* sp.
Onion	Yellow dwarf	*Aphis gossypii*
		A. craccivora
		Rhopalosiphum maidis
Pulses and Legumes Bean (*Phaseolus vulgaris*)	Mosaic	*Aphis gossypii*
		A. craccivora
		Myzus persicae
***Mung* (*Phaseolus aureus*)**	Yellow mosaic	*Bemisia tabaci*
Urd (*Phaseolus mungo*)	Leaf crinkle	*Aphis craccivora*

Table 2.7 (contd.)

Host plants	Disease	Vector
Double bean (*Phaseolus lunatus*)	Yellow mosaic	*Bemisia tabaci*
Common bean (*Dolichos lablab*)	Yellow mosaic	*Bemisia tabaci*
Soybean	Mosaic	*Myzus persicae* *Aphis gossypii* *A. craccivora* *Lipaphis erysimi*
Pigeon pea (*Cajanus indica*)	Sterility mosaic	*Aceria cajani*
Cowpea	Mosaic	*Aphis gossypii* *A. craccivora* *Myzus persicae*
Pea	Mosaic	*Aphis gossypii* *Myzus persicae* *Rhopalosiphum pseudobrassicae*
Broadbean (*Vicia faba*)	Mosaic	*Aphis craccivora* *A. rumicis* *Macrosiphoniella sanborini* *Myzus persicae*
Onion	Yellow dwarf	*Aphis gossypii* *A. craccivora* *Rhopalosiphum maidis* *Myzus persicae*
Lettuce	Ordinary mosaic	*Aphis gossypii* *Myzus persicae*
Groundnut	Mosaic	*Aphis gossypii* *A. craccivora* *Bemisia tabaci*
	Chlorosis	*A. craccivora*
	Clump disease	*A. craccivora*
Mustard (*Brassica juncea*)	Mosaic	*Aphis gossypii* *Myzus persicae* *Brevicoryne brassicae*

Table 2.7 (contd.)

Host plants	Disease	Vector
Rai (*Brassica compestris*)	Mosaic	*Myzus persicae* *Lipaphis erysimi*
Safflower	Mosaic	*Aphis gossypii* *Myzus persicae*
Sesamum	Leaf curl	*Bemisia tabaci*
Mulberry	Mosaic	*Rhopalosiphum nraidis* *Myzus persicae* *Aphis gossypii*
	Yellow vein	*Bemisia tabaci*
Large cardamon (*Ammomum subulatum*)	Foorkey	*Pentalonia nigronervosa*
Small cardamon (*Elettaria cardamomum*)	Katte, Mosaic or Marble	*Pentanolia nigronervosa*
Sugar cane	Mosaic	*Schizaphis gramineum* *Aphis maidis*
Coconut	Root (Wilt) Cadany cadong	*Stephanitis typicus* *Aphis gossypii*
Tobacco	Leaf curl Yellow net vein Broken ring spot	*Bemisia tabaci* *Bemisia tabaci* *Myzus persicae* *Aphis gossypii* *A. craccivora* *Bemisia tabaci*
	Ring spot	*Xyphinema americanum*
Banana	Bunchy top	*Pentalonia nigronervosa*
	Mosaic	*Aphis gossypii*
Papaya	Mosaic	*Aphis gossypii* *A. medicaginis* *Myzus pericae*
	Leaf curl	*Bemisia tabaci*
Citrus	Tristeza and Quick decline	*Toxoptera citricidus* *T. aurantii*

Japanese scientists (Ashuyama and his co-workers) in 1967, discovered in plants organisms resembling Mycoplasma. It is now known that certain diseases which have been previously attributed to infection of viruses are due to Mycoplasma. Mycoplasma differ from viruses inasmuch as they are pleomorphic, i.e., can exist in different forms and have cell membranes. They are particularly sensitive to tetracyclines. Some Mycoplasma species from animals can be grown in cell-free media. It is now claimed that Mycoplasma-like organisms isolated from plants can similarly be grown in the laboratory. In pathogenesis they show similarities to viruses.

TABLE 2.8

Common Mycoplasma Diseases of Crop Plants

Host	Disease	Vector
Rice	Yellow dwarf	*Nephotettix cincticeps* *N. nigropictus* *N. virescens*
Sandalwood	spike	*Jassus indicus*
Safflower	phyllody	*Neoaliturus fenestratus*
Sesamum	phyllody	*Orosis albicinctus*
Eggplant	little leaf	*Hishimonus phycitus*
Citrus	greening	*Diaphorina citri*
Potato	purple top witches broom	*Orosius albicintus* "
Sugar cane	grassy shoot	*Aphis idiosacchari* *A. sacchari* *A. maidis*
	white leaf	*Epitettix hiroglyphicus*
Coconut	lethal yellowing	*slow flying insect*

Angiospermic plants may be parasitic on other important hosts and can cause damages at economic levels. These plants can parasitise on roots or stems of plants and may be wholly or partially parasitic depending whether they can have a part of their food of their own or not. Examples of wholly stem parasites dodder (*Cuscuta* spp.), partially stem parasites *Loranthus, Arcethobium,* wholly root parasite *Orobanche,* partial root parasite sandal tree (*Santalum album*).

There are a few instances where algae may cause a disease, e.g., *Cephaluros* species causing red rust of tea

Diseases or abnormalities in plants may appear due to unfavourable environmental conditions, viz. unfavourable temperature and water relations, excess or deficiency of nutrients, lack of aeration, presence of toxicants or pollutants in the environment, smog, injudicious uses of chemicals including pesticides.

QUESTIONS FOR DISCUSSION

1. In what stages are insects usually harmless to crops ? Why ?
2. What insects are usually harmless to plants in the adult stage ?
3. How can a knowledge of mouth parts be helpful in chemical control of insects ?
4. What type mouth part does a boring insect have ? Why ?
5. Of what importance is it to the grower to know the life history of crop pests ?
6. Why is accurate timing of control measures, dependent on knowledge of the life history of pest ?
7. Why is the classification of insects important to those who deal with insect control ?
8. Why are nematodes called disease-producing organism ?
9. What are the harmful groups in the animal kingdom ?
10. What is biotic potential ? In which way is the knowledge of biotic potential useful for the control of pests ?-
11. Name the order of insects which include plant pests. Give their diagnostic features.
12. As mites are wingless, how do they lisseminate themselves in nature ?
13. What are the symptoms of nematode infestation ?
14. Why are mites considered important plant pests ?

15. Give a classified account of different symptoms found in diseased plants. In which groups of diseases, are floral organs modified ?
16. How does rotting take place in plant tissues ? Is it possible that rotting can take place without pathogenic organism ?
17. Can you name diseases which are systemic ?
18. What are the principal symptoms produced by bacteria on plants ?
19. What are the sequences in the infection process of plants ?
20. When does an epidemic take place ? What factors are important in the development of an epidemic in plants ? How do you plan disease control in the light of the knowledge of epidemics ?
21. Can a virus multiply outside the host cell ? If not, why not ?
22. Plant pathogenic bacteria do not form resting structures or spores. How do they perennate in nature between two successive crops ?
23. How do environmental factors influence the development of plant disease ?
24. Is correct timing more essential for the control of diseases, or insect control ?
25. Can you mention the fundamental difference between attack of a disease-producing organism and an insect ?
26. In which way are insects important in virus disease ?
27. Mention different levels of parasitism found in plants.
28. What are angiospermic parasites ? How do you classify them ?

CHAPTER 3

CULTURAL METHODS OF CONTROL

Cultural methods of control aim at either reducing insect population or inoculum potential of pathogens, or preventing damages due to pests ; either encouraging a healthy growth of plants, or circumventing the attack by changing various agronomic practices. The importance of these methods lies on the basic fact that usually comparatively little additional expenditure—capital or recurring, or effort is involved except in adjustments in the cropping system. Operations under this category are directed towards field sanitation ; clean cultivation ; crop rotation ; adjustments in dates of sowing or planting ; adjustment of sowing or planting distances ; water and fertiliser management ; tillage practices, etc. These methods are considered to be particularly valuable where other control measures are not economical or feasible. They can be easily integrated with others. Cultural methods of control are not entirely new in concept and adoption. Probably before the advent of chemicals in the post-World War II and resistant varieties in the post-World War I era, the cultural methods constituted important available means of keeping the infection or infestation in check.

Normally good farmers adopt such agronomic practices from the sowing to harvesting for the efficient management of inputs as well as crop growth so as to obtain high yield. By trial and error as well as experience over a considerable number of years, beneficial effects of these methods, have been quite evident in the securing of higher yield. The adoption of these methods has not encouraged incidence of diseases and pests to a great extent. Normally vigorously growing healthy plants can withstand the attack of pests and build-up a defence mechanism. For the adoption of successful cultural methods of control, it is necessary to have a fairly good knowledge of the biology and ecology of the pathogenic organisms or insect pests, including methods of perennation in relation to passing through uncongenial conditions in the absence of host plants. Information on the effect of environmental and biotic factors in the build-up of the pests and infection or infestation is also required.

Field Sanitation

Field sanitation is an essential prerequisite to reduce the inoculum of the pathogen or insect population, to minimise or defer the possibilities of the appearances of epiphytotics or eipzootics. Measures adopted for field sanitation include : (a) destruction of crop residues, stubbles and self-sown tillers, etc. ; (b) use of eradicant sprays where complete destruction of crop residues may not be possible ; (c) eradication of affected plants and plant parts ; (d) eradication of weeds or other plants serving as alternate, alternative or collateral hosts harbouring pathogens or insects in the off-season and growing season ; and (e) tillage practices which will reduce the inoculum or lead to damage and destruction of resting stages of pathogens or eggs of insects.

(a) Destruction of crop residues

Destruction of crop residues, like dry leaves, sticks, stubbles, earheads or other plant parts results, to a large extent, in the elimination of sites of hibernation and shelters for insects in the off-season and carry over of the pathogens in the absence of the host. Quite often these plant parts are infected or infested with pests. It has been observed that leaf blight diseases of rice particularly one caused by *Heliminthosporium oryzae* is carried over in the stubbles and primary infection is evident in the self-sown tillers arising from these stubbles. Infection of *Sclerotium rolfsii* on jute is carried over in the foot and root regions in the stubbles left over after harvest of the jute plants. Sugar cane stubbles left over in the field help to carry over red rot fungus *Glomerella tucumanensis*, as would be evident from the infection of young plants in disease-free sugar cane fields receiving irrigation water through the harvested field of sugar cane with left over stubbles. Similar dissemination has been noticed in rice bacterial leaf blight pathogen, *Xanthomonas oryzae*, which is capable of surviving for some time in rice stubbles. Rice stem borer insects *Scirpophaga incertulas* and others are supposed to hibernate in the rice stubbles. Tree stumps left after felling or damage due to storm, etc., in plantations serve as sites of colonisation of pathogens which later infect healthy plants. To eliminate this eventuality, special treatment of tree stumps by surrounding them with a trench filled with lime has been suggested.

In many cases, diseased planting materials left in the field after discarding them, serve as sources of infection as in the case of late blight of potato where piles of refuses of rejected tubers later become an important source of infection. Left over plant parts of corn infected with smut *Ustilago zeae* constitute an important source of infection later.

The proper disposal of straw of *toria* (*Brassica napus*) early in spring results in the reduction in the infestation of painted bug. Pink boll worm of cotton *Pectinophora gossypiella* hibernates in the debris of cotton fields and their timely destruction reduces their population. Destruction of stubbles of sugar cane, corn and rice results in lesser incidence of stem and root borers in the subsequent season. Abandoned fields of egg plant, and many other plants are found to harbour borers, and scale insects which are often oligophagus. Numerous such examples can be cited which will point out the need and efficacy of these measures. It may be emphasised that where the climate is comparatively mild and temperatures normally do not attain too high or too low a level, weather permits cultivation of more than one crop and the pathogens have a wider host range and insect pests are oligophagus, the crop residues constitute important sources of infection and need to be eliminated.

(b) Use of eradicant sprays

In the floor of orchards there may be substantial number of infected leaves. Effective disposal of them is not feasible. They constitute sources of infection in the next season. In such cases the use of an eradicant spray has been found to be useful, as in the scab disease of apple, phenyl mercuric chloride (25 gms of mercury/400 litres of water) is sprayed on the fallen leaves to reduce infection of ascospores.

(c) Eradication of affected plant parts and plants

Eradication schemes have been undertaken in a number of cases to tackle the sudden appearance of a disease in an area. Elimination of citrus canker (*Xanthomomas citri*) from the U.S.A. after its introduction from Japan by removal and destruction of 20 million trees between 1914 and 1934 is an example of the success of an eradication scheme. Such total successes have however not been achieved in the eradication of the Dutch elm disease (*Ceratocystis ulmi*) in the U.S.A. and the fireblight of pear (*Erwinia amylovora*) from England.

Rogueing of infected plants in the early stages constitutes an important measure of control of potato viruses and the production of certified seed tubers. Mass eradication of downy mildew (*Sclerospora graminicola*) affected corn plants results in reduction of the spread of the diseases.

Physical removal of the entire diseasesd plant is not the only method of eradication. Removal of affected plant parts may also reduce the inoculum as in canker of apple caused by *Nectria* spp. and in similar cases including citrus canker. Removal of smutted inflorences constitutes an important method of control in whip smut of sugar cane (*Ustilago scitaminea*) and is also recommended in corn smut (*Ustilago maydis*). The removal of branches and twigs of mango infected with angiospermic partial parasite *Loranthus* gives good results. Systematic destruction of the affected plant or parts in the proper manner to keep down the population is resorted to reduce the damages caused by fruit flies infesting cucurbits, mango, guava, peach, etc., and many tissue borers of plants.

(d) Destruction of weeds, alternative, alternate or collateral hosts

It is well known that weeds or uneconomic unrelated plants harbour the pathogens and insects in the off-season when appropriate host plants are not available. They constitute important sources of infection. Alternate hosts are necessary for completion of the life cycle of a number of fungi. In many cases destruction of alternate hosts e.g. barberry in cereal stem rust (*Puccinia graminis* f. *tritici*) however has not reduced infection, because the pathogen has other means of perennation and spread. It may not be possible in all cases to get rid of such uneconomic plants growing as weeds and harbouring the pests, nevertheless in some cases good results may be obtained. Destruction of *Malva parviflora, Althea rosea, Malvastrum* spp. during April-June reduces carry over of spotted boll worm of cotton from one season to another. Eradication of *Sorghum haplense* keeps the population of sugar cane mite low. Similarly destruction of *Heliotropium supinum* in fields reduces the incidence of red hairy caterpillars.

Beet leaf hopper *Circulifer tenellus* has successfully been reduced in population with much less incidence of curly top by the destruction of weeds harbouring them and replacing the barren lands adjacent to the cultivated ones by suitable non-host perennial grasses. One of the recommended measures for control of sorghum midge (*Contarinia sorghicola*) is to destroy by burning or other suitable means the Johnson grass *Sorghum haplense* on which the midge overwinters. In India and Pakistan, wild grasses and hedges near the sugar cane fields serve as alternative or collateral hosts of various species of sugar cane borers. Mechanical removal of such hosts, particularly wild sorghum early in the season brings about effective reduction in incidence up to 50 per cent. In some tropical countries, legislative measures have been enforced for the destruction of alternative or collateral hosts during a specific period, often termed as "dry season". In Sudan and Lesser Antilles. no alternative or collateral hosts of cotton pests are allowed to grow during the dry season.

It has been reported that the influx of swarming caterpillars (*Spodoptera mauritia*) on rice fields with first heavy monsoons takes place from the graminaceous weeds in the bunds and nearby areas.

In many cases, weeds are perennial and too many weeds may be involved and it may not be possible to achieve any significant result, nevertheless a clean cultivation should be aimed at to reduce inoculum and insect population and keep down chances of infection.

(e) Tillage

Summer ploughing and upturning of the top layer of soil exposes the soil to the summer heat with the result that the fungi and insects in the soil are destroyed to a considerable extent. This is particularly stressed because minimum tillage is being advocated as a major agronomic practice in multiple cropping programme.

Use of clean planting material

Use of disease and pest-free planting material is also an important prerequisite for clean cultivations, as many pathogens, nematodes and some insect pests are carried over in the seeds and planting materials. In virus diseases, the most important practical measure is the use of virus-free planting material. The absence or presence of a very low level of initial inoculum is definitely helpful in delaying or suppressing the incidence of pests. This can be easily achieved by use of clean planting materials which may be considered as a major sanitation measure.

Crop rotation

Rotation of crops or change in sequence of cropping pattern is in use for a very long time. Apart from the advantages normally derived in greater yield, this system in many cases, results in much less incidence of pests. It is most useful against diseases caused by fungi and nematodes and damage against insects with restricted host range and limited power of dispersal. Crop rotation is essentially a preventive measure and has its effect mainly on the succeeding crop.

The main principle, as far control of pests are concerned is to disrupt the continuity in the readily available food supply or host plant with the result that the organisms will face starvation and there will be a consequential decline in population. Crop rotation is a very effective method of control of root diseases in field crops where other methods are not either readily available or cannot be adopted. In many cases a break for one year by a non-susceptible host may be sufficient, particularly where the pathogens are soil inhabitants. Incidence of potato tuber moth (*Gnorimoschema operculella*) can be effectively reduced by suspending the cultivation of potato crop in the affected areas for two to three years. In white ant infested lands, sugar cane, wheat or chillies which cannot be grown may be replaced by onion and tobacco.

Crop rotation has its own limitations. For adoption of crop rotation as a method of control of plant pests, knowledge of the life history with reference to host-range, mode of perennation, longevity of the resting structures, etc., is essential. The organisms should not be capable of remaining alive in the soil for a number of years and rotation should not include any susceptible crop. It cannot be practised where pathogens are typical soil inhabitants and can remain alive in the soil for a long time in the absence of any host or which have a very wide host range or the overwintering and oversummering resting bodies may remain alive for a number of years, e.g., resting spores of clubroot organism (*Plasmodiophora brassicae*, or wart (*Synchytrium endobioticum*). A six years' rotation in clubroot, with the absence of any cruciferous host including weeds is necessary to have the soil free from the pathogen. This method cannot be practised where vegetable cultivation is undertaken in a very intensive manner throughout the year. Rotation will not be of use in many vascular wilts caused by *Fusarium* spp. where pathogens can remain alive for a long time in the soil or *Rhizoctonia solani, Macrophomina phaseolina, Sclerotium rolfsii* which have a very wide host range. It may, however, be stated that apart from plant protection considerations, rotation as desired may not be adopted because of economic feasibility.

It has been claimed that in shifting or *jhoom* cultivation which is practised in the Northeastern hill areas in India particularly by the local people, population of pests and disease producing organisms is often at a low level. In such cases forest trees are felled down and the area is burnt and is put under crop for a number of years, then the area is put under afforestation programme. Ecological diversity due to diverse mixed cultures of crops with neighbouring forest plantations at different stages of growth acts as a check against any serious outbreak of pest.

Trap crops or secondary crops

In many cases an early cultivation in small area of a susceptible crop ahead of the main crop to draw the insects and to destroy them to reduce the damage is practised. For successful employment of this method, a crop often termed as a 'trap' or secondary crop must be highly susceptible to the pest attack and should be destroyed before the main crop. Early planting of a few rows of cotton has been advocated to reduce the population of overwintering bollworm (*Anthonomus*).

Instead of growing ahead, they may be grown along with main crop and must be destroyed in time. *Bhendi* (*Abelmoschus esculentus*) is often sown with 'cotton to attract cotton jassid and spotted bollworm and the plants should be destroyed before the insects migrate to the cotton. *Arhar* (*Cajanus indica*) may be used as a trap crop in mixed cropping with cotton to reduce attack of cotton grey weevil which shows preference to *arhar* in relation to cotton. Planting of wild beets with sugar beets is useful in reducing attack of beet nematole *Heterodera rostochiensis*.

In plant diseases, this method, is however, not useful, as in some cases it may aggravate the disease. In black stem rust of wheat, off season cultivation of wheat has been ascribed to provide for the inoculum. Growing of susceptible crops in between has been found to increase infection in air borne foliar diseases.

Adjustment in sowing or planting

Many plants are susceptible to attack of pests during a limited period in their life and serious attack will result if the population or inoculum build-up takes place during that period which is again conditioned by ecological factors. Adjustments in date of sowing may be profitably practised to circumvent attack of pest by avoiding the peak period of attack. To put this concept into practice, as in previous cases, the knowledge of the conditions favouring appearance of pest and vulnerable age of the plants at which they are susceptible to attack is essential.

This circumvention can be effected in relation to space also. Plants may be grown in areas, where vectors are either absent or not active, to avoid virus diseases for having disease free seeds or planting materials. This is taken advantage of in obtaining tubers or seed potatoes free from virus diseases in growing potato in aphid-free areas. Similarly seed multiplication in dry areas is often practised to have a low level of seed-borne infection by pathogenic fungi which are more active in humid areas.

Soil factors

Various factors pertaining to soil namely moisture, texture, pH and organic matter often influence the development of many soil-borne diseases.

The avoidance of such conditions as far as practicable has beneficial effect on the control of the disease. These measures may prove to be more useful as chemical control in such cases is not possible owing to : heavy cost ; crop rotation not being practicable ; and resistant varieties not being available.

Diseases normally favoured by high temperature are : cabbage yellow *Fusarium oxysporum* f. *conglutinans* ; flax wilt (*F. oxysporum* f. *lini*) ; and tomato wilt (*F. oxysporum* f. *lycopersici*) ; while low temperature favours attack of *Rhizoctonia solani* ; tobacco root rot (*Thielaviopsis basicola*), and onion smut (*Urocystis cepulae*).

Use of a cover crop of *Phaseolus aconitifolius* between rows of cotton has been found to be beneficial in reducing attack of root rot by *Rhizoctonia bataticola* (*Macrophomina phaseolina*) in the hot summer months in the Punjab, by lowering soil temperature.

High moisture content favours : club root of crucifer (*Plasmodiophora brassicae*) ; most diseases incited by *Pythium* spp. *Phytophthora* spp. and *Fusarium* spp. ; and violet root rot of of tea (*Sphaerostilbe repens*). Usually smuts are favoured by low moisture content. Potato scab (*Streptomyces scabies*) is also severe under conditions of low soil moisture. Most soil-borne diseases are not favoured by heavy soils and light soils particularly favour wilt caused by *Fusarium* spp. Diseases caused by *Fusarium* spp ; club root, of crucifers, powdery scab of potatoes (*Spongospora subterranea*) are favoured by acid soils, whereas bacterial diseases including potato scab, and wilts caused by *Pseudomonas solanacearum* are favoured by alkaline soils.

A large number of diseases may be kept under check by the application of organic matter, but there are a few diseases favoured by organic matter which include bunt caused by *Tilletia* spp., flag smut (*Urocystis agropyri*) of wheat, and root rot of tea caused by *Rosellinia* spp.

Water management of the soil also constitutes a very important cultural method, though most farmers may have little control over the amount of water which comes to the stand. Flooding of rice fields at the proper time has been found to have a beneficial effect in suppression of several species of rice borers. Similar measures may be taken against borers of sugar cane, namely, *Diatraea saccharalis,* and *Castnia licoides.* It has been stressed that too many irrigations in closely spread rice fields are responsible for the creation of a proper environment of epizootics of brown plant hopper (*Nilaparvata lugens*). Flood fallowing may be adopted to make the soil free from soil-borne infection in a few cases, e.g., organism (*F. oxysporum* f. *cubense*) causing Panama disease of banana.

Application of fertilisers particularly nitrogen, phosphorus and potassium has been claimed to have an effect on diseases and insect pests. It is known that plants receiving an adequate level of balanced nutrients have good growth. Vigorously growing and healthy plants generally can withstand the attack of pests. The application of high doses of nitrogenous fertilisers in general predisposes the plants to attack of diseases and insects. Such vegetative growth with much succulence in the aerial parts makes the plants more attractive to the attack of insect pests. There are some evidences to indicate that the application of potassium and phosphorus reduces the incidence of pests. In rice tungro virus, the application of high doses of nitrogen is recommended to mitigate the symptoms and effects of virus. In this context it may be worthwhile to note whether direct application of anhydrous ammonia into the soil as fertiliser

5

may have an effect on soil microflora in general and soil pathogens and insect pests in particular. Sometimes the application of fertilisers which helps in having a good stand may reduce the attack of some pests which colonise more in thin patches of the crop.

Planting distance

Maximisation of yield necessitates a definite number of plant population per unit area. Very often planting has to be done at a very close spacing for this purpose. This results in the dense crowding of the plants.' In fibre crops like jute (*Corchorus capsularis* and *C. olitorius*), the close spacing is taken recourse to discourage branching and encourage linear growth for better production of fibres. These agronomic practices often have repurcussions on the incidence of pests. Very close spacing of rice plants results in a more humid microclimatic condition which favours incidence of foliage blights and brown plant hopper. Spacing at a wider inetrval has been found to be beneficial for avoiding attack of pests. Early spread of black rot of cabbage (*Xanthomonas campestris*) which takes place by plant to plant contact may be checked by avoiding planting at close distance. In respect of insect attack it has been claimed that in some cases close spacing may be beneficial, as insects may favour to colonise patches of crop, whereas in many cases dense crowding may help in the easy spread of the pests. A balance has to be maintained between planting distance for maximisation of yield and the consequent effect on microclimatic conditions favouring pests.

In this context comparative merits of direct seeding and transplanting of rice in relation to incidence of diseases and pests deserves consideration. While direct seeding of rice has certain advantages, under such conditions attack due to diseases and insect pests is likely to be more severe.

Tillage practices

The depth of seeding sometimes affects seedling blight and damping off. Deep planting may cause delay in the emergence of seedlings, which may be vulnerable to pre-emergence damping off. Early emergence results in early lignification of tissues which become resistant to attack of soil-borne pathogens. On the other hand deep ploughing has been claimed to be beneficial for control of gram wilt (*Fusarium oxysporum f. ciceri arietini*).

Deep ploughing may bury insects too deep in the soil for emergence. Many insect pests, namely corn earworm (*Heliothis zea*), European corn borer (*Ostrinia nubilalis*), wheat stem sawfly (*Cephus cinctus*) have been reported to be controlled to a large extent in many countries by deep ploughing.

There should be encouragement of parasites and predators by suitable manipulation of cultural methods. Measures taken to keep down population of insect pests should not interfere adversely with parasites and predators, on the other hand, they should be helpful to beneficial insects which may be of value in the suppression of pests.

QUESTIONS FOR DISCUSSION

1. What is the aim of cultural methods of pest control ? Why is it considered important ?
2. What information is needed for the adoption of successful cultural control measures ?
3. What are the important operations under cultural methods of control ?
4. Why is field sanitation considered important ? What measures are undertaken in field sanitation ?
5. Why is the destruction of crop residues considered beneficial ?
6. Can you cite examples of usefulness of removal of affected plants or plant parts in pest control ?
7. What is the utility of crop rotation from the stand point of pest control ?
8. How can trap crops be useful in pest control ?
9. Why is adjustment in sowing or planting date recommended for the control of pests ?
0. How is water management useful for control of plant diseases ?

11. Why should a balance be maintained between planting distance and the consequent effect on microclimatic conditions favouring pests ?

12. How can tillage practices be useful in reducing the incidence of pests ?

CHAPTER 4

CHEMICAL CONTROL OF PLANT PESTS

Historically pest control is very ancient. Reference to blasting, mildew, locusts, caterpillars, etc., as causing damages to the crops is found in the Bible and were considered to be act of displeasure or punishment by God to evildoers. In Greece, farmers in ancient times eloquently sought the help of the gods to spare their fields from destruction. Romans to protect their wheat crop, presumably from rust, used to hold an annual festival—the Robigalia in honour of God Robigo or Robigus. Apart from worship of gods, other measures were adopted, but they had no rational basis, excepting mechanical methods of controlling insects in cases where the attack was obvious, selection of plants which looked apparently healthy and much less affected in a pest-affected field.

The history of crop protection is closely linked with scientific study of the organisms which have been designated as pests as well as epiphytotics. Earliest record of the use of fungicides, however, is by Homer, 3000 years ago, who referred to the pest-averting properties of sulphur, which is still used as the basis of some modern fungicides. Crop protection had its scientific footing based on the following discoveries : that of Anton de Bary in 1853, that fungi are capable of causing infection in plants and inciting parasitic diseases ; work of Louis Pasteur on bacteria ; enunciation of Koch's postulates (1881) ; existence of bacterial diseases in plants by Thomas Burrill (1879-81) ; and infectious nature of the mosaic disease of tobacco by Iwansoswki in 1882.

Use of pesticides in griculture in a technicab sense dates back to the middle of the nineteenth century and prominent events which may be regarded as milestones have been observed to coincide with the occurence of particular epiphytotics. The appearance of powdery mildew of grapevine in Europe (1845) was followed by the use of sulphur as a fungicide for the control of mildew (1848). The appearance of downy mildew in a very severe form in Southern France eventually led to the discovery of Bordeaux Mixture by Millardet in France (1882). The spread of Colorado beetle (1850-59) was sought to be checked by the introduction of an arsenical compound—Paris Green (1867) which was later replaced by lead arsenate previously used for control of gypsy moth (1889). For the control and eradication of scale insects that appeared in virulent form in California (1886), the tent method of fumigation with hydrocyanic acid was adopted.

Prior to World War II, most pesticides were inorganic chemicals which are simple in nature and a few insecticides of plant origin. The discovery of DDT in 1939 revolutionised the concept of chemical method of pest control. This was soon followed by the discovery of gamma-isomer of BHC. Schrader's discovery of organophosphorus, materials of which became available after World War II and chlorinated hydrocarbons by Diels-Alder reaction greatly stimulated chemical methods of pest control. Gradually other groups of insecticides were developed for commercial use and newer products are continually appearing in the market.

In the field of fungicides, apart from replacement of Bordeaux Mixture by fixed copper fungicides, thiocarbamates and quinone compounds which came into the market were found to be very effective for control of diseases in plants. They have been followed by pthalimides and other groups of chemicals. Examples may also be cited of the development of organomercuric compounds as seed treating chemicals following Hiltner's discovery that *Fusarium* disease of rye could be controlled by mercuric chloride, and first record of use of chlorophenol mercury in 1914 for the control of bunt.

It may be stated that the tremendous developments in the range of chemicals to be used as pesticides have made a definite impact on pest control. The rapidity and effectiveness with which the pests can be eliminated by the use of such chemicals have revolutionised the concept of pest control. It may also be pointed out that the chemical industry has been gradually attracted towards enterprise of development of new chemicals for the protection of crops. During World War II, two synthetic weed killers, namely, 2,4-D, MCPA were discovered, but they appeared in the market after the War. Following these compounds, a large number of diverse chemicals, often selective in action are available for use as herbicides. Other chemicals, commonly known as "third generation pesticides", which have been tried on a small-scale include chemosterilants, sexattractants, and juvenile hormones. Majority of these chemicals are yet to be exploited for a large-scale commercial use. In spite of limitations of cost, degradation, toxicological problems, interest in these substances is growing because they seem to have the answer for pollution and toxic hazards caused by the use of common pesticides.

Along with the development of chemical pesticides, there has been an almost simultaneous development of machineries for the application of pesticides. From conventional hand operated dusters and high volume sprayers, there has been gradual development to more sophisticated power operated application machiners operated by oil engines or with power take off from tractors to present day low volume or ultra low volume sprayers and aircraft for the application of pesticides.

The efficiency of a chemical to be used as a pesticide apart from its inherent toxicity against the target organism or organisms is determined by the manner in which it is applied ; spray on the foliage, or application on the surface of the seed or in the soil ; the stage of growth of the crop and the pest ; its physical form and associated diluents powder or liquid. The inherent toxicity or technological possibility of the pesticide can be fully exploited if the conditions are optimal or near optimal.

Pesticides can be effective only when they are applied in time. Hence survey and surveillance and warning systems need to be developed for advising the growers to take up measures at the appropriate time. It is also necessary that a schedule of pesticide application has to be developed keeping in view the biology of a pest, vulnerable stages in the host plants when they are susceptible to the attack of pests, and environmental conditions favouring easy multiplication and dispersal of pests.

With the developing chemical technology and progressive use of greater quantity of pesticides, various pesticides belonging to diverse chemical groups with varying structures are now available for pest control. The main requirements of a satisfactory application of pesticides are : (a) inherent toxicity and easy availability of active constituent for pest control ; (b) low phytotoxicity ; (c) low toxicity to man and animals ; (d) stability in storage as concentrate ; and (e) stability when diluted to prepare active spray strength.

Pesticides do not increase yields. Any increased yields recorded after the application of pesticides is due to control of losses which might have taken place due to attack of pest, had they not been controlled in time. Hence a basic difference in the use of fertilisers and pesticides has to be recognised. Fertilisers even when applied at a dose below the recommended one may show some effect in augmenting yield ; similarly variations in timings of the application, may not produce optimal results, but good response is likely to be noted. In the case of pesticides, the application at suboptimal doses or curtailing the number of applications may be entirely useless for the control of pests, and in fact may be responsible for the appearance of strains of pests which may be resistant to pesticides.

Plant protection chemicals are generally costly, but the benefits that are likely to be derived due to the control of pests and assured yield are greater in relation to the cost. The benefit-cost ratio varies from 1.5 to 16, even 100 (in case of seed treatment).

Toxicity of pesticides and their residues often pose serious problems. Before a chemical is released for use in the field, tolerance limits are to be fixed. In case of inherent toxic hazards, safety measures have to be taken. Besides wrong handling during preparation, use, storage and transport may result in danger to men and animals. The usefulness of chemicals also depends on the quality of the chemicals. To ensure this rigorous checking is necessary to prevent the sale of substandard articles. These aspects are looked after in India through legal provisions in the Insecticide Act (important provisions given in the appendix).

In the following chapters, an account of major chemicals, thier method of application and toxicity hazards are presented.

QUESTIONS FOR DISCUSSION

1. How is the history of plant protection closely linked with the study of the organism ?
2. What factors determine the efficiency of chemicals as pesticides ?
3. How is the efficacy of a pesticide influenced by the manner of application ?
4. What should be the minimum qualities of a pesticide ?
5. How are pesticides useful in agricultural production ?
6. Why safety precautions are to be used in the handling of pesticides ?

CHAPTER 5

INSECTICIDES

The main plant insecticides

Substance	Chemical Group	Originating Plant (organ, content)	Effect
1. Nicotine	Alkaloid	*Nicotiana tabacum N. rustica* (leaves 5-14%) and a further 9 alkaloids	Food, contact and respiratory poison. Fatal dose for man 60 mg. while 4 mg. produces serious illness
2. Anabasine	Alkaloid	*Anabasis aphylla* (leaves 1-2.6%)	Cf. nicotine
3. Piperine	Alkaloid	*Piper nigrum* (seeds) and other plants	*Synergist*, practically inocuous to vertebrates
4. Veratrine Alkaloids (effective as cevadine and veratridine)	Alkaloid	*Schoenocaulon officinale* (seeds 2-4%) *Veratrum album* (V. viride) (root)	Selective contact and food insecticide (ten times as effective as DDT), and also poisonous to man
5. Ryanodine	Alkaloid	*Ryania speciosa* (wood 0.16-0.2%)	Oral food poison (stomach poison) Selective effect, Contact poison. Very low toxicity to vertebrates
6. Wilfordine	Alkaloid (a mixture of 5 alkaloids)	*Tripterygium wilfordii* (root)	Selective food poison, for example for pests within stores; low toxicity to vertebrates
7. Quassin, Neoquassin Picrasmin	Diterpenoids Lactones	*Quassia amara Picrasma excelsa* (wood)	Selective action. No toxic effect on vertebrates

Substance	Chemical Group	Originating Plant (organ, content)	Effect
8. Sesamin	The crystallinefraction of sesame oil (0.25%)	*Sesamum indicum* (seeds)	*Synergist.* Very low toxicity to vertebrates
9. Rotenone (ellipton, sumatrol, malaccol, deguelin, a-toxicarol)	Rotenoids	*Derris (Deguelia) elliptica* (root)	Contact poison, food poison (fatal dose for man 2000-3000 mg. a person)
10. Pyrethrin I Pyrethrin II Cinerin I Cinerin II Jasmolin I Jasmolin II	Pyrethrins	*Chrysanthemum cinerariaefolium C.roseum, C.carneum* (flowers,0.7-3%)	Contact poison, practically non-toxic to man and domestic animals

Long before the development of synthetic organic insecticides, natural substances derived from plants were successfully employed in pest control. However at present the use of insecticides of vegetable origin is extremely limited and is on rapid decline. The main vegetable insecticides are given in the table below :

The best known vegetable insecticides have the most varied effects :

(a) pronounced deterrent or repellent effects of essential oils of some unbelliferous plants, or neem *(Melia azedarch)* :

(b) food, contact and respiratory poisons of alkaloids like nicotine, anabasine ;

(c) food and contact poisons containing rotenone (*Deris elliptica* from Indonesia, and *Lonchocarpus* species from South America) ;

(d) purely contact insecticides namely pyrethrins of chrysanthemum sp. ;

(e) highly active synergists e.g. piperine from seeds of *Piper nigrum, sesamin* from seeds of *Sesamum indicum.*

Of the various insecticides use of the alkaloids nicotine and anabasine is on rapid decline, because of their extreme toxicity, when used as a tobacco infusion. Synergists derived from pepper or sesame seeds etc, are of less importance than chemically related cheaper synthetic products like piperonyl butoxide. Insecticides with a certain limited and selective effect from Ryania, Quassia etc. are still used in specific instances, but they are not substances of international significance. Pyrethrum and derris preparations are, by far, the most commonly used vegetable insecticides and are important in their countries of origin.

A. Naturally occurring contact insecticides of plant origin

(1) Nicotine [l-3 (1-methyl-2-pyrrolidyl) pyridine]
That tobacco leaves possess property of killing insects probably become known sometime after their introduction into the European countries. Washes prepared by soaking tobacco leaves overnight in water were used in the eighteenth century in England. Commercial extracts were in the market by late nineteenth century. Posslet and Reimann in 1828 traced this property of tobacco leaves to the alkaloid nicotine, and its structure was determined in 1893. Synthesis of nicotine was achieved in 1904, but the main source is still the two species of genus *Nicotiana*, namely *tabacum* and *rustica*.

Nicotine which has a comparatively high vapour pressure (volatility mgHg. 0.0425 at 25°C) mainly acts in vapour phase. It can be used as dust, spray and fumigant. Nicotine vapour penetrates through the cuticle and gut wall of the insects and paralyses the nervous system by blocking the motor nerves in the ventral nerve cords. It is particularly effective against most soft bodied insects particularly aphids. Fifty years ago it was the only known aphicide.

Both commercial nicotine and its sulphate have excellent keeping properties. On ageing the colour depends to brown to browinsh-black without any deterioration in its toxicity.

Nicotine has virtually no phytotoxic effect. It leaves no toxic residue. Edible crops may be harvested two days after spraying—a much shorter interval than required by most other insecticides.

Formulations include dusts usually containing 3 percent nicotine with a mixture of dolomite (calcium and magnesium carbonate) which acts as chemical accelarator. A number of liquid preparations containing 95 percent nicotine, or nicotine sulphate with 40 percent nicotine are available. Because of the high rate of volatilisation, it can be used for fumigation in glasshouses.

In spite of its effectiveness use of nicotine is falling off rapidly because of high mammalian toxicity (50-60 mg/kg— acute oral does LD_{50} for rats—single dose) and ineffectiveness in cold weather. It is rapidly absorbed through the skin and splashes on the skin must be washed immediately.

(2) Derris (Rotenone)

Oxley in 1848 suggested tuba root (*Derris elliptica*), a fish poison, for the control of insect pests of nutmeg. Hooker in 1877 recorded its use as an insecticide by the Chinese in Singapore. The use of ground roots of certain species of *Derris* was patented in England in 1912. The active ingredients are a series of compounds known as rotenoids, of which rotenone constitutes the main insecticide. Apart from different species of *Derris*, roots of *Lonchocarpus, Tephrosia (Cracea)* spp. have been found to possess similar insecticidal properties.

The exact mode of the action of rotenoids is not yet known ; but the respiratory system and heart beat of insects are depressed and slowly paralysed resulting in death. Rotenoids are highly toxic to fish, but they are virtually nontoxic to warm blooded animals. Normally powdered roots are used in formulations though extracts are used in some. Rotenone is sparingly volatile. Dust fomulations contain 0.2-0.5 percent rotenone, whereas water dispersible and emulsified concentrates usually have 1-6 percent rotenone. It is active against a wide range of insects and leaves no toxic residue.

(3) *Pyrethrum*

The flower heads of certain species of *Pyrethrum* particularly *Chrysanthemum cinerariaefolium* have been used in a finely ground form as an insecticide for household purposes for more than a century. Jumtikoff, an Armenian discovered this effective powder of ground flower heads of *Chrysanthemum* spp. being used as an insecticide by tribes of the *Caucasus*. His son, in 1828 manufactured this power in a large-scale and in 1850 this product was introduced into France. The active ingredients are a group of compounds, commonly known as pyrethroids or pyrethrins which are mixed esters of pyrethrolone and cinerolone with chrysanthemic and pyrethric acids and are present in amount from 0.7 to 3 percent. Since these substances are never separated in commercial products, they are collectively known as pyrethrins.

The characteristic action of pyrethroids is the quickness with which the insects, particularly the housefly are paralysed ("knock down" effect). The rapidity of action indicates an immediate penetration and a very fast spread along the nerve membrane and an adverse effect on the nervous system.

The esters which are the main constituents are extremely unstable and rapidly lose their toxicity or potency during storage especially in sunlight. For this reason they are of little value for use in field crops. It has also been observed that insects are able to detoxify these compounds which are decomposed in the insect body. Effectiveness of pyrethrum can be improved upon by the addition of a number of compounds which act as synergists by intervening in the process of detoxification. For use as dust, ground flower heads are mixed with a nonalkaline carrier or diluent; and for spray or aerosol in oil emulsion, solubility in water being negligible. A synergist is usually added often with DDT to increase effectiveness. Because of the negligible vapour pressure pyrethrum cannot be used as a fumigant.

Pyrethrins have one particular important feature namely, insects have not shown any resistance towards it. Pyrethrum and its synergists have low mammalian toxicity.

A number of synthetic relatives of pyrethrins have been prepared and tested. Of them allethrin, dimethrin and barthrin are of promise because of their stability, effectiveness and low toxicity to warmblooded animals.

B. Hydrocarbon oils

Distillates from petroleum often known as mineral oils have been in use as insecticide since the days when kerosene was introduced for illumination purposes. These hydrocarbon oils are particularly useful in the control of scale insects which are not easily amenable to treatment by other insecticides. It has been found that hydrocarbon oils should be of a minimum boiling range or viscosity and should be applied as emulsions of great stability without affecting their insecticidal efficiency. In dormant trees, success of tar oil washes is dependent on preponderance of aromatic hydrocarbons particularly when when used as ovicides. Hydrocarbon content of the oils is of importance in determining which insects are to be controlled. Some insects are more susceptible to oils with high contents of aliphatic hydrocarbons, while others are more easily controlled by aromatic constituents. Hydrocarbon oils are generally applied as winter washes and they are normally phytotoxic to foliage, unless special formulations are used. Insecticidal efficiency of hydrocarbon oils can be enhanced by dinitro-O-cresol (DNOC). These oils are widely used as winter and sometimes summer washes in orchards in temperate countries and control of scale insects in *Citrus* and other plants.

C. Inorganic salts

With the successful introduction of synthetic organic compounds, which are highly toxic to insects by contact and ingestion, the popularity of inorganic salts which are stomach poisons has diminished. Nevertheless some arsenic compounds, e.g. Paris Green (copper arsenate), lead arsenate, calcium arsenate are still in use. They are mainly employed for the control of slugs and soil insects being incorporated in the baits of dried blood or ground offal or rice bran with molasses. These compounds are strong poisons to warm blooded animals. Hence precautions are to be taken in keeping away the domestic animals from the bait. These insecticides must not come in contact with injured skins or wounds.

D. Chlorinated hydrocarbons

The chlorinated hydrocarbons include substances varying in their chemical structure, but because of their common properties like high insecticidal activity and chemical and biological persistence, they are grouped into one.

All organochlorine insecticides are poorly soluble in water and well soluble in organic solvents, including fats. Many of them are quite volatile. They are thermally and chemically stable substances withstanding the action of various factors of the environment namely temperature, solar radiation, moisture etc. This underlies the prolonged protective action of these substances against pests, but at the same times creates danger of contaminating the environment and agricultural products. Organochlorine compounds incorporated into the soil in large doses may inhibit nitrification processes for 1-8 weeks and suppress the general microbiological activity for a short period. They do not however affect substantially the properties of the soil. All organochlorine compounds usually do not have any adverse effect on growth of plants when applied in recommended doses.

The representatives of this group are mainly contact insecticides with prolonged residual effects and a broad spectrum of action. Organochlorine insecticides upon entering on insect body act upon the nervous system, disturbing, as is assumed the lipid equilibrium of the nerve cell membranes and preventing the transmission of nervous impulses. Insects perish as a result of malfunctioning of their nervous system.

The metabolism of chlorinated hydrocarbon derivatives in an insect body and other living beings occurs in three main directions:

(1) dehydrochlorination with the detachment of one or more HCl molecules and formation of products of low toxicity,

(2) oxidation to epoxides and other no less toxic metabolites,

(3) hydrolysis to water soluble substances that are easily excreted by the organism.

Most substances of this group are moderately toxic to men and animals, only a few are highly toxic.

(1) DDT

The name DDT stands for an abbreviation for dichloro diphenyl trichloroethane which is more accurately described chemically as to p2-bis (p-chlorophenyl) 1,1,1-trichloroethane. It was discovered in 1939 and patented by the Swiss Firm of J.R. Geigy A.G. for insecticidal purpose in 1942. The insecticide was first used by the Allied troops in 1942 in World War II and in 1945 the production figure reached 14 million kg, because it was the first protective insecticide to be discovered. Since then its use has increased gradually and now it is the most used insecticide. Because of environmental pollution and toxic hazards to human health, some countries such as Sweden, Great Britain, Canada and the U.S.A. have banned the use of DDT. The World Health Organisation, however, is in disfavour of the total ban on DDT, as no satisfactory substitute is yet known particularly for the control of insect vectors of human diseases in the tropics.

The technical product is a mixture of compounds and may contain up to 30 per cent OP' isomer. The universally accepted and WHO approved specification calls for at least 70 per cent PP' isomer, which has more potency as an insecticide.

DDT is both stomach as well as contact insecticide. It has no action as an ovicide and little effect on pupal stages. Nevertheless its persistence is sufficient for the newly hatched larvae and emerging adults to acquire a lethal dose.

DDT may be termed as a broad spectrum insecticide, as it has its effect on a large number of species of insects. Contact and stomach insecticide highly persistent on solid surface with little action on phytophagous mites.

It has been mainly used against biting and chewing insect pests, domestic insects and mosquitoes. Many species have developed marked resistance towards DDT. There is also wide variation in susceptibility among the various groups of insects, even where resistance has not been acquired.

DDT has been valued as a protective pesticide because of its low volatility, water solubility and good stability in solution. In aqueous solution, it is readily decomposed by alkalies with loss of hydrochloric acid and toxicity against insects. Similar detoxification of DDT by its dechlorination by an enzyme is considered to take place in insects which have acquired resistance against DDT as a result of continuous use.

Although a considerable quantity is being used annually for more than forty years and DDT is a very widely known insecticide, the mechanism of action is not fully known. It is generally agreed that it is a nerve poison and causes violent tremors by the multiplicative effect on a nerve impulse, as a result nervous and muscular activities become uncoordinated.

DDT is non-phytotoxic except on cucurbits and many varieties of barley. DDT is toxic to warm blooded animals in the sense it is not degraded in the body and may persist in the body fat. Acute oral LD_{50} for rats is 113-118 mg/kg.

DDT is used for agricultural purposes as a dust with 1-10 per cent content, usual being 5 per cent. Water dispersible formulations range from 20-50 per cent usual being 50 per cent. Emulsified concentrates with varying contents of DDT are also used for aerosol. Heat stability of DDT permits it to be used as smoke either by itself or in combination with gamma-BHC.

Of the different cyclopropane relatives of DDT, two closely related to DDT are DDD or TDE (tetrachlorodiphenylethane) 1,1, dichloro, 2-2-bis (p-methoxyphenylethane) have shown promise as contact insecticides, but they are effective only against a narrow range of insects. Both these compounds have low mammalian toxicity and low level of accumulation in the fatty tissues and milk of mammals.

(2) Gamma-BHC

Benzene hexachloride was first put into use as an insecticide in the early 1940's. Crude preparation of BHC contains several isomers of hexachlorocyclohexane—a chlorinated hydrocarbon of which the gamma isomer is the most active as insecticide. BHC, an abbreviation of benzene hexachloride is a misleading-term as the carbon atoms of cyclohexane unlike benzene do not lie in one plane. Non-planarity of carbon ring makes possible existence of eight isomers.

Gamma-BHC commonly known as lindane contains not less than ninety-nine per cent of this compound. It is expensive when used as lindane. Crude BHC has an offensive musty odour. When used on vegetables, root and tuber crops, and fruits it imparts an unpleasant taint, consequently its use on them is limited.

Like, DDT, the mode of action of gamma-BHC is not known. It is capable of rapid penetration through the cuticle and acts as a poison on the nervous system, namely tremor, ataxia, convulsion and prostration. Nevertheless some striking differences exist which suggest that the mode of action is not identical with that of DDT.

It is persistent stomach poison and contact insecticide with fumigant action against gundhi, hispa, caterpillar, caseworm, whorl maggot, brown plant hopper of rice, stemborer, cutworms, armyworms, thrips, aphids, grasshoppers of wheat, jowar, maize, ragi, beetle, grey weevils of soyabean and fruit crops, beetles, sphinx moth, leaf caterpillars of pulses. Also used in baits as a rodenticide. Used as a foliar spray or dust, soil application and seed treatment with fungicides.

Gamma-BHC is practically insoluble in water and of low volability. It is toxic against a wide range of insects and is sufficiently insecticidal to act in a pseudo systemic manner on insects within the plant tissue adjacent to the deposit. In solution, it is dechlorinated even by mild alkali. It is particularly effective against ectoparasites and soil insects. It is known to persist in the soil for a long period. It is compatible with a number of insecticides including DDT. Gamma-BHC has an appreciable vapour pressure. It is stable to heat and can be volatilised, a property which has been utilised for the preparation of smoke generators.

Dust formulations include 0.20-0.65 per cent Gamma-BHC while emulsified concentrates range from 12-20 per cent. Gamma-BHC smoke generators are prepared in small pellets with a pyrotechnic mixture. Gamma-BHC is also incorporated in seed dressings in 20-40 per cent concentrations.

It is not phytotoxic, though damages to curcurbits particularly in the early stages of growth have been reported. An interval of two weeks should normally elapse between application and harvest and permitting the access of livestock or poultry to the treated area. Acute oral LD_{50} for rats is 88-91 mg/kg, acute dermal LD_{50} for rats is 900-1000 mg/kg.

(3) Cyclodiene insecticide

With hexachlorocyclopentadiene as the starting point, a series of hydrocarbons are prepared by Diels-Alder reaction, a well known process in Organic Chemistry. Of them the following have been utilised as insecticides for commercial purposes.

(a) Aldrin

It is a common name given to a product containing at least 95 per cent of hexachlorodimethanonaphthalene (1, 2, 3, 4, 10, 10-hexachlori-1, 4, 4a, 5, 8, 8a-hexahydro-exo-1, 4-endo-5, 8-dimethanonaphthalene). It is an effective stomach poison and contact insecticide and particularly effective against soil insects. It has low solubility in water, is somewhat volatile (volatility mmHg 2.31×10^{-5} at 20°C). By itself it is not persistent, but by biological oxidation, it is epoxidised to dieldrin which is very persistent. It is compatible with almost all other insecticides. Certain minerals used as diluents induce catalytic dechlorination and consequent detoxification which may be inhibited by the addition of specific inhibitor.

Aldrin may be used as dust or spray or may be incorporated in seed dressings. Dust formulations usually contain $1\frac{1}{2}$-2 per cent aldrin and emulsified concentrate for spray preparation 30 per cent. In seed dressings, normally 30 per cent concentration of aldrin is used. It is effective against soil insects at the rates of 0.5 to 5 kg a.i. per hectare.

Aldrin has comparatively high mammalian toxicity (67 mg/kg for rats—oral administration single dose). Because of its conversion to persistent dieldrin and uncertainty of effects of small quantities in animal body, uses of aldrin has been recommended to be limited to specific doses of active ingredient both as dust and spray—quantity of active ingredient as dust depending on the crop and pest. In the United Kingdom under the Pesticides Safety Precaution Scheme its use has been limited due to its conversion to dieldrin and uncertainty of long term effects of its persistence in the body.

(b) *Dieldrin*

It is the common name of a product containing at least 85 per cent HEOD (1, 2, 3, 4, 10, 10-hexachloro-6, 7-epoxy-1, 4, 4a, 5, 6, 7, 8, 8a-octahydro-exo-1, 4-endo 5-8 dimethano-naphthalene). It is an epoxide of aldrin. The product is of high chemical stability and compatible with all other pesticides. It is persistent, nonsystemic. Dieldrin is widely used by WHO for control of mosquitoes, certain strains of which have rapidly built up resistance. It is non-phytotoxic.

Water dispersible formulations usually contain 50 per cent dieldrin and emulsified concentrate 15 per cent. It is incorporated in seed dressings the content varying from 40-80 per cent depending on the material in which it is incorporated. As with aldrin, its usage is limited in the U.K. under Pesticides Safety Precaution Scheme. In India, it is permitted to be used only for locust control. The acute oral LD_{50} for rats is 46 mg/kg, and acute dermal LD_{50} for rats is 10-120 mg/kg.

(c) *Endrin*

This insecticide is an isomer of dieldrin. It is mainly used on foliage as spray. Mammalian toxicity (17.5 mg/kg for rats—oral single dose) is very high. Hence it is recommended for limited use in food crops. It is sold in the liquid formulation with 20 per cent emulsified concentration. Like dieldrin it is a very potent insecticide with persistent effect. It is compatible with many other insecticides. Its use is at present banned in India. It is non-systemic, persistent. The acute oral LD_{50} for rats is 7.5-17.5 mg/kg and acute dermal LD_{50} for rats is 15 mg/kg.

(d) *Endosulfan*

This cyclodiene derivate, though included in chlorinated hydrocarbons contains a sulphite grouping. The active component is 1, 2, 3, 4, 7, 7-hexachlorobicyclo (2, 2, 1)-2-heptan-5-6-bis (hydroxymethylene) sulphite. The technical product contains ninety per cent endosulfan which is a mixture of two isomers with different boiling points, but of same insecticidal ability. It is highly persistent, practically insoluble in water, non-volatile and stable, but its use is being recommended for restriction both in application and dosage because of high mammalian toxicity (40-50 mg/kg for rats—single oral dose). It is used as spray, usual formulation being 20 per cent emulsified concentration. Because of high toxicity, at least six weeks should elapse between its last application and harvesting of food crops and livestock and poultry should be kept away from sprayed fields at least for three weeks. It is toxic to fish. Acute oral LD_{50} for rats is 80-110 mg/kg. Contact and stomach insecticide, controls aphids, thrips, beetles, foliar feeding larvae, mites, borers, cutworms, bollworms, whiteflies, leafhoppers, slugs, termites, also control certain mites.

Other products which have been of interest are chlordane, heptachlor, etc. Chlordane has low mammalian toxicity, but insects easily build resistance to it.

(e) Chlordane

Technical chlordane is the name given to a product containing not less than 60 per cent 1, 2, 4, 5, 6, 7, 8, 8-octachloro-4, 7-methano-3a, 4,7,7a-tetra hydroindane with the remainder. It occurs as two structural isomers, the cis and trans forms which cannot be separated easily. Trans form is ten times as toxic as cis-form. The technical chlordane is a viscous liquid, which on refinement becomes pale amber with an aromatic odour. It is miscible with polar solvents including aliphatic and aromatic hydrocarbons, alcohols, ethers. It is insoluble in water and stable in acid. It is readily detoxified by alkaline dehydrochlorination, hence cannot be stored in galvanised containers. It is compatible with dinitrocompounds, arsenicals, other chlorinated hydrocarbons, dithiocarbamates and sulphur. It acts both as stomach and contact poison. It is non-systemic and persistent. It has a certain amount of fumigant activity. It is active against a wide range of insects and is particularly effective against termites and soil insects. It should not be used on leafy vegetables, as it tends to impart a taint. It has low mammalian toxicity. It is marketed as: (i) dust with 5-10 per cent active ingredient, (ii) 40-50 per cent wettable powder, (iii) 40-75 per cent emulsified concentrate, or (iv) 20 per cent oil solution for use with petroleum oil derivatives. Acute oral LD_{50} for rats is 457-590 mg/kg. Its use is restricted in India.

(f) Heptachlor

Isolated from technical chlordane with empirical from C_{10} $H_{15}Cl_7$ 1, 4, 5, 6, 7, 8, 8-heptachlori-4, 7 methanoindene. Commercial preparations contain 70 per cent heptachlor and not more than 30 per cent of related compounds. It is residual type fast acting insecticide which acts both as contact and stomach poison. It is considered four to five times more effective than chlordane. It is used in a number of ways including seed treatment, soil treatment. It leaves no toxic residue.

Acute oral LD_{50} for rats is 100-162 mg/kg.

It is effective against a wide range of insects occurring as pests on foliage or diverse agricultural and horticultural crops. It is compatible with most insecticides and fungicides. It leaves no taint on foliage unlike chlordane. It is marketed as: (i) dust with $2\frac{1}{2}$-5 per cent active ingredient, (ii) 25 per cent wettable power, (iii) 40 per cent emulsified concentrate, and (iv) 30-35 per cent oil solution.

Like other chlorinated hydrocarbons namely DDT and gamma-BHC, the exact mode of action of cylodiene insecticides is not known. As with DDT and gamma-BHC, these insecticides can easily penetrate through the cuticle and have lipid solubility. They act on the nervous system. Symptoms produced on insects have a close resemblance to those produced by gamma-BHC.

(g) Chlorinated terpenes

Toxaphene (Campheclor)—technical product of chlorinated camphene containing 67-69 per cent chlorine, with empirical formula $C_{10}H_{10}Cl_8$. It is primarily a stomach poison, and slower in action than gamma-BHC and DDT. It is active against a wide range of insects. One of the advantages is its harmlessness to pollinating bees. It is four times more toxic to mammals than DDT. Acute oral LD_{50} for rats is 80-90 mg/kg. A time interval of four to five weeks should elapse between last application and harvest. It is compatible with most common insecticides and fungicides. It is not normally phytotoxic except to cucurbits. Toxaphene is marketed as: (i) 10-20 per cent dusts, (ii) 40-50 per cent wettable, power, (iii) 25-60 per cent emulsified concentrate, and (iv) 80 per cent oil solution.

A product similar to toxaphene was introduced in 1951 as "strobane".

E. Organo-phosphorus insecticides

Organo phosphorus compounds are one of the most important groups of modern pesticides. The widespread use of these compounds is due to (i) high insecticidal and acaricidal activity, (ii) broad spectrum, (iii) rapidity (high initial toxicity) of action on pests, (iv) low stability in biological media, (v) decomposition with the formation of products nontoxic to men and animal, (vi) absence of an ability to be deposited inside the the body of men, animals or plants, (vii) systemic action of a number of these compounds, (viii) low rate of use per unit area treated, and (ix) rapid decomposition in soil and water. Prominent harmful features of these compounds are high toxicity to men and animals and relatively rapid appearance of resistant population of pests after long usage of these compounds. Specific resistance to organophosphorus compounds appears more rapidly in the species of insects that produce several generations in a single season. Resistance is known to have appeared in aphids, houseflies, cockroaches, mosquitoes and some other species of insects.

Modern organophosphorus insecticides and acaricides are compounds of pentavalent phosphorus of the general formula.

$$\begin{array}{c} O \ (or \ S) \\ R_1 \diagdown \ \| \\ P \longrightarrow X \\ R_2 \diagup \end{array}$$

R_1 and R_2 are alkyl, alkoxyl, or aryl radicals combined with alkoxyl ones, or dimethylamine groups and X is a residue of weak acid (central P atom).

The bond of X with phosphorus acts as a phospholyrating agent and phospholyrates vitally important substrate, namely, the active site of acetyl cholineesterase, an enzyme which plays a very vital role in the transmission of nerve impulse. The phospholyrated enzyme is irreversibly inhibited and is no longer able to carry on normal function of rapid removal and destruction of neurochrome (acetylcholine).

The fundamental structural element of an animal nervous system is a nerve cell (neuron) whose designation is to receive, interpret, and transmit information in the form of nerve signals (impulse). The impulse in the form of an original electrical signal travels along a neuron to the nerve ending of another cell or to a muscle fibre. The ending of a nerve and membrane of another cell or muscle are separated by a synaptic cleft of 30 to 50 nm wide. The cleft is filled with a gel-like substance with a tremendous electrical capacitance, therefore an electrical signal cannot pass through it. Hence a nerve impulse has to be transmitted through synaptic cleft with the aid of chemical substance (mediators or transmitters) acetylcholine. The process of synaptic transmission is an involved biochemical cycle or acetylcholine exchange. Acetylcholine esterase has an exceedingly important role in this cycle, because the inhibition of activity of acetylcholine esterase leads to the accumulation of free acetylcholine in the synaptic cleft. As a result the normal passing of nerve impulse is disrupted, convulsive activity of the muscle sets in that transforms into paralysis, and other features of self-poisoning of the organism by surplus acetylcholine appear.

The toxic action or biological activity of organophosphorus compounds on insect is due to inhibition of the activity of acetylcholine esterase in the synapses of the nervous system. Upon penetrating into an organism through the integuments of an insect or through the alimentary or the respiratory tracts, the insecticide reaches the nervous system. The activity of the acetylcholine esterase system rapidly drops. Indications of poisoning appear very rapidly and expresses themselves in hypersenstivity, hyperactivation of insects and tremors of the appendages followed by convulsions. Next paralysis sets in ending in death. Most organophosphorus insecticides and acaricides have a high initial toxicity and pests perish during the first few hours after application of the insecticide.

Organophosphorus compounds are effective for controlling larvae and adult, but ovicidal action is weak, as it cannot penetrate egg shells. Oil solutions of some organophosphorus insecticides can enter inside the eggs and cause death.

The biological activity of the organophosphorus compounds is due to the capacity of the central P-atom to phospholyrate the active site of the enzyme cholinesterase an essential constituent of the nervous system, not only of insects, but also of higher animals. The phospholyrated enzyme is irreversibly inhibited and is therefore no longer able to carry out its normal functions of rapid removal and destruction of neurochrome (acetylcholine) from the nervous synapse. This results in accumulation of acetylcholine with consequent disruption of the normal functioning of the nervous system giving rise to typical cholinergic symptoms associated in insects with organophosphorus poisoning like hypersensitivity, hyperactivity, tremor, convulsions, paralysis and death.

Organophosphorus compounds are often acutely toxic to many forms of animal life, e.g. parathion and related compounds. However it is also possible to develop such organophosphorus compounds such as malathion which has a very low degree of toxicity to mammals.

It is customary to employ the term organophosphate as a generic term to include all the toxic organic compounds containing phosphorus. Normally the nomenclature is based on consideration of the atoms attached to phosphorus. When the compound contains (I) it is called a phosphate, with (II) it is a phosphonate with (III) it is a phosphorothionate, with (IV) it is a phosphorothiolate, and with (V) it is a phosphorodithiolate, and with (VI) it is a phosphoramidate.

Most organophosphates are regarded as esters of alcohols with a phosphorus acid or as anhydrides of phosphorus acid with some other acid. Compounds of this category with insecticidal action may be grouped as follows:

(a) Derivatives of phosphoric acid e.g. DDVP, phosphamidon,

(b) Derivatives of thiophosphoric acid e.g. parathion, demeton.

(c) Derivatives of dithiophosphoric acid e.g. malathion,

(d) Derivatives of pyrophosphoric acid e.g. schardan, TEPP

(e) Derivatives of phosphonic acid e.g. EPN

(f) Others_____e.g. isopestox

A. Important derivatives of phosphoric acid are dichlorovos, phosdrin, phosphamidon, chlorfenvinphos, demefox, dithiolane.

$$\overset{O}{\underset{\parallel}{}}$$

1. **Dichlorovos (DDVP, Nuvan)** $(CH_3O)_2$ $POCH=CCl_2$; 0,0-dimethyl 0-2, 2-dichlorovinyl phosphate. Dichlorovos is a colourless liquid, highly soluble in most of the organic solvents, and the solubility is about 1% in water. In the presence of traces of moisture, dichlorovos on standing, breaks down with the formation of acidic products which further catalyse the decomposition. To counteract this rapid decomposition and stabilise the technical material 2 to 4% epichlorohydrin is added to tie up the acidic substances and improve conditions of storage. Dichlorovos is hydrolysed to dimethylphosphoric acid which is harmless and dichloroacetaldehyde which rapidly decomposes and evaporates. Hence on application on plants no toxic residue is left.

Dichlorovos is a contact and stomach insecticide with fumigant and penetrant action. It is used as a household and public health fumigant, especially against diptera and mosquitoes, for the protection of stored products at 0.5-1.09 a.i./100 m³ for crop protection against sucking and chewing insects at 300-1000 g a.i./ha. It is non-phytotoxic and non-systemic.

The acute oral LD_{50} for rats is 56-108 mg/kg; the acute dermal LD_{50} for rats is 75-210 mg/kg.

$$\overset{O}{\underset{\parallel}{}} \quad \overset{CH_3}{\underset{\mid}{}}$$

2. **Phosdrin (Mevinphos)** $(CH_3O)_2$ P O C=CHCOOCH₃; 0, 0-dimethyl-0-(1-methyl-2-carbomethoxyvinyl) phosphate.

Phosdrin is a colourless liquid, slightly soluble in petroleum ether, but highly soluble in water, acetone and benzene, resistant to hydrolysis in neutral medium, but in alkaline medium highly hydrolysable. Usually phosdrin breaks down in plants in 0.8 to 4.2 days. It exists in the form of 'cis' and 'trans' isomers of which 'trans' isomers are more stable, but 'cis' is 100 times more powerful than 'trans' isomers. Technical grade product contains 60 per cent 'cis' and 40 per cent 'trans' isomers. Phosdrin is toxic to warmblooded animals. It is a good insecticide with short term action for sucking and chewing insects. It is usually marketed in the form of 50 per cent emulsion concentrate.

It is a contact and systemic insecticide and acaricide with short-residual activity. The technical product (60 per cent E isomer) is effective against sap feeding insects at 125-250 g/ha, mites and beetles at 200-300 g/ha, caterpillars 250-500 g/ha. It is nonphytotoxic. The acute oral LD_{50} for rats is 3-12 mg tech/ kg. It is not approved for registration in India.

3. **Phosphamidon (Dimecron)** 0-0-dimethyl-0 (2-chloro-N, N-dimethyl-carbamyl)-methylvinylphosphate $(CH_3O)_2$ P O C= CClCON $(C_2H_5)_2$.

$$(CH_3O)_2 \overset{O}{\overset{\|}{P}} O \overset{CH_3}{\overset{|}{C}}=$$

Pure phosphamidon is a colourless to pale yellow liquid. The commercial formulation which is usually 85% w/w is bright violet in colour due to the presence of a dye. It is highly soluble in water, alcohol, acetone. It is stable in neutral and weakly acidic aqueous solutions, but is rapidly hydrolysed in alkaline medium. It is compatible with most pesticides except alkaline ones like Bordeaux mixture, lime sulphur, nicotine sulphate and copper oxychloride based fungicides. Being soluble in water, it is suitable for low volume and ultra low volume sprays.

Phosphamidon is systemic insecticide. It gets absorbed into the plant tissue within one to three hours and is translocated more towards the top. The active ingredients get degraded into less stable metabolities having low mammalian toxicity. Insecticidal property is reduced when phosphamidon is mixed with copper oxychloride fungicides. It is nonphytotoxic for all practical purposes. It is however not advisable to spray, like any other systemic organophosphorus insecticides when there is lack of adequate moisture in the soil, as marginal burn or necrosis in the leaves may result due to accumulation of the active ingredient along the margin. Foliar or surface application will not be effective against insects like cutworms. In spite of its toxicity for mammals it is used for spraying against sucking insects. Acute oral LD_{50} for rats is 17-30 mg/kg.

4. **Dicrotophos (Bidrin)** (3-dimethoxphosphinyloxy)-N-N dimethyl crotonamide with dimethyl phosphate of 3-hydroxy N-

$$N\text{-dimethyl cis-crotonamide}; \quad (CH_3O)_2 \overset{\overset{O}{\|}}{P}\text{-O-C}\overset{\overset{CH_3}{|}}{=}CHCON(CH_3)_2.$$

Dicrotophos is a persistent systemic vinyl phosphate insecticide.

Bidrin is the name applied to pure cis-isomer of the compound. It is yellow to brown liquid, miscible with water, and various organic solvents. The insecticide is highly toxic to human beings and animals and may be absorbed by inhalation or through unbroken skin. Its mechanism of action is like other organophosphorus insecticides by inhibiting the enzyme cholinesterase.

Dicrotophos is a systemic insecticide and acaricide of moderate persistence. It is effective against sucking, boring and chewing pests at doses of 300-600 g a.i./ha and is recommended for use on rice, cotton, coffee and other crops. It is nonphytotoxic except to certain varieties of fruit under some conditions.

The acute oral LD_{50} for rats is 12.8-30 mg/kg and acute dermal LD_{50} is for rats 141-181 mg/kg. It is not approved for registration in India.

5. **Chlorfenvinphos (Birlane, Sapecron)**
2-chloro-1-(2,4-dichlorophenyl) vinyl-diethyl phosphate)

$$(C_2H_5O)_2 \overset{\overset{O}{\|}}{P}OC = CHCl$$

Chlorfenvinphos in a moderately persistent contact and stomach insecticide. It is sparingly soluble in water, but miscible with a number of organic solvents, due to relative stability in alkaline and acidic conditions. It has relatively low vapour pressure in comparison to other organophosphorus compounds and extremely stable to thermal decomposition and relatively stable in water. It breaks down in soil and plants, and the breakdown products are more persistent than parent compounds. Chlorfenvinphos is used in the form of emulsions, wettable powders and granules.

Chlorfenvinphos is effective for use in soil to control rootflies, root worms and cutworms 2-4 kg a.i./ha. As a foliage insecticide, it is recommended for the control of Colorado beetle on potato and scale insects on citrus at 200-400 g/ha and for stemborers on maize, sugarcane and rice at 550-2200 g/ha. It is nonphytotoxic at these concentrations; 50 per cent loss from soil normally occurs in a few weeks.

The acute oral LD_{50} for rats is 10-39 mg/kg.

6. Monocrotophos (nuvacron, corophos)

3-hydroxyl-N-methyl-ciscrotonamide dimethyl phosphate.

Monocrotophos is highly effective organophosphorus compound which provides both systemic and contact control and is an active foliage insecticide. Besides it works also as acaricide.

It is water soluble and is available in the form of water soluble concentrate containing 36% of the active ingredient. It is absorbed inside the plant through both young and mature leaves within 12 hours of application. It provides excellent control of chewing, boring and sucking insects and at the same time acts also as a contact one. It is effective against mites.

Monocrotophos is nonphytotoxic when used in the recommended doses. Symptoms of necrosis, corkiness, in apple and scorching of leaves of sorghum have been reported. It is not compatible with alkaline pesticides.

The acute oral LD_{50} for rats is 14-23 mg/kg.

B. Derivatives of thiophosphoric acid:

1. Parathion (Thiophos) or Ethylparathion
   0,0-diethyl-0-4-nitrophenyl thiophosphate

$$(C_2H_5)_2 \; P \overset{\overset{\displaystyle S}{\|}}{-} O -\!\!\!\left\langle\!\!\!\bigcirc\!\!\!\right\rangle\!\!- NO_2$$

One of the earliest known organophosphorus insecticide in use. Insecticidal properties of this compound was first discovered by Schrader in 1944. It is slightly soluble in water, sufficiently volatile to exert a pseudosystemic action in insects, moderately persistent, being stable for a short time at higher temperatures. It is an excellent contact insecticide and is rapid in its action. It has some nematicidal properties, but as acaricide its performance is poor. It is not compatible with alkaline insecticides, as it is degraded by alkali. It is widely used in the tropical countries for effective control of a number of insects, in spite of its high mammalian toxicity (LD_{50} for rats—oral 6-4 mg/kg). Because of high toxicity, safety precautions are to be taken during its application. The compound being somewhat relatively stable, spray and dust residues remain effective for several days, but the residues on plants disappear sufficiently to leave no detectable parathion after 30 days. Hence there should at least be a gap of four weeks between the last application and the use of food crops when used as spray, two days when used as aerosol and 24 hours when used as smoke. It causes little phytocidal damage. Formulations include 15, 25, 50 per cent wettable powders, 25 and 50 per cent emulsified concentrates, dusts, aerosoles. Ethylparathion is phased out of use in India.

2. Methyl Parathion (Metacide).
   0,0-dimethyl 0-4-nitrophenylthiophosphate

$$(CH_3O)_2 \; P \overset{\overset{\displaystyle S}{\|}}{-} O -\!\!\!\left\langle\!\!\!\bigcirc\!\!\!\right\rangle\!\!- NO_2$$

Methyl parathion is a white crystalline substance slightly more soluble than parathion, sparingly soluble in paraffinic hydrocarbons, but soluble in aromatic hydrocarbons and most organic solvents. Rate of hydrolysis of methyl parathion is considerably higher than that of parathion and is relatively unstable thermally. Other chemical properties of methyl parathion are similar to those of parathion. Toxicity of methyl parathion is much lower. It penetrates the skin with much greater difficulty than parathion. Acute oral LD_{50} for rats is 14 mg/kg; acute dermal LD_{50} for rats is 67 mg/kg.

Methyl parathion is usually marketed in the form of emulsion, wettable powder and dusts. It is gradually replacing parathion on account of lower mammalian toxicity. It is a good contact and stomach insecticide with some fumigant action especially effective against boll weevil, aphids, armyworms, leaf hoppers, mealy bugs and thrips. It is non-systemic and non-phytotoxic.

3. Fenitrothion (Sumithion, folithion, accothion etc.)

0,0-dimethyl-0, 4-nitro-3-methyl phenyl thiophosphate

$$(CH_3O)_2 \ \overset{\overset{\textstyle S}{\|}}{P} - O - \underset{}{\bigcirc}{\overset{(CH_3)}{}} - NO_2$$

Fenitrothion is a clear liquid, sparingly soluble in water, but highly soluble in most organic solvents.

It is a non-systemic contact insecticide. In chemical properties, fenitrothion does not differ much from methyl parathion, except that its rate of hydrolysis in water and alkalies is lower, thermal stability is also lower. Iron promotes decomposition of this compound as with most organophosphorus compounds. It is compatible with other pesticides except those of alkaline nature, as the active material is relatively unstable in alkali.

0,0-dimethyl-0, 6-nitro-3-methyl phenyl thiophosphate (commonly called methyl nitrophos) is an active synergist for many organophosphorus insecticides including fenitrothion. For use in agriculture mixture of fenitrothion and methyl nitrophos in the ratio of 1 : 1 is recommended. This mixture does not differ from fenitrothion in pesticidal properties. Mammalian toxicity is comparatively lower. Acute oral LD_{50} for rats is 250-500 mg. tech/kg. Acute dermal LD_{50} for rats >3000 mg/kg.

4. Fenthion (Lebaycid)
0,0-dimethyl-0-(4-methyl mercapto-3-methyl phenyl) thiophosphate.

Fenthion is a fast killing contact general purpose insecticide with long residual action. It is sparingly soluble in water, but soluble in most organic solvents. Fenthion is more resistant to hydrolysis and heating than methyl parathion. It may be mixed with other insecticides and fungicides. Its high alkaline stability permits it to be combined with Bordeaux mixture also. It is marketed in the form of 50% emulsified concentrate, 25% wettable powder and 3% dust. Contact and stomach insecticide with a penetrative action, effective against mosquitoes, roaches, flies, fruit fly, leaf hoppers, cereal bugs and rice stem borers, should not be used at temperatures exceeding 36°C. Toxic to aquatic life, bees, fowl, dogs and poultry. Acute oral LD_{50} for rats is 190-315 mg/kg, acute dermal LD_{50} for rats is 330-500 mg/kg.

5. Fensulphothion (Dasanit, Terracur P)
0,0-diethyl-0. p-(Methyl sulphinyl) phenyl thiophosphate.

Dasanit is an oily liquid, soluble in most organic solvents, but slightly soluble in water. It is highly toxic (acute oral LD_{50} for rats is 10.5 mg/kg). It is used in the form of emulsion concentrate, wettable powders and dusts.

Fensulphothion is an insecticide and nematicide active against free-living cyst-forming and root-knot nematodes. It is recommended for soil treatment, has long persistence and some systemic activity.

6. Diazinon (Basudin, 0,0-diethyl-0-[2 isopropyl-4-methylpyrimidyl-6] thiophosphate). It was introduced in 1952.

$$(C_2H_5)_2P(S)O- \text{(ring)} --CH_3(CH_3)_2 \quad Or$$

$$(CH_3)_2-CH-C- \text{(ring)} -OP \begin{array}{c} S \\ OC_2H_5 \\ OC_2H_5 \end{array}$$

Diazinon is a contact nonsystemic insecticide used widely against aphids, thrips, mites and flies. It is also effective against soil insects. It is slightly soluble in water, and slightly volatile. It has much less toxic hazards to men. It is compatible with other fungicides excepting copper fungicides. Two weeks should normally elapse between its last application and harvest. It is not phytotoxic, but may cause damage to tomato and cucumber at lower temperature. It is marketed as 5 per cent dust, emulsified concentrate 20 per cent and wettable powder 40 per cent particularly for use in soil as drench.

Diazinon is a non-systemic insecticide with some acaricidal action. Main applications are in rice, fruit trees, vine yards, sugarcane, corn, tobacco, potatoes and horticultural crops for a wide range of sucking and leaf eating insects. Used also against flies in glass houses, mushroom houses.

The acute oral LD_{50} for rats is 300-850 mg/kg and acute dermal LD_{50} for rats is > 2150 mg/kg.

7. Chlorpyriphos (Dursban, Coroban)

0,0-diethyl-0-3,5,6-trichloropyridylthiophosphate.

$$(C_2H_5O)_2 \ \overset{\overset{\displaystyle S}{\|}}{P} - O - \underset{N}{\overset{Cl}{\underset{}{\diagdown}}}\hspace{-0.2em} \begin{array}{c} Cl \\ \diagdown \\ Cl \end{array}$$

Dursban is an active insecticide for control of sucking and chewing plant pests, soil inhabiliting pests and household pests. It is highly soluble in organic solvents, but almost insoluble in water. In acid or alkaline media, the compound is slowly hydrolysed by water forming diethylthiophosphoric acid, and ethylthiophosphoric acids and trichlorodihydroxypyridine. Duration of action of dursban, which is nonsystemic in nature, on different surfaces is 6 to 11 weeks, but on leaves of plants, it is short acting due to hydrolysis. The compound is active on grain crops for several weeks. The acute oral LD_{50} for rats is 135-163 mg/kg for chickens 32 mg/kg. Toxic to fish.

It is marketed in the form of emulsified concentrate.

8. Thionazin (Zinophos)

0,0-diethyl-0-pyrazinyl thiophosphate was first introduced in 1959 as a systemic pesticide against sap feeding insects. Thionazin is a soil insecticide and nematicide effective against a number of plant parasitic as well as free living nematodes, including those attacking bulbs, buds, leaves and roots as well as soil pests such as symphylids and root maggots and foliar insects such as aphids and leaf miners. It is of short persistence.

The acute oral LD_{50} and acute dermal oral LD_{50} for rats are 12 mg/kg and 11 mg/kg respectively.

It is used also as soil drench. The available formulations are 46 per cent emulsified concentrate and 10 per cent granules.

9. Demeton (Systox)

0,0-diethyl-2-ethyl mercapto ethylthiophosphate.

Schrdan, Octamethylpyrophosphoroamide abbreviated as OMPA was one of the first organophosphorus insecticides discovered in 1941-42 by Schrader, the pioneer worker on these group of insecticides. The compound is of historic interest as it was found to be translocated within the tissues of the plant rendering it toxic to sucking insects.

Property of schradan was noticed in the esters of dialkyl-phosphoric acids and glycol ethers or thioethers. Such an ester of 2-ethylthioethanol is demeton which is 0,0-diethyl-2-ethyl mercapto ethylthiophosphate (demeton 0) which spontaneously isomerises into thiolo isomers 0-0-diethyl-S-ethylmercapto-ethylthiophosphate (demeton S). A mixture of demeton-0 and demeton-S is marketed under the name 'Systox'.

Demeton has powerful systemic effect and is used for the control of many sucking plant pests. It has also some contact and fumigant action and is also used for seed treatments. Duration of action of the compound under field conditions depends on the crop, the period varying from 4 to 6 weeks. It is marketed in the form of emulsified concentrate.

10. Methyldemeton (Metasystox)

Dimethyl homologue demeton-0-methyl (0-0-dimethyl-2-ethyl mercaptoethylthiophosphate) in mixture with its thiol isomer demeton-S. Methyl was put in the market under the name 'metasystox'. Methyl demeton is an improvement over demeton in its action on insects. The conversion and decomposition of methyl demeton in plants takes place relatively quickly. It is a very active systemic insecticide, but the duration of insecticidal activity does not exceed three weeks. Demeton-S-methyl has fairly moderate mammalian toxicity, hence requires adoption of safety precaution measures. At least three weeks should elapse between the last application and harvest.

Demeton and methyl demeton, within the tissues of the host plant, are partially oxidised to more active compounds or partially broken down to inactive compounds. Hence it was considered appropriate to use oxidised compounds. Consequently the next step in the development of these group of insecticides was the introduction of oxydemeton methyl, sulphoxide of demeton-S-methyl and demeton-S-methyl sulphone. Both these compounds are efficient insecticides for control of aphids and red spider mites. These insecticides have been successful for control of vectors in some virus diseases.

11. Quinalphos (Ekalux, Bayrusil)
0,0-diethyl-0-0 (quinoxalinyl-2) thionophosphate.

It is a potent contact insecticide with powerful stomach action against chewing insects, particularly caterpillars. It has no phytotoxic effect. It has a good penetration and quick knockdown effect. It is compatible with most non-alkaline insecticides and fungicides. It is marketed in the form of 25 per cent emulsified concentrate.

Quinalphos is a contact and stomach insecticide and acaricide with good penetrative properties. It is used at 190-500 g a.i. (as e.c.)/ha against caterpillars on vegetables, groundnut, cotton, also scales and caterpillars on fruit trees at 250-500 g a.i. (as e.c.)/ha or 0.75-1.0 kg a.i. (as granules)/ha against pest complex on rice. It is degraded in plants within a few days.

The acute oral LD_{50} for rats is 62-137 mg/kg and the acute dermal LD_{50} for rats 1250-1400 mg/kg. It is dangerous to honeybees, the topical LC_{50} (24 hours) is 1.6 mg/kg.

C. Derivatives of Dithiophosphoric acid

A large number of different derivatives of dithiophosphoric acid are now used in agriculture. In most cases, the toxicity of the derivatives of dithiophosphoric acid is less than the corresponding compound of thiophosphoric acid. But the chemical stability of the compound is increased in dithiophosphoric acid derivatives.

1. Malathion
0,0-dimethyl-S-1,2,dicarboethoxyethyl-dithiophosphate.

$$(CH_3O)_2 \ P(S) \ CH\,COO\,C_2H_5$$
$$CH_2\,COOC_2H_5 \ or$$

Malathion was first introduced in 1950 because of low mammalian toxicity (2800 mg/kg-acute, oral LD_{50} for rats). It is slightly soluble in water and of short to moderate persistence. It is a general purpose nonsystemic insecticide particularly for control of household, home garden, vegetable and fruit pests.

It is active against a wide range of sucking insects, including aphids, scale insects and mealy bugs, beetles, thrips and mites. It is used for control of stored grain pests. It is not phytotoxic except on ornamentals and green house crops. It is. not compatible with alkaline insecticides. Formulations include both dust and spray as emulsified concentrate. Though toxic hazards is low, nevertheless four days should elapse between its last application and harvest to eliminate toxic effect if any and possible taints. Acute oral LD_{50} for rats is 280 mg/kg. It is extremely toxic to honey bee.

2. Dimethoate (Rogor)
   0,0-dimethyl-S-(N-methyl-carbamoylethyl) dithiophosphate.

$$\underset{(CH_3O)_2\ PSCH_2CONHCH_3}{\overset{\displaystyle S}{\overset{\displaystyle \|}{}}}$$

It is a systemic insecticide first introduced in 1966. It is used against sap feeding insects. It is highly soluble in water and most organic solvents. It is degraded by alkalies and is effective for a short period. Depending on the dosage of dimethoate applied it is usually broken down in plants in 15-20 days. While applying dimethoate care should be taken to see that there is no water stress, as it will cause injury to the plants. In storage dimethoate is relatively unstable and breaks down particularly at higher temperature. It is used as spray, usual formulation being forty per cent emulsified concentrate.

Dimethoate is a contact and systemic insecticide and acaricide effective against a wide range of insects and mites on different crop and horticultural plants, the dosage recommended being 0.3-0.7 kg a.i./ha. It is nonphytotoxic at recommended doses except to a few olive, citrus, fig and nut varieties.

The acute oral LD_{50} for rats is 320-380 mg a.i./kg.

3. Morphothion (Ekatin M, Ekatin F)
0,0-dimethyl-S-(morpholinocarbamoylmethyl) dithiophosphate.

$$(CH_3O)_2 - PSCH_2CON$$

It is similar to dimethoate. Chemical properties of morphothion are similar to those of other organophosphorus compounds of this group. It is marketed as emulsified concentrate.

4. Formothion (Anthio)
0,0-dimethyl-S-(N-methyl-N-formylcarbamoylmethyl) dithiophosphate.

$$(CH_3O)_2 - PSCH_2CON - CHO$$

Formothion is another systemic compound similar to dimethoate. It is poorly soluble in water, but highly soluble in organic solvents. Its chemical properties are similar to those of dimethoate, but is more stable on storage and on heating. It is marketed as 25% emulsion concentrate.

Formothion is a contact and systemic insecticide effective against a wide range of sucking insects, thrips, jassids, aphids, psyllids, scales, bugs, whiteflies, fruitflies and some chewing insects, epilachna beetles and dipterous mining larvae on a variety of crops including fruit trees, cotton, tobacco, vegetables, ornamentals. The normal dosage is 175 g a.i./ha.

The acute oral LD_{50} for rats is 360-500 mg/kg.

5. Disulphoton (disyston, dithiosystox)
Thiolothionate of the ethyl homologue of demeton-methyl-diethyl-S [2-(ethylthio) ethyl] phosphorothiolothionate or 0,0-diethyl-S-(2-ethylmercaptoethyl) dithiophosphate. It was first introduced in 1956. It is almost insoluble in water and stable except to strong alkalies. Being extremely toxic to mammals, it is used in the form of granules usual formulations containing

7.5 per cent active ingredient. It is an excellent aphicide. It is widely used for the control of sucking pests of cotton and other crops through soil application. As safety precaution measure, its dosage and application are kept within certain ranges. There should be an interval of six weeks between the last application and harvest of the crop. It is systemic insecticide and acaricide. The acute oral LD_{50} for rats is 2.6-8.6 mg/kg. It is not approved for registration in India.

6. Phorate (Thimet)

Thiomethyl derivative of disulphoton **0,0-diethyl-S-(ethyl thiomethyl) dithiophosphate.**

$$(C_2H_5O)_2 \overset{\overset{\displaystyle S}{\|}}{P} S\ CH_2\ SC_2H_5$$

It is a systemic insecticide of low water solubility and is unstable in solution. Because of high level of toxicity it is used for soil treatment for protection of plants in the form of granules containing 10 per cent active ingredient. In the plant sap, it is oxidised to stable sulphone. In many cases it gives a complete protection to plants for 20-25 days. It is very toxic compound and same precautions as in case of disulphoton are to be used. Acute oral LD_{50} for rats is 1.6-3.7 mg/kg.

7. Carbophenothion (Trithion)

0,0-diethyl-s-(4-chlorophenylthiomethyl) dithiosphate.

It is powerful contact insecticide and acaricide, used for the control of various sucking pests. It is almost insoluble in water but highly soluble in organic solvents. Its LD_{50} for rats is 10-30 mg/kg.

Carbophenothion has a long residual action. In combination with petroleum oil, it is used as a spray to control overwintering mites, aphids, scale insects on dormant deciduous fruit trees. It is phytotoxic at high concentrations on some plants. It is not approved for registration in India.

8. Ethion

0,0,0,0-tetraethyl-s-methylene bis (dithiophosphate).

Ethion is used for control of aphids and plant feeding mites. It is practically insoluble in water, but highly soluble in aromatic hydrocarbons and their halogen derivatives. It is marketed in the form of wettable powder (25%) and as emulsified concentrate. It is non-systemic. The acute oral LD_{50} for rats is 280 mg a.ı./kg.

9. Thiometon (Ekatin)
0,0-dimethyl-S-(2-ethyl mercaptoethyl) dithiophosphate.

Thiometon is a good systemic insecticide of the same duration of action as methyl demeton, but its contact insecticidal properties are somewhat less. It is sparingly soluble in water, highly soluble in most organic solvents. Thiometon is stable at normal temperature, but breaks down on heating. Thiometon is marketed in the form of 25-50 per cent emulsified concentrate and in granulated form. Thiometon is a systemic insecticide and acaricide which controls sucking insects mainly aphids and mites on most crops. Acute oral LD_{50} for rats is 120-130 mg/kg.

10. Phosmet (Imidan)
0,0-dimethyl-S-(naphthalimidomethyl) dithiophosphate.

Imidan is used for the control of various pests of cotton fruits and other crops. It breaks down easily and does not leave any toxic residue on fruits. It is practically insoluble in water, but highly soluble in a large number of organic solvents. Imidan is marketed in the form of 2 per cent emulsified concentrate, 50 per cent wettable powder and 10 per cent granulated formulation. It is safe for wide range of predators of mites and therefore useful in integrated programme.

11. Phosalone (Zolone)
0,0-diethyl-S-(6-chlorobenzolinyl-3-methyl) dithiophosphate.

It was first introduced in 1963 and has been found to be effective against aphids, larvae, flies, beetles and lepidopterous insects. It is considered to be a promising substitute for DDT and other persistent organochlorine insecticides. Though it has low mammalian toxicity, yet safety precautionary measures are applied by fixing a time interval of three weeks between last application and harvest. Acute oral LD_{50} for rats in 120-170 mg/kg.

12. Azinophos methyl (Guthion)
    0,0-dimethyl-S-(3,4-dichloro-4-keto-1,2,3-benzotriazinyl   3-methyl) dithiophosphate.

Azinophos methyl and azinophos ethyl was first introduced in 1954 and are used as an insecticide and acaricide. It is moderately persistent and relatively stable. It is used as spray, the usual formulation being 22 per cent w/v emulsified concentrate and wettable powder formulation containing 25 per cent active ingredient. It has fairly high mammalian toxicity. The least permissible time interval between last application and harvest varies from 2 to 3 weeks depending on the crop.

Azinophos methyl and Azinophos ethyl are non systemic insecticide and acaricide of long persistence. They are mainly effective against chewing and sucking insect pests of wide range of crops.

Acute oral LD_{50} for rats for Azinophos ethyl is 12.5-17.5 mg/kg and Azinophos methyl is 16.5 mg/kg.

Both Azinophos ethyl and Azinophos methyl are not approved for registration in India.

13. Menazon (Sayfos)
    0,-dimethyl-S-(4, 6-diamino-1,3,5-triazinyl-2-methyl) dithiophosphate.

Menazon is a systemic insecticide mainly aphicide with low toxicity to mammals. It is capable of penetrating into the plants through root system and imparting an insecticidal effect of long duration. It is used in the control of aphids of potato viruses. It is marketed as wettable powder and also as granules. The acute oral LD_{50} for rats is 1950 mg/kg.

D. Derivatives of pyrophosphoric acid
    1. HETP and TEPP. The name HETP (hexaethyl tetraphosphate) was given to a compound with empirical formula $C_{12}H_{30}O_{18}P_4$. It was marketed in Germany under trade name "Bladan" as a substitute for nicotine. In reality the product is a mixture of linear pyrophosphates, the main insecticidal principle being tetraethylpyrophosphate (TEPP) with empirical formula $C_8H_{20}O_7P_2$).

TEPP is readily soluble in water, but is easily hydrolysed to non-toxic diethyl phosphoric acid. It is miscible with organic solvents and aromatic oils, but not with kerosene or other paraffin oils or petroleum ether. It is incompatible with alkaline pesticides, but compatible with chlorinated hydrocarbons and sulphur. It is particularly effective against aphids, thrips, mealy bugs and red spider mites. The insecticidal activity of TEPP is about thirty-three per cent of parathion and its use is gradually decreasing. It is marketed in the form of 40 per cent anhydrous concentrate with a surface active agent. The acute oral LD_{50} for rats is 1-12 mg/kg.

2. Schradan (OMPA)
Octamethylpyrophosphoramide

Schrader first noted that certain toxic compounds could be absorbed through the leaves or roots of plants and could thus protect the plant from insect attack. Octamethyl pyrophosphoramide was one such compound. An aqueous solution of this compound in pure form is neutral and can be stored for unlimited time. It is relatively stable in neutral and alkaline aqueous media, but is rapidly hydrolysed under acid conditions.

It is not phytotoxic to most plants. It is not highly toxic as a contact insecticide when applied directly, but is readily absorbed by roots and leaves of plants and translocated with the result that the plant sap becomes toxic to many insects and mites feeding on plant sap. In the plant 90 per cent of the material disappears in 40 days in mid summer probably by decomposition in the plant system. The chemical is rapidly translocated in plants. The compound is highly toxic to mammals. It is effective against sap feeding insects and mites. The acute oral LD_{50} for rats is 9.1 mg/kg.

3. Miscellaneous organophosphatic compounds
1. Trichlorophon (Diptrex)
0,0-dimethyl-(1-hydroxy-2,2,2-trichloroethyl) phosphonate.

Trichlorophon is a systemic insecticide used for the control of chewing insects. It is soluble in water, but sparingly so in paraffinic hydrocarbons. It is stable in acid medium, but rapidly hydrolysed in alkaline medium. It is a safe insecticide. Trichlorophon is a contact and stomach insecticide with penetrant action. It is effective against lepidopterous larvae and fruitflies at 75-120 g a.i./ha. Its activity is attributed to its metabolic conversion to Dichlorovos.

The acute oral LD_{50} for rats is 560-630 mg/kg.

2. EPN (0-ethyl-0-(4-nitrophenyl) benzene thiophosphate.

EPN is a powerful acaricide besides being toxic to some insects. It is practically insoluble in water, but highly soluble in most organic solvents. It is relatively hydrolysed in alkaline medium, but not in acid and neutral media, EPN is more stable. It is relatively non phytotoxic except for a few varieties of apples. It is highly toxic to warm blooded animals. EPN is a non-systemic insecticide and acaricide with contact and stomach action against lepidopterous larvae especially bollworm and leafworm of cotton, rice stemborer and leafeating larvae on fruit and vegetables. It is non-phytotoxic. It is not approved for registration in India.

The acute oral LD_{50} for rats is 33-42 mg/kg.

It is marketed as dusts or wettable powders.

3. Mephosfolan (Cytrolane)

Diethyl (4-methyl-1,3-dithiolan-2-Ylidane) phosphoroamidate. It was introduced in 1963 by the American Cyanamide Co. under the name 'Cytrolane'. It is yellow to amber colour liquid, which is stable in water under neutral conditions, but is hydrolysed by acid or alkali. Mephosfolan is a contact and stomach insecticide with systemic activity following root or foliar absorption. It is effective against borers, bollworms of major crops. It is extremely toxic to mammals.

The acute oral LD_{50} for rats is 3.9-8.9 mg/kg.

It is marketed as e. c. or granules. It is not approved for registration in India.

4. Methomyl (Lannate)

S-Methyl N-(methylcarbamoyloxy) thioacetamide

$$\underset{MeS}{\overset{MeS}{>}}C=N-\overset{\overset{O}{\|}}{O}C\,NH\,Me$$

Methomyl was introduced in 1966 as an experimental insecticide and nematicide. It is a colourless crystalline solid with slight sulphurous odour. It is stable as solid and in aqueous solution under normal conditions. Methomyl is used as a foliar treatment for control of many insects. In cases of soil treatment, the chemical is taken up by roots and translocated and thus control insects attacking above ground parts.

It is fairly toxic. The acute oral LD_{50} for rats is 17-24 mg /kg.

5. *Leptophos (Phosvel)*

O-4-bromo-2, 5-dichlorophenyl O-methyl phenylphosphonothionate. The insecticide was first introduced by Veliscol Chemical Corp. in 1967 with trade mark 'Phosvel'. It is a non-systemic insecticide particularly effective against lepidopterous pests on cotton, vegetables and fruit. As the insecticide has positive temperature coefficient against lepidoptera and therefore be more useful in warmer climates. It is moderately persistent. It is moderately toxic. The acute oral LD_{50} for rats is 50 mg/kg.

It is marketed as e.c., wettable powder or granules. It is not approved for registration in India.

F. Carbamate insecticides

The fact that organophosphorus compounds are highly effective against a wide range of insects due to anticholinesterase activity led to the search for insecticides among carbamate compounds, which have been in use in human medicine for such activities, namely phytostigmine or ersine. Investigations in this direction produced a number of compound like pyrolan, isolan and eventually to carbaryl and carbofuran compounds.

N-methyl and N, N-dimethyl carbamic esters of phenols and heterocyclic enols possess useful insecticidal properties. Among the derivatives of carbamic acid, the aryl esters of N-methyl carbamic acid are used for control of insects pests. All the insecticidal esters of N-alkyl carbamic acids cause inhibition of cholinesterase. Carbamates behave as competitive inhibitors of the cholinesterase, the relationship between insecticidal structures and anticholinesterase is somewhat complicated.

1. *Carbaryl (Sevin)*

1-napthyl-N-methyl carbamate

Carbaryl is a good contact insecticide with occasional systemic action in certain cases though not of much practical value. It is of broad range activity. It is sparingly soluble in water, but highly soluble in organic solvents. It is moderately resistant to the action of water, light and oxygen in the air at room temperature. It is effective against leaf-eating caterpillars, beetle, larvae, some aphids, boll weevils of cotton and many lepidopterous insects, but is ineffective against mites. It is ineffective against houseflies though isolan and pyrolan are.

It is not compatible with compounds of alkaline nature. It is weakly phytotoxic. It is a fairly safe insecticide, LD_{50} value for rats being 85 mg/kg but it is toxic to bees and fish. It is marketed in the form of dusts and wettable powders.

2. *Propoxur (Baygon)*

2-Isopropoxy phenyl-N-methyl carbamate

Baygon is a broad spectrum non-systemic insecticide with essentially contact action. It produces rapid knockdown effect and has an extremely long residual action. It is effective against jassids, bugs, aphids and domestic pests.

Baygon is a white to cream-coloured crystalline compound with mild phenolic odour. It is slightly soluble in water, but highly soluble in organic solvents. LD_{50} for rat is 100 mg/kg. It is formulated as emulsion concentrate (20%), wettable powder (50%) and dusts. It is extensively used by WHO for mosquito control. It is claimed to have an antifeedant effect.

3. *Carbofuran (Furadan)*

2, 3-dihydro-2, 2-dimethyl-7-benzofuranyl methyl carbamate.

Carbofuran is a broad spectrum systemic insecticide and acaricide and nematicide to a large extent. It is stable in acid and neutral media, but unstable in alkaline medium. It is sparingly soluble in water, slightly soluble in organic solvents. It is compatible with all non-alkaline pesticides and fertilizers. Carbofuran is not phytotoxic. Carbofuran applied to the soil in which the plants are growing is rapidly absorbed by the roots of the plants and distributed to stems and leaves, where it is slowly converted into non-toxic compounds over a period of 30 days. The quantity not absorbed by the roots is rapidly degraded in the soils, rate of degradation being dependent on pH and clay structure of the soil.

Toxic residues do not remain in the soil from one season to the next. It is marketed in the form of granules (3%). Oral LD_{50} for rats is 11 mg/kg.

4. Aldicarb (Temik)

2-methyl-2-methylthiopropyonal-oxime-O, N-methyl carbamate

Aldicarb is a systemic nematicide. It is extremely toxic to mammals LD_{50} value for rats is 0.93 mg/kg. It is insoluble in paraffinic hydrocarbons, sparingly soluble in water, acetone, ethyl alcohol, chloroform, toluene and chlorobenzene.

Other compounds of promise are zectran 4-dimethyl amino 3, 4 xylyl N-methyl carbamate which is systemic with both insecticidal and acaricidal activity and toxic to snails and slugs.

5. Bendiocarb (Ficam)

2,3-isopropylidenedioxyphenyl methyl carbamate

Bendiocarb is contact and stomach insecticide effective against wide range of household, and storage pests. In agriculture, it is effective against lepidoptera, coleoptera, collembola, specially soil pests.

Available as wettable powder, granules or dust. The acute oral LD_{50} for most mammals is 34-64 mg/kg.

G. Synthetic Pyrethroids

In view of the high production costs of pyrethrum, it was considered expedient to search for compounds which would be synthesized by chemical modification of the natural structures that would possess the same properties as the natural product namely good knockdown effect, low mammalian toxicity. The first synthetic pyrethroid allethrin appeared in the market as long as 1950 and is still in use to-day. By 1964, tetramethrin, a good knockdown effect was discovered, a few years later, resmethrin and bioresmethrin with greater insecticidal activity and lower mammalian toxicity, but they are unstable in sunlight. The photostable pyrethroids discovered through the work of Elliot *et al.* in Rothamsted in England and Ohno *et al.* Sumitomo, Japan, are the most interesting new products in the insecticial field at the present time. They are characterised by much lower application rates—application rates being one-fifth to one-tenth of those of classical organophosphorus and carbamate insecticides. While allethrin, resmethrin, bioresmethrin, permethrin are mainly used as insecticides in enclosed spaces e.g. houses, hospitals, food processing plants, also against mosquito, cockroach and for control of vectors of insect borne communicable diseases, fenvalerate, cypermethrin, decamethrin are mainly used in crop protection, permethrin is however useful against crop pests and decamethrin is used for control of tsetse fly.

Synthetic pyrethroids which are in use in agriculture in India are as follows:

Scientific name	Commercial trade name/formulation		Dosage per hectare
1. Permethrin	Permasect	25 EC	200—300 ml.
2. Cypermethrin	Cymbush	25 EC	200—300 ml.
	Ripcord	10 EC	1000—1500 ml.
	Cyperkill	25 EC	200—300 ml.
3. Fenvalerate	Sumicidin	20 EC	500—750 m.
	Agrofen	20 EC	
	Fenval	20 EC	
4. Decamethrin	Decis	2.4 EC	500—750 ml.

The natural pyrethroids are purely contact poisons that penetrate rapidly into the nervous system and cause the characteristic symptom in the insect. A phase of exceptional excitation is followed by a disturbance in the coordination of the movement, paralysis and finally death. The initial knockdown effect is very quick, but insufficient to be lethal, because the natural pyrethroids are rapidly detoxified in the insect by enzymatic action.

Precise mode of action of synthetic pyrethroids is not known. They are recognised as nerve poisons, but do not interact with acetylcholinesterase. In this respect their action differs from that of organophosphates and carbamates. Synthetic pyrethroids developed so far have very little solubility in water and therefore little systemic or translaminar activity. They mainly act as stomach or contact poisons. These chemicals initially destroy adult insects and eggs and later caterpillars. At a later stage, they act as repugnant and antifeedant against insects.

Synthetic pyrethroids are fully inactivated in most soil types within 2-4 weeks, hence there is very little likelihood of accumulation of toxic residues in soil at normal rates of application.

Generally synthetic pyrethroids are of low toxicity to mammalian metabolism before they can reach sensitive regions. Decamethrin is however moderately toxic to mammals. Synthetic pyrethroids are toxic to fish, but toxicity of pyrethroids to a given species of fish parallels insecticidal activity.

Use of synthetic pyrethroids results in increased incidence of phytophagous mites. Resurgence of pests particularly aphids and scale insects has been reported. Resistance to synthetic pyrethroids has been reported in the aphid *Myzus persicae*, brown plant hopper *Nilaparvata lugens*.

Permethrin (Permasect, Pounce, Ambush*)*,
(3-phenoxyphenyl) methyl 3-(2, 2-dichloroethenyl)-2, 2-dimethylcyclopropane carboxylate.

It is a good contact insecticide effective against broad range of pests. It controls leaf and fruit-eating lepidopterous and coleopterous pests in cotton, vegetables, vines, tobacco and other crops. It has good residual activity on treated plants.

Acute LD_{50} for rats 430-4000 mg/kg, for mice, 540-2690 mg/kg, chickens >3000 mg/kg (LD_{50} value varies with cis/trans ratio).

Cypermethrin (Cymbush, Ripcord, Cyperkill).
Cyano (3-phenoxyphenyl) methyl 3-(2, 2-dichloroethenyl)-2, 2-dimethyl cyclopropanecarboxylate.

Cypermethrin is a stomach and contact insecticide effective against a wide range of insect pests, particularly leaf- and fruit-eating lepidopterous and coleopterous pests in cotton, fruit, vines, vegetables, tobacco and other crops at 25-150 g a.i./ha. If applied before infestations become well established, it will also give protection against hemiptera in most crops. It has

good residual activity on treated plants and no case of phytotoxicity has been reported.

Acute oral LD_{50} is for rats 303-4123 mg/kg depending on the
carriers and conditions used, for chickens >2000 mg/kg.

Fenvalerate (Sumicidin, Agrofen, Fenval, Belmark, Pydrin)
Cyano (3-phenoxyphenyl) methyl 3, 3-dimethyl-spiro (cyclo-
propane-1, 1' (IH) indene)]-2-carboxylate.

Highly active contact insecticide effective against broad range
of pests including strains resistant to organochlorine, organophosphorus and carbamate insecticides. It gives good control
of leaf- and fruit-eating insects on a wide range of crops including cotton, fruit, vegetables, vines. Aphid *Myzus persicae* is
controlled by sprays at 2-5 g a.i./100 1 and cutworms by soil
surface sprays at 50-70 g a.i./ha. Its stability to sunlight and
wash off by rains ensure good residual activity on treated
plants.

Acute oral LD_{50} for rats 300-630 mg/kg, for domestic fowl
1600 mg/kg, 200-300 mg/kg for mice.

Decamethrin (Decis)
Cyano (3-phenoxyphenyl) methyl 3 -(2, 2-dibromoethenyl)-2,
2-dimethyl cyclopropane carboxylate.

Very potent insecticide effective as a contact and stomach
poison against wide range of insects. It controls numerous insect pests of field crops at 11 g a.i/ha.

Acute oral LD_{50} for rats 135 mg/kg.

H. Acaricides

The bridged diphenyl group. An examination of the structure of these compounds would tend to attribute insecticidal
properties to them because of their closeness to DDT. Surpris-
ingly enough, these compounds do not have insecticidal properties, on the other hand they have been found to be toxic to
Acarinae. Exact reason to their effectiveness as acaricides is
not known, as the physiology of mites has not been fully investigated yet. Some of the important compounds are described below:

(i) *Azobenzene* (Diphenyldiazine $C_{12}H_{10}N_2$)
On account of its fairly high volatility it has been used as

smoke or aerosol for control of mites particularly in green houses, at the rate of 2.5 gm/5 m^3. However to some extent, it causes damage to foliage and bloom in the effective concentration.

(ii) *Chlorobenzilate* (Akar, Foibex)

Ethyl 4, 4', dichlorobenzilate is an effective acaricide without any insecticidal effect so that it may be safely used on honey bee for control of tracheal mites. Chemically, it is an ester and has close resemblance to DDT. It is susceptible to degradation by alkali. It is formulated as a water dispersible powder or emulsified concentrate both at 25 per cent concentration of active ingredient. The acute oral LD_{50} for rats is 700-3100 mg/kg.

(iii) *Chlorofenson*

4-(chlorophenyl 4-chlorobenzenesulphonate). It was first introduced in 1949. It is toxic to eggs and immature stages of sap-feeding mites. It is of high persistence due to low volatility and high chemical stability. It is, however, broken down by alkali being an ester. It has very little toxic hazards to warm blooded animals. It is usually available as 50 per cent wettable powder for spray formulations. The acute oral LD_{50} for rats is 2000 mg/kg.

(iv) *Fenson*

4-chlorophenyl benzene sulphonate.

Closely related to chlorofenson with the similar properties. But it causes russetting of apples and damages to young cucurbits. It is marketed as 20 per cent wettable powder and emulsified concentrate.

(v) *Tetradifon*

2,4,5, 4-tetrachlorodiphenylsulphone

This acaricide was first introduced in 1954—virtually insoluble in water, non-volatile of high stability with a good persistence with pseudosystemic effect. It is effective against eggs and non adult stages of mites. It is harmless to warm blooded animals and plants except young cucurbits. It is available in formulations of 8 per cent ingradient and smoke generators.

(vi) *Aramite*

2-chloroethyl 2-(p-tert-butyl phenoxy) isopropyl sulphite is in use in commercial scale since 1950 for control of various mites. It has been claimed to be an ideal acaricide, while controlling effectively various mites it is not toxic to men and animals and predators, but it may cause irritation to skin. It is not normally phytotoxic except on pears and hops. A period of four weeks should elapse between last application and harvest. It is available as three or four per cent dusts, 15 per cent wettable powder or 25 per cent emulsified concentrate.

(vii) *Dicofol* (Kelthane)

2, 2, 2-trichloro-1, 1-bis (4-chlorophenyl) ethanol.

The chemical was first introduced in 1955 by Rohm and Hass Co. under the name 'Kelthane'. Though chemically it resembles DDT, nevertheless it has no insecticidal property. It is a non-systemic acaricide effective against wide range of mites. It has rapid killing action and long residual effect. It has a low mammalian toxicity. The acute oral LD_{50} for rats is 668-842 mg/kg. It is available as 25 per cent wettable powder or 18.5 per cent emulsified concentrate.

QUESTIONS FOR DISCUSSION

1. How the different insecticides are classified into different groups?
2. Which insecticides of plant origin are still in use and why? Why is the use of nicotine sulphate falling off in spite of its effectiveness?
3. How do synergists help in the effectiveness of pyrethrum?
4. In which way are hydrocarbon oils used for pest control? What specifications should be present for the effectiveness?
5. Are inorganic salts still in use for control of insects? What type of poisons are they?
6. Why is there an attempt to ban DDT? Which isomer of DDT is more potent? Is there any specification regarding the content of the specific isomer?
7. Why has DDT been valued as a protective pesticide? What are the closely related compounds of DDT which have shown promise?

8. Why is gamma-BHC widely used? In which forms is it used?
9. What are the different cyclodiene derivatives? How are they used?
10. What are the different categories of organophosphorus insecticides? Are all organophosphorus compounds systemic?
11. Are all organophosphorus compounds equally toxic to warm blooded animals?
12. How do systemic insecticides differ from non-systemic ones? In which way are systemic insecticides of value? How do organophosphorus insecticides act?
13. Why is attention being given to carbamate insecticides? How do they act? What are the important insecticides of this group of compounds?
14. What are acaricides? Have acaricides any insecticidal properties or vice versa? Is there any systemic acaricide?
15. What are synthetic pyrethroids? What are the principal advantages of synthetic pyrethroids?

CHAPTER 6

FUNGICIDES

Fungicides may be broadly classified on the basis of their chemical composition, or mode of action. On the basis of their mode of action, they are classified as (i) protectants—effective only when they are applied before fungal invasion as a protective measure, (ii) eradicants which are aimed to kill the fungal pathogens in a substrate or infection court before infection can possibly take place and (iii) therapeutants which are capable of killing the fungal pathogens after establishment of infection inside the host tissue. Most therapeutants, to be effective, enter inside the host tissue and eventually the system of the plant and are thus called systemic fungicides.

An ideal fungicide should have the following properties, namely,

1. High field performance which is determined by (a) innate fungitoxicity, (b) easy availability, of the fungicidal substance or active ingradient, (c) good coverage of the surface on which they are applied, (d) tenacity, i.e., retentivity of the fungicide and its active ingradient on the host surface;
2. Should not be damaging or harmful to the host plants;
3. Can be kept in storage for a long time without losing its effectiveness;
4. Easy to prepare and apply;
5. Low toxicity to men and animals;
6. Not easily degraded or decomposed in nature;

7. Good benefit/cost ratio or economically feasible and
8. Low effect on environment.

A. Sulphur fungicides

(1) *Elemental sulphur*

Sulphur is known as fungicide since ancient times even before the knowledge of the exact nature of the fungi causing plant disease was available. The earliest scientific record on its use dates back to 1821 A.D., when a paper was read before the London Horticultural Society on the value of sulphur for the control of peach mildew. Sulphur gradually became recognised as a valuable fungicide for the control of powdery mildews and it remains so to date as one of the most effective and widely used chemical for the purpose. It is also an effective acaricide. It is insoluble in water and almost non-volatile though it can be substituted for fumigation. It is used as dust, spray and fumigant. Chemically it is stable and compatible with other pesticides. It is non-toxic to man and warm blooded animals. It can however cause damage to plants particularly in hot weather by producing both chronic and acute injuries on foliages and fruits.

The efficiency of sulphur as fungicide depends on the particle size which has a direct relation to its toxicity. Dusting formulations normally contain 95 per cent sulphur with particles passing through 200-300 mesh sieve (47-74 microns) or still finer particles. Aggregation of finer particles is prevented by the addition of a small amount of kaolin or similar materials. Spray formulations are prepared by rendering sulphur wettable or water dispersible through the addition of appropriate chemicals. Spray formulations according to the U.K. Ministry of Agriculture, Forests and Fisheries should have 70 per cent sulphur with such a particle size that at least 40 per cent should be of 6μ or less and not less than 9 per cent should be of 2μ. Paste formulations are also available containing not less than 40 per cent sulphur of which 90 per cent should be 6μ or less in diameter with 55 per cent of them less than 2μ in diameter. Dry sulphur dust has been used, in the past, for treatment of seeds against externally seed-borne smuts and bunts.

Mode of action of sulphur is not fully understood. It is believed that it enters or permeates the fungal spores or cells as such and compete with oxygen as hydrogen acceptor and disrupts the normal hydrogenation and dehydrogenation with the formation of hydrogen sulphide which inactivates vitally important enzymes like catalase, cytochromoxidase, laccase. Thus a disturbance is caused in Kreb's cycle. Formulations of elemental sulphur have low toxicity to men and warm blooded animals. No tolerance levels have been established for sulphur content in food products.

(2) *Lime sulphur*

Lime sulphur has been in use for more than a century. It is prepared by boiling sulphur and lime water, the resultant product being calcium polysulphide ($CaS.S_x$) with a small quantity of calcium thiosulphate. On exposure to atmosphere, acidification by action of carbon dioxide or action of water soluble salt of a metal exuding or depositing on the plant surface polysulphide sulphur is precipitated as an elmental sulphur. The amount of deposition of sulphur depends on the content of calcium polysulphide of the spray. Lime sulphur solution should be clear and free from sludge and have an apparent density of 1.30 ± 0.1 at $20°C$ add polysulphide (S_x) content should be 24 per cent w/v. Action of lime sulphur is due to elemental sulphur, hence the mode of action is the same as that of sulphur. It is capable of causing both chronic and acute damages to the sprayed plants particularly in the sunny weather and at the fruit formation stage. It is effective against mildew, mites and scale insects.

One of the disadvantages of lime sulphur as compared with sulphur is its bulk adding cost to transport and spraying. The second disadvantage is that being alkaline in reaction it is incompatible with a large number of pesticides which are hydrolysed by alkali.

B. Copper fungicides

(1) *Bordeaux mixture*

Copper sulphate was first recognised as a fungicide in 1807 by Prevost for treatment of wheat seeds for control of bunt.

Millardet following a slender clue that the leaves of grape-vines pasted with mixture of lime and copper sulphate showed freedom from attack of downy mildew discovered a fungicidal preparation which has been named as Bordeaux Mixture, name being derived from the locality from which it originated.

Bordeaux Mixture is prepared by mixing a solution of 4.5 kg copper sulphate ($CuSO_4. 5H_2O$) of 98 per cent purity, 4.5 kg hydrated lime and 450 litres of water. Copper sulphate should be dissolved either in wooden barrels or plastic containers (not in metal containers as they would receive a deposit of copper). As dissolving of copper sulphate is time-consuming, it should be done well in advance, preferably overnight. Lime used should always be fresh. Both the solutions should be sieved. Care should be taken to see that there is no free copper in the solution. To check for the same, a few drops of 10 per cent potassium ferrocyanide solution may be added to a small quantity of the mixture to see whether there is any reddish brown deposit or in the absence of ferrocyanide, a clean knife blade may be kept in the mixture for a few minutes to see whether any deposit of copper takes place on the knife blade. If the reaction is found to be positive in either case, more lime should be added. The mixture should be slightly alkaline in reaction. Bordeaux Mixture should always be prepared fresh. On standing, the precipitate forms small spherocrystals or crystallites and loses the fungicidal efficiency. The stability of the mixture may be increased for a comparatively short period with the addition of sugar or jaggery molasses or sulphite lye at the rate of 1 kg for 1000 litres of water.

In spite of the difficulty that the preparation has always to be made fresh, Bordeaux Mixture is still a popular and highly effective fungicide, qualities of which have not yet been surpassed by its substitutes.

It is an excellent fungicide for basically two reasons: (a) inherent fungitoxicity and (b) tenacity. On weathering its fungicidal efficiency improves further instead of deteriorating. Adherence of the precipitate, however, deteriorates with time, requiring further application for the protection of the sprayed surface. Bordeaux Mixture is usually alkaline in reaction, hence it is not compatible with insecticides which are hydrolysed by alkali.

Fungicidal activity of Bordeaux Mixture is associated with the gradual and slow formation, from the deposit, of soluble copper compounds from which ultimately toxic cupric ions are liberated. These ions gradually permeate the spores and cells of the pathogen and bring about disruption of normal activities, and eventual death by poison.

Bordeaux Mixture is effective against leaf blights and fruit falls or rots associated with large number of diseases, notable exceptions being apple scab and powdery mildews. Bordeaux Mixture has always to be used in high volume sprays because of lime. In order to make it amenable to low volume spraying, the quantity of lime has to be reduced (8:4:20), by replacement of lime by washing soda or soda ash (8:10:80), the preparation being known as Burgundy Mixture. Bordeaux paste can be prepared for treatment of wounds (1:1:1).

Bordeaux Mixture can be phytotoxic to many plants particularly in cool cloudy weather when the stomata remain open and the dissolved copper salts have a chance to enter inside the plant tissues. Burgundy Mixture is likely to cause more phytocidal damage.

Bordeaux Mixture is only slightly toxic to men and warm blooded animals. The tolerance limit in fruits and vegetables is 5 mg/kg and in meats and eggs 2 mg/kg. The maximum tolerated concentration in the air of working zone is 0.3 mg/m^3 and in water basins is 0.1 mg/litre.

(2) Other copper compounds

To avoid the cumbersome preparation of Bordeaux Mixture which has always to be used fresh, many fixed copper compounds have been tried as substitutes. The most important and widely used of them are copper oxychloride formulations which are formed by aeration of scrap copper in copper chloride solution the resultant product closely approaching 3 Cu (OH)$_2$, CuCl$_2$. Copper oxychloride formulations can be used both as dust and spray. Dust formulations may contain 6-35 per cent copper. For spraying, colloidal formulations may be prepared by intensive grinding of particles to such a size that the ground particles remain permanently in paste or in solution in the desired strength. Copper oxychloride is basic salt of copper chloride. It is a solid crystalline substance insoluble in water and organic substances. It is stable to sunlight, moisture and an elevated temperature, but is decomposed by alkalis. It acts as a protoplasmic toxin. It is moderately toxic to men and warm blooded animals (LD_{50} for mice is 470 mg/kg). When injected into an organism, it causes inflammation of gastro intestinal tract.

When molasses or sulphite lye is added to maintain fungicidal efficiency of Bordeaux Mixture there is a gradual reduction of precipitate to cuprous oxide without deterioration in fungitoxicity. Dispersible cuprous oxide powders are now in wide use as foliage protectants, the familiar example being Perenox. Cuprous oxide has also been used a seed treating fungicide.

(3) Cheshunt compound

It is cuprammonium fungicide introduced in 1922 specifically for control of damping off of seedlings. It is prepared by intimate mixing of two parts by weight of crystalline copper sulphate and 11 parts of ammonium carbonate. Solution of this mixture is used for watering of soil and seed boxes for control of damping off.

Basic copper carbonate. $Cu(OH)_2$, $CuCO_3$ has been in use for treatment of seeds particularly against externally seed borne smuts.

The fungicidal or more accurately fungistatic action of copper arises through non specific destruction of proteins and enzymes which is due to non selective and non specific affinity of the cupric ion for certain isogenic groups such as imidazole, carboxyl, phosphate or sulphahydryl.

C. Thiocarbamate fungicides

Discovery of thiocarbamate fungicides is compared to that of DDT in the field of insecticides. Though spergon was the first organic fungicide to be commercially used, nevertheless these compounds opened a new vista in the arena of protective fungicidal control. This group of fungicides derived from dithiocarbamic acid are at present the most important, most versatile and most widely used ones. They are used as protectants, seed dressing chemicals, preservatives in textiles, paints, etc.

Though the use of these compounds as fungicides, is comparatively recent, they have been in use in rubber industry for acceleration of vulcanisation process. Tisdale in 1931 demonstrated fungicidal possibilities of thiocarbamates, though they have been put into use much later.

These fungicides are derived from dithiocarbamic acid NH_2. CS_2 H-represented structurally.

$$
\begin{array}{c}
H \\
\diagdown \\
N-\overset{\displaystyle S}{\overset{\|}{C}}-S-H \\
\diagup \\
H
\end{array}
$$

They may be classified into three groups.

(1) *Thiuram disulphides*

$$
\begin{array}{c}
N \\
\diagdown \\
N-\overset{\displaystyle S}{\overset{\|}{C}}-S-S-\overset{\displaystyle S}{\overset{\|}{C}}-N \\
\diagup \qquad\qquad\qquad\qquad \diagdown \\
R \qquad\qquad\qquad\qquad\qquad R
\end{array}
$$

(2) *Metallic dithiocarbamates*

$$
\begin{array}{c}
R \\
\diagdown \\
N-C-S-Metal \\
\diagup \\
R
\end{array}
$$

(3) *Ethylene bisdithiocarbamates*

$$
\begin{array}{c}
R \\
\diagdown \\
C-N-\overset{\displaystyle S}{\overset{|}{C}}-S-Metal \\
\diagup \\
R
\end{array}
$$

$$
\begin{array}{c}
R-\overset{\diagup R}{\underset{\diagdown R}{C}}-N-\overset{|}{C}-S-Metal \\
\underset{S}{|}
\end{array}
$$

R = alkylradical

(1) *Thiuram disulphide*

Thiuram disulphides are formed by joining two molecules of dithiocarbamic acid through the 'S' atom. Of the different compounds that may be formed, only tetramethyl thiuram disulphide has been in use as fungicide for control of plant diseases.

$$CH_3\diagdown \atop CH_3\diagup N-\overset{\overset{\displaystyle S}{\|}}{C}-S-S-\overset{\overset{\displaystyle S}{\|}}{C}-N\diagup ^{CH_3} _{CH_3}$$

This product is named as "Thiuram", though various trade names have been assigned to it by different manufacturers. It was introduced in 1934, primarily as a protectant of seeds, but later it has also been used for foliage protectant and in some cases for soil treatment. It is practically insoluble in water, nonvolatile and is compatible with most other pesticides. It is not phytotoxic at the concentrations used and has low toxicity to warm blooded animals except poultry. It can cause, however, irritation to skin and dermatitis in some cases. It is not advisable to spray fruits before harvest which are intended for canning or deep freezing because of taint and damage to containers. The usual formulations for spray are 80 per cent wettable powder and a 50 per cent colloidal suspension. For seed treatment 50 per cent powder is normally used. Better results are obtained with soak treatment in a suspension of thiuram containing 250 gm of active ingredient in 100 litres of water. For soil treatment 10-15 kg of active ingredient per hectare is usually recommended. Though this compound has been in use of seed treatment, it is not as effective as organo-mercurials.

(2) *Metallic dithiocarbamates*

(a) *Ferbam*—Ferric dimethyl dithiocarbamate, $C_9H_{18}FeN_3S_6$

$$CH_3\diagdown \atop CH_3\diagup N-\overset{\overset{\displaystyle S}{\|}}{C}-S-Fe-S-\overset{\overset{\displaystyle S}{\|}}{C}-N\diagup ^{CH_3} _{CH_3}$$

$$\underset{\underset{\underset{\underset{CH_3 \qquad CH_2}{N}}{\|}}{C=S}}{S}$$

Of the numerous metal dialkyl dithiocarbamates tried, only iron and zinc dimethyl dithiocarbamates have been found to be of practical value and have been developed commercially as fungicides.

Though developed in 1931 by Tisdale and Williams, ferbam was introduced as a commercial fungicide in 1943. It is somewhat unstable to heat, light and liable to deterioration in storage. It is somewhat soluble in water, and is completely soluble in many organic solvents. It has been successfully used as a protectant against a number of diseases of horticultural crops and vegetables. In temperate countries it has been used to control a number of diseases of apples and other fruits. It is used as spray or dust, the usual formulations being 80 per cent active ingredient as dust or 50 per cent wettable powder as spray. Usually 0.2-0.3 per cent concentration is used in spray. It has virtually no phytotoxicity and low mammalian toxicity. It can be used with other pesticides, but copper, lime and mercury compounds reduce its effectiveness.

(b) *Ziram*—Zinc dimethyl dithiocarbamate $C_6H_{12}N_2S_4Zn$

It is the most stable of dithiocarbamate fungicides, of comparatively low solubility in water and insoluble in alcohol or ether, but soluble in chloroform. It is of low toxicity to warm blooded animals, but it may be toxic if orally administered, but may cause irritation to skin and dermatitis to some individuals. Because of effect on skin, it is not favoured by many as a spray fungicide. It is effective against early blight of potato and a number of diseases of fruits and vegetables caused by Fungi Imperfecti. It has been found to have additional beneficial effect on plants in areas with deficiency in zinc.

It is used as a spray in 0.2-0.3 per cent concentration of active ingredient. It is marketed in the formulation of 50 per cent wettable powder.

Zinc-manganese-copper, iron dithiocarbamate complex has been introduced in 1967 for control of a large number of foliar diseases. It is compatible with most insecticides and copper oxychloride, but should not be mixed with lime or lime sulphur.

(3) *Ethylene bisdithiocarbamates*
  (a)*Nabam*—sodium ethylene bisdithiocarbamate $C_4H_6N_2Na_2S_4$

It was the first of ethylene bisdithiocarbamate compounds to be marketed in 1943. It is soluble in water and the solution is fairly stable. It decomposes on exposure. In the field, nabam was found to give unsatisfactory performance because of phytotoxicity and instability. It was later on found that addition of zinc sulphate and lime stabilises its performance.

700 gm of nabam, 100 gm of zinc sulphate and 100 gm of lime in 100 litres of water are to be used when nabam is to be used in the field as a spray. It can also be successfully applied in the field in soil for control of seedling diseases of a number of crops at the rate of 5·6 kg per hectare of active ingredient. Nabam is available both as powder containing at least 93 per cent of anhydrous salt and a stock solution containing 20 per cent anhydrous salt. It is of low toxicity to warm blooded animals when used alone or in conjunction of zinc sulphate.

(b) *Zineb*—Zinc ethylene bisdithiocarbamate.

When it was found that addition of zinc sulphate improves efficiency of nabam, zineb was prepared in the field, but now ready-mix zineb is available in the form of water dispersible powder of 75 per cent active ingredient. It is practically insoluble in water, non-compatible with fungicides or insecticides containing lime but compatible with other pesticides. It is unstable in light, heat or moisture and may deteriorate in storage. Its mode of action is the same as that of nabam. It is used for control of a large number of diseases affecting foliage and fruit including blight diseases of potato. It is not phytotoxic except on zinc-sensitive plants. It may also be used for soil treatment for control of damping off and other soil-borne diseases. It also shows nematicidal properties against *Meliodogyne* spp. Zineb in mixture with copper oxychloride preparations is also marketed. It has low mammalian toxicity, but can cause irritation to the skin in some cases.

(c) *Maneb*—*Manganous ethylene bisdithiocarbamate*. Maneb is similar to zineb, Mn^+ replacing Zn. This fungicide has been introduced in 1950. It shows properties similar to zineb. It decomposes in storage in excessive heat or in presence of moisture. It is applied in the form of spray and the usual formulation are wettable powder with 80 per cent active ingredient. It is also compatible with most pesticides except fixed copper fungicides and Bordeaux Mixture, and some growth promotion substances like gibberellic acid. It is of low toxicity to warm blooded animals, but may cause injuries to skin. It is not phytotoxic but excessive application may cause damage to cauliflower, tomato, pepper, lettuce, etc. It is a very effective fungicide for control of foliar diseases particularly blight diseases of potato and tomato. Maneb is now followed by a coordinate complex 'Mancozeb' of zinc ions and manganous salts.

(d) *Metham-sodium or Vapam.* Sodium N-methyl-dithiocarbamate

$$CH_3-N-C-S-Na$$
$$\quad\;\; |\quad |$$
$$\quad\;\; H\quad S$$

It is a soil fumigant introduced in 1954. It has been found to be useful for partial sterilisation of the soil against fungi and nematodes. It also acts as a herbicide. It is readily soluble in water. Though stable in aqueous solution it is decomposed in the soil to form methyl isothiocyanate which is the active component. Temperature, moisture and physical conditions of the soil influence decomposition of vapam. It is an irritant to eyes and skin and safety precautions are to be used for the purpose. It is strongly phytotoxic and treated soils should not be planted until decomposition is complete and all smell of the chemical is gone. The compound is sold as a stock solution containing 33 per cent anhydrous salt. Its mode of action is the same as that of other bisdithiocarbamates.

Mode of action of Thiocarbamate fungicides.

Precise mode of action in either case is not known. It is considered that antifungal activity of thiuram disulphides and dialkyl dithiocarbamates is due to formation of "thioureide" ion.

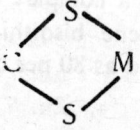

They are also strong chelating agents. They disrupt the activity of cell besides acting as chelating agents by attachment to 1 : 1 complex metal and dithiocarbamate ions to enzymes, attachment of dithiocarbamate ions to metals bound to proteins and reaction of free radical intermediates with cellular components and lethal catalysis. Bisdithiocarbamate groups act by production of diisothiocyanate groups ultimately, ethylene thiuram monosulphide acting as an intermediate product. Isothiocyanate reacts with SH groups essential for functioning of SH bearing enzymes. Isothiocyanates react with thiolo group of proteins and enzymes. However there is evidence to indicate that isothiocyanate may not be entirely responsible for toxicity of bisdithiocarbamates. N-methyl-dithiocarbamate has been found to cause alteration of permeability of the mycelium of *Rhizoctonia solani*.

There are several combinations of fungicides of which at least one component is a dithiocarbamate. Commercially more important ones are:

1. Metiram

This is a combined formulation containing zinc ethylene bisdithiocarbamate and ethylene bisthiuram polysulphide. It is light yellow solid

$$[-CH_2-NH-C(S)-S-S-C(S)-NH-CH_2]n$$
$$[-CH_2-NH-C(S)-S-Zn-S-C(S)-NH-CH_2]m$$
$$n : m = 1 : 3$$

It acts as a contact poison. LD_{50} for rat is 6100 mg/kg. The tolerance limit in food products of vegetable origin is 1 mg/kg. The maximum tolerated concentration in the air of the working zone is 0.1 mg/m^3.

2. Mancozeb (Manzeb, Dithane-M-45)

The active ingradient is a complex of zinc ions (2%), manganese (16%) and ethylene bisdithiocarbamate (62%). It is grayish yellow powder sold as 80 per cent w.p.

3. Cuprosan

The active ingradient is a mixture of Dithane-Z-78 and copper oxychloride. It is a bluish green odourless powder. It is formulated as 80 per cent w.p. (65% copper oxychloride and 15% Dithane-Z-78).

It is moderately toxic, LD_{50} for rats is 400 mg/kg. The maximum tolerated concentration in the air of the working zone is 0.1 mg/m^3 and in water basins 0.1 mg/litre. The tolerance level in vegetable and fruits is 0.1 mg/kg.

4. Miltox special

Another mixture of copper oxychloride and zineb. It is marketed in the form of 57 per cent wp containing 37 per cent copper oxychloride and 20 per cent zineb.

5. Polychom

It is a mixture of metiram with copper oxychloride. It is marketed in the form of 80 per cent wp containing 3 parts of 75 per cent metiram and 1 part of 90 per cent copper oxychloride.

D. Other organic fungicides

(1) Glyodin

Immediate success of thiocarbamates as commercial fruit protectants stimulated investigation on other groups of organic chemicals as safer protective fungicides with no phytocidal damage. Wellman and McCallan in 1946 discovered fungicidal properties of the derivatives of glyoxalidine which resulted in the introduction of 2-heptadecyl 2-imidazoline acetate (2-hepadecyl glyoxalidine) as a commercially acceptable fungicide named "Glyodin". It is almost insoluble in water, soluble in sopranol up to 39 per cent. It has been found to be effective against apple scab and a few other fruit diseases in temperate countries. It is marketed as 70 per cent wettable powder (glyoxide dry) or 34 per cent solution in isopropanol. Mixtures with hydrocarbon oils, parathion, malathion, dieldrin, etc. may be injurious to foliage.

Fungitoxicity of glyodin is considered to be due to interference with biosynthesis of purines which are constituents of nucleic acids.

(2) Cyprex or Dodine

n-dodecylguanidine acetate $C_{12}H_{25}NH.C(=NH.).NH_2.CH_3.$ COOH is a protectant fungicide commercially marketed in 1959. It is soluble in hot water, alcohol, but insoluble in other solvents. It is stable even under moderately alkaline or acid conditions, but degraded by strong alkali. It has low toxicity to warm blooded animals, but may cause irritation to skin and eyes on contact. It is harmless to bees. Dodine is extremely surface reactive (cationic wetter) with unusual tenacity. It has also eradicant properties to some extent. It can be safely mixed with most of the commonly used fungicides and insecticides. The usual formulations for spray are 65 per cent wettable powder and 20 per cent liquid.

It has been found to be effective against apple scab and a number of foliar and fruit diseases.

Effect on cellular membranes with alteration of permeability and inhibition of intracellular enzymes are considered to be the mechanism of fungitoxicity of dodine.

(3) Quinone compounds

(a) *Chloranil.* 2, 3, 5, 6-tetrachloro-1, 4-benzoquinone, was released in 1940 under the trade name "Spergon" as a seed treating fungicide. It has been mainly used as a seed treating fungicide. Foliar application has not been found to be successful due to its decomposition in light. It is usually compatible with other seed treating chemicals. It has low mammalian toxicity. Dosage for seed treatment is 2-3 gm per kg of seed depending on the size and weight of the seeds.

(b) *Dichlone.* 2, 3-dichloro-1, 4-napthaquinone, was introduced in 1943, following the success of chloranil, as a seed treating fungicide under the trade name "Phygon". It is practically insoluble in water, stable to heat, light and water, but degraded by alkali. It is compatible with dry or wettable powder formulations of other seed treating fungicides, but not with oils or oil emulsions. It has low toxicity to warm blooded animals, but may cause irritation to skin. Besides seed protecting fungicide it can be used as a foliage protectant. It is nonphytotoxic.

Quinone compounds are believed to act through their interference in the oxidation-reduction mechanism of the pathogens.

(c) Dithianon (Delan, Thynon)

The active ingradient is 5, 10-dihydro-5, 10-dioxonaptho (2, 3b)-1, 4-dithin-2, 3, dicarbonitrile or 2, 3-dicyano-1, 4-dithia-anthraquinone. It forms brown odourless crystals melting at 230°C. It is used as a protective fungicide against many foliar diseases. Sold as 75% w.p. dosage being 0.2% active ingradient 2-4 kg/ha.

(4) *Captan*

N-(trichloromethylthio), cyclohex-4-ene 1, 2, dicarboximide was introduced in 1949 as a protective fungicide. The compound is virtually insoluble in water, stable, practically nonvolatile and of pungent odour. It is of low toxicity to warm blooded animals, harmful to fish. Since its introduction, this fungicide has proved its efficacy against a large number of fungi inciting diseases of fruits, vegetables and ornamentals. It is not phytotoxic. Captan has been found to be useful as seed treating fungicide as well as for reducing post harvest losses in storage caused by a number of fungi. It is incompatible with all alkaline materials, lime sulphur, Bordeaux Mixture. It can, however, be used with other pesticides. Common formulation for spray preparation is 50 per cent wettable powder and for seed treatment 75 per cent dressing. It can also be used for drenching of soil in nurseries for protection against damping off at 0.5 per cent concentration.

(5) *Folpet*

N-(trichloromethylthio) phthalimide closely related to captan recently introduced and commonly sold under the name "Phaltan". It resembles captan in physical and biological properties, but is better for certain purposes. Nevertheless its use so far has been comparatively limited. It is available as wettable powder.

(6) *Difoltan*

Closely related to captan and folpet is difoltan N-1, 1, 2, 2-tetrachloroethyl) sulphenyl-cis-4-cyclohexene-1,2-dicarboximide. It is marketed as 80 per cent wettable powder or emulsified concentrate containing 80 per cent active ingredient.

Mechanism of action of captan and related compounds is due to—$SCCl_3$ group which causes interference with sulphahydryl enzymes and oxidative processes particularly activity of co-carboxylase. It is also considered that the action may be due to unspecific mechanisms to toxicity.

(7) *Karathane* (*Dinocap*)

Dinitro (1-methylheptyl) phenyl crotonate (Capryl dinitrophenyl crotonate) was first introduced in 1954 specifically for the control of powdery mildew. It is an ester, and unstable in the presence of alkali and should not be mixed with lime-sulphur or other pesticides giving an alkaline reaction. It is soluble in oil hence it cannot be used as oil based spray. It has proved to be an excellent substitute of sulphur for control of powdery mildew on sulphur-shy or sulphur-sensitive plants. Besides giving a protective action, it also acts as an eradicant but no systemic. It also controls mites to some extent. It is of comparatively low toxicity to warm blooded animals, but inhalation should be avoided and hair and skin should be protected from its staining properties. It is toxic to fish. Usual formulations are 25 per cent wettable powder or 50 per cent emulsified concentrate for sprays. The acute oral LD_{50} for rats is 980-1190 mg/kg.

(8) *Pentachloronitrobenzene (quintozene) and*

teenazene-Tetrachloronitrobenzene

Chlorinated nitrobenzenes were introduced in 1930 and its efficacy against soil borne pathogens and dry rot of potato were demonstrated in 40's, but it has been recently widely used for control of *Rhizoctonia*, *Sclerotium* and other soil fungi. It is practically insoluble in water, of low volatility and stable in soil. It has a long persistent effect in soil. It is of low toxicity to warm blooded animals. Cucurbits and tomatoes are sensitive to it. PCNB has nematicidal properties also. Usual formulation is 20 per cent dust. These compounds mainly act as antisporulants and fungistatic rather than fungicidal. The acute oral LD_{50} for rats is >12000 mg/kg.

(9) Organic tin compounds

Inorganic compounds of tin do not have any fungicidal value, but a large number of organic tin compounds were found to have fungicidal properties. Majority of them have phytotoxic properties only fentin acetate (triphenyltin acetate [TPTA] and triphenyl tin hydroxide-[fentin hydroxide (TPTH)] have been found to be useful fungicides. TPTA is sold under the name of brestan and TPTA and TPTH are sold in the name of deuter. TPTA has also insecticidal properties particularly against surface feeding insects. It was introduced in 1954 for the control of potato blight. Since then it has also been found to be effective against a number of foliar diseases. The rate of each application should not exceed 112 gm of active ingredient per acre 156 gm of 60 per cent wettable powder in TPTA or 20 per cent wettable powder in TPTH (in which formulations they are marketed). These compounds should not be mixed with oil based formulations. Organic tin compounds are often mixed with maneb and marketed as mixed formulation (60 per cent fentin, 20 per cent maneb). Use of these compounds is banned in India at present.

(10) Oils

Oils have attracted attention for fungicidal purposes since the beginning of the twentieth century. Preliminary works by earlier workers with 2 per cent kerosene water emulsion and later highly refined oil suggested feasibility of control of powdery mildews in some shrubby crops.

Practical use of petroleum oil as fungicide was rediscovered, in 1955 when it was observed that oil used as carrier of copper or zineb sprays for control of sigatoka (*Cercospora musae*; *Mycosphaerella musicola*) disease of banana was equally effective against the disease even in the absence of the fungicide. It was, however, noticed that there was reduction in yield in banana plants sprayed with oil alone. It was found that this effect would be eliminated by change in application methods. Nowadays oil-mist sprays either with oil alone or oil-water emulsion are widely used, for control of sigatoka disease of banana. It has been considered that the fungicidal action of oil is a purely physical one. Oil which penetrates into the leaf tissue and is not translocated from the same prevents induction of disease symptoms. Phytotoxicity of mineral oils has been reported in citrus, banana and a few other crops due to inefficient spraying leading to accumulation in concentrated areas whereby necrotic flecks arise.

(11) *Salicyanilide* (Salicylamide).
2-hydroxy-N-phenylbenzamide.
It is a broad spectrum non systemic fungicide mainly used in glass houses. It is toxic to oomycetes and ascomycetes. If 0-hydroxyl group is replaced by a chlorine or bromine it loses its toxicity to above fungi and becomes specifically toxic to basidiomycetes. 0-toluanilide is marketed as mebenil and recommended against rusts.

(12) *Anilazine* (-dyrene).
2,4-dichloro-6 (-2-chloroanilino)-1,3,5,-S-triazine or 4, 6-dichloro-N-(2-chlorophenyl)-1,3,5-triazin-2-amine.
It is used against the leaf spots of grasses due to *Dreschlera* spp. and rots due to *Fusarium* and *Rhizoctonia* spp. It is also effective against *Botrytis, Septoria, Colletotrichum* spp. as a protectant when applied at 170-250 g a.i./100 1 and is obtained as 50% w.p. (This formulation is defunct).

(13) *Dicloran* (DCNA, allisan or botran).
2-6,-dichloro-4-nitroaniline is a protective fungicide. It has a limited area of use against rots and post-harvest decays due to *Botrytis* spp., *Molinia* spp., *Rhizopus* spp., and *Sclerotinia* spp.

E. Antibiotics
The chief merits of antibiotics as potent agents in plant disease are: (i) selectivity between target and non-target organisms except in a few cases, (ii) easy decomposition, (iii) low dose, (iv) easy and non-hazardous manufacture, and (v) no operational hazard.

With the development and successful application of antibiotics for control of diseases in men and animals, there has been a continuous search for antibiotics for the control of plant disease. According to Martin, more than 340 antibiotics have been enlisted for the purpose but majority of them have been found to be of no practical value because of their instability, and one or other undesirable properties. Only a few have been found to be of promise.

(1) Streptomycin

It is obtained from culture filtrates of certain strains of *Streptomyces griseus*. Chemical structure of streptomycin is that of a glycoside in which aglycone streptidine is linked to N-methyl glucosamine through an unusual sugar streptose. It is strongly basic in character and is marketed either as sulphate or hydrochloride. A formulation of streptomycin mixed with oxytetracycline (terramycin) is marketed under the name Agrimycin. Streptomycin is effective against both Gram-positive and Gram-negative bacterial plant pathogens, but they do not show any fungitoxicity against true fungi. It is systemic in nature, but because of its possible phytotoxicity, application in the desirable concentration is limited. Streptomycin has been successfully used against bacterial seed-borne pathogens at 100 ppm or more. There is possibility of development of strains of bacteria resistant to streptomycin. Addition of oxytetracycline retards the development of resistant strains, hence the use of "Agrimycin" is advocated. Streptomycin and Agrimycin are compatible with lead arsenate, DDT, aldrin, fixed copper fungicides, ferbam, parathion, wettable sulphur and zineb but not with gamma-BHC, chlordane or glyodin.

(2) Cycloheximide

Besides streptomycin, from the culture filtrates of *Streptomyces griseus* an antifungal antibiotic was obtained named cycloheximide-popularly marketed under the trade name "Actidione". It is chemically -[2-(3,5-dimethyl-2-oxycyclohexyl) -2-hydroxyethyl] glutarimide. It is slightly soluble in water, stable in neutral and acid solutions. Cycloheximide is obtained as a byproduct in streptomycin manufacture and is remarkably active against a number of fungi including Phycomycete *Pythium debaryanum* but inactive against bacteria. It is systemic in nature, easily taken up by roots and leaves are rendered highly antifungal. Fungitoxicity is reported to persist for a period of five weeks after spraying. In spite of its having a wide spectrum antifungal activity including that on wheat rust, it has not been able to make an impact in the field of disease control because of the high cost of production and a narrow margin between the effective and phytotoxic concentration. Phytotoxicity has been ascribed to mutagenic effect. Concentration of 5-100 ppm depending on the disease and host plant has been found to be effective. Cycloheximide can be mixed with lead arsenate, DDT, dieldrin, malathion, parathion, methoxychlor, ferbam, glyodin, captan, thiuram and organomercurials, but not with chlordane or pesticides containing lime or alkaline in reaction.

It is of comparatively low mammalian toxicity, though toxicity level is much higher than streptomycin or griseofulvin.

Mode of action of cycloheximide is due to its interference with RNA synthesis.

(3) *Griseofulvin*

7-chloro-4,6-dimethoxycoumarin-3-one-2-spiro-1′-(2′-methoxy-6′-methylcyclohex-2′-en-4′-one) was first isolated from *Penicillium griseofulvum* in 1939. More recently it is recognised as a metabolic product of many *Penicillium* spp. It is neutral in reaction and systemic in nature. Uptake through transpiration stream is rapid.

Griseofulvin is not a true fungicide in the strict sense of the term. It does not prevent germination of spores but has a remarkable effect on the development of hyphae. Even at a very low concentration 10μg/ml it causes curling stunting, abnormal branching with distortion and loss of apical dominance in hyphae or germ tubes. Fungi with chitinous cell walls are affected, but not all fungi, even a few with chitinous cell wall have not shown any response. Bacteria are unaffected. Its activity is now attributed to interference with nucleic acid. It is relatively non-phytotoxic, and toxicity to warm blooded animals is extremely low ; but it is slowly degraded in plant tissue and more quickly in soil, hence it is of very limited practical utility in the field. It has been found to be effective against a number of powdery mildews, and a few foliar diseases.

Dosage for foliar sprays varies between 100 and 1000 ppm.

(4) *Aureofungin*

Produced in submerged cultures of *Streptomyces cinnamomeus* var. *terricola* in the laboratory of Hindusthan Antibiotics, India (1964). Chemically it belongs to a new aromatic subgroup among heptaene—the aromatic moiety being N-methyl-p-amino aceptophenone and mycosamine.

It is insoluble in water, but soluble in alkaline solutions, hence soap solutions containing this antibiotic (1 gm in 10 ml of water) can easily be prepared. It is unstable in the presence of moisture and light. It has been claimed to be a broad-spectrum antifungal antibiotic systemic in nature. It is virtually non-toxic to warm blooded animals. It has been found to be effective under experimental field conditions in controlling several fungal diseases. Its effectiveness as a post-harvest dip for the prevention of rot in storage and seed treating chemical in seed-borne infections has also been observed. Normal concentration for use is 5-10 mg/1000 ml.

(5) *Blasticidin-S*

It is produced by *Streptomyces griseochromogenes*. It contains a nucleoside, cytosinine, and an acid blastidic acid. It has been found to be effective against many species of bacteria and a few fungi including *Pyricularia oryzae*. This antibiotic has also the property of inhibiting growth, hence can be applied after infection has been established as a chemotherapeutant. It is slightly phytotoxic on rice plants, and is toxic to mammals. It is used as a wettable powder. Effective dosage of active ingredient is 50-100 mcg/ml.

(6) *Kasugamycin*

Obtained from the culture broth of *Streptomyces kasugaensis* in 1965. It is a water soluble base. It is selectively effective against *Pyricularia oryzae* and *Pseudomonas* spp. It is effective in low pH in the same concentration as that of blasticidin-S against blast disease.

It is more effective as an eradicant than a protectant.

It appears to be inactive against rice blast cultures in vitro, but vivo substances in juice of rice plant active it in some way.

Strains of *Pyricularia oryzae* resistant to blasticidin-S and kasugamycin have been reported.

F. Systemic fungicides

Success of systemic insecticides in the control of insect pests prompted the search for chemicals other than antibiotics to be used as systemic fungicides which will act as direct chemotherapeutants particularly for vascular pathogens. Interest in chemotherapy and use of systemic fungicides was stimulated by the availability of diverse organic compounds possessing specificity for action as fungicides, chemotherapy of human diseases and discovery of antibiotics. Hundreds of chemicals have been tried and periodic reviews on the progress of development of systemic fungicides and chemotherapy, have been published. Compounds which have shown promise for commercial application are as follows.

(1) *Oxathins*

Two oxathiin compounds developed by Von Schmeling and Kulka in 1966; (a) 2,3-dihydro-5-carboxianilido-6-methyl-1,4, oxathiin—now changed to 5,6-dihydro-2-methyl 1,4 oxathiin-3-carboxianilide (DMOC, vitavax) and its oxide (b) 2,3-dihydro, 5-carboxianilido-6-methyl-1,4. oxathiin-4,4 dioxide (DCMOD, plantvax). Both these compounds are water soluble and do not show any phytotoxic properties on crops used. Both these compounds, however, display a remarkable selectivity in the antifungal spectrum. Fungi belonging to Basidiomycetes are particularly sensitive to DMOC. Of the fungi belonging to other groups only *Verticillium albo-atrum* and *Monilia cinerea f. americana* have been found to be sensitive to DMOC. DCMOD has, however, been found to be active against a number of pathogens belonging to Fungi Imperfecti. Selectivity in action of these compounds has been attributed to the degree of uptake and binding. DMOC is not very stable when applied to the soil and complete degradation may take place in 10-30 days. DMOC (Vitavax) has shown promise as a seed dressing or soil drench for the treatment of cereals against loose smuts (*Ustilago nuda, U. avenae*) which are internally seed borne and *Rhizoctonia* disease of cotton and sugar beet. Both these have been employed against control of smuts, and there are indications that they may be of value in the control of rusts of cereals and vegetables. The mode of action of these compounds is considered to be the interference with synthesis of protein, RNA and DNA in rapidly metabolising cells. They have been found to inhibit mitochondrial respiration. Rate of use is 50-250 gm/50 kg of seed in seed treatment and 10 per cent dust in soil treatment.

2. Pyracarbolid (Sicarol)

3,4-dihydro-6-methyl-2H-pyran-5-carboxamilide.

It is a systemic fungicide controlling basidiomycetes such as rusts, smuts, Rhizoctonic solani, blister blight of tea. It is recommended at 0.28 kg/ha against blister blight of tea in Sri Lanka.

3. Benodanil (Calirus)

2-iodobenzoic acid anilide. It has been developed by BASF (West Germany) available as 5 per cent w.p. and 20 per cent e.c. It is systemic, prophylactic as well as as curative fungicide. The recommended dose is 1.1 kg a.i./ha. It is more effective against rust diseases of cereals.

4. Pyrimidine derivatives

(a) Methyrimol. 5-n-butyl-2-demethylamino-4-hydroxy-6-methyl pyrimidine has been shown to be systemically active against powdery mildew of cucurbits, cinarea, chrysanthemums and sugar beet. Its fungitoxicity is much less against powdery mildews of apples, pears, roses and cereals, and it has no effect against mildew of grapes.

It is stable to heat, acid and alkali and sparingly soluble in water. It is quickly degraded inside the plant, though it is persistent in the soil. Because of the possibility of loose fixation in the soil, the chemical has to be applied near the root zone as a drench as soon as symptoms of powdery mildew appear. The rate of application will vary from 0.1 to 1 gm of active ingredient per plant.

Sphaerotheca fuligena has shown resistance to methyrimol.

(b) Ethirimol. It is closely related to methyrimol, (5-n-butyl-2-ethylamino-4-hydroxy-6-methyl pyrimidine) (substitution by ethylamino group of dimethyl amino group). This compound gives good control of powdery mildew of cereals, when applied as a seed treatment or as a combined drill granule at the rate of 3 kg and 1 kg per hectare results in complete control.

Exact mode of action of these two fungicides is not known, but evidences point to an interference with tetrahydrofolic-acid-directed C-1 transfer reactions required for wide range of cellular reactions and inhibition of some pyridoxal-requiring enzymes. Some strains of *Erysiphe graminis* have shown resistance to these chemicals.

(c) *Triarimol.* (4-chlorophenyl)-bis-pyrimidine-yl methanol.

This compound is active at a very low concentration against many fungi. Triarimol has much broader spectrum of action than any other in the group, being active at 30-40 ppm against the apple scab, powdery mildews and cereal rusts as foliar sprays. There are also reports of action against cereal smuts as soil drench, even against lentil wilt caused by *Fusarium* sp.

(d) *Bupirimate.* 5-butyl-2-ethyl-amino-6-methyl pyrimidin-4-yl-dimethylsulphamate) is the sulphamate ester of ethirimol. It is marketed by I.C.I. as 'Nimrod'. It is specifically used for control of powdery mildew in apples, as spray, because of intralaminar movement in leaves.

(3) *Benzimidazole*

Benomyl or Benlate methyl (1-butylcarbamoyl) benzimidazol-2-yl-carbamate is an excellent systemic fungicide which shows good eradicant and protectant activities and has mite ovicidal action. It has been found to be effective against a large number of fungal diseases of crops including rice blast, disease caused by *Cercospora*, *Verticillium* on cotton, powdery mildews and black spot or roses, but it has little effect on diseases caused by *Helminthosporium* spp. and Phycomycetous fungi. It is not phytotoxic and has a very low toxicity to warm blooded animals. It is not a skin-irritant. It is nonvolatile.

Addition of surfactant is beneficial in dispersion of the chemical. Normal dosage for use varies from 0.025 to 0.1 per cent depending on the disease and host plant.

In aqueous solution benomyl is hydrolysed to MBC (methyl benzimidazol-2-yl-carbamate) and this breakdown product is responsible for fungitoxicity. MBC is considered to interfere with DNA synthesis or some closely related process namely nuclear or cell division.

Resistance against benomyl has been reported in a large number of fungi.

(4) *Chloroneb*

1,4-dichloro-2,5-dimethoxybenzene was introduced as an experimental fungicide in 1967. This compound is almost exclusively fungistatic and relatively non-fungicidal. It is easily taken up by root system by absorption and is mainly concentrated in roots and lower stem, so it is useful as a supplemental seed treatment and soil treatment at the planting time to control seedling disease. It is effective against *Rhizoctonia* spp. moderately so to *Pythium* spp. but gives poor control of *Fusarium* spp. and is inactive against *Trichoderma* spp. It has been found effective against *Sclerotium rolfsii*. It has a low solubility in water and is compatible with most pesticides and can be mixed with organomercurials for seed treatment. The seed treatment is effected as an overcoat of 0.6 per cent active ingredient through slurry. Soil treatment can be applied as a spray in between furrows (2-3 kg of 65 per cent wettable powder in 100 litres of water) or granular at the 15 kg/ha. It is of low toxicity to warm blooded animals but causes mild irritation to skin. Strains of *Ustilago maydis* resistant to chlorneb have been reported.

(5) *Carbendazim* (Bavistin, Derosal)

Chemically it is methyl benzimidazol-2-yl-carbamate.

Broad spectrum fungicide systemically active against ascomycetes, fungi imperfecti and some basidiomycetes but not against phycomycetes. It is both prophylactic and curative, commercially formulated at 50 per cent w.p. It is recommended for seed treatment 100-200 g/100 kg against seed borne diseases of cereals, and root rot of different crops caused by a number of fungi, also as foliar spray against leafspot, anthracnose, boll rots, powdery mildews.

Its action is like that of benomyl-interference with DNA synthesis.

(6) *Thiabendazole* (TBZ) 2-(4'-Thiazolyl) benzimidazole.

Thiabendazole was originally introduced as anthelminthic, but it has been found to be a broad spectrum systemic fungicide effective against many pathogens. It is translocated from the roots to leaves, but also in the less usual reverse direction and it is not degraded or broken down inside the plant tissue. The mode of action is similar to that of benomyl or MBC. Its effectiveness has been demonstrated against blue and green moulds of citrus fruits. Thiabendazole has been reported to give good control of seed-borne bunt, *Fusarium nivale*, *Septoria nodorum* when applied at the rate of 120 g/100 kg. Some workers have stated that a low dose of 20 g/100 kg also gives good results. Two sprays at one month interval at the rate of 60 gm/100 litres of water can give good control of leaf spot of sugar beet (*Cercospora beticola*). Verticillium wilt of cotton and potato have been found to be controlled by soil application of 30 kg/ha.

It is marketed as Thiabendazole W-7 containing 60 per cent of the active ingredient by Merck & Co., U.S.A.

(7) *Terrazole*

5-ethoxy-3-(trichloromethyl) - 1,2,4 - thiadiazole, commonly known as 'terrazole' or OM-2424 has been found to be effective in control of seed and seedling diseases of maize, cotton, sorghum, soybean, bean, potato, tomato and cucumber. It has been observed to give good control of seedling blight of ground nut against *Rhizoctonia*, *Fusarium* and *Thielaviopsis*, when the seed was treated with a mixture of terrazole (6.1 per cent) and quintozene (23 per cent) marketed as Terrachlor super X in the U.S.A. the dosage being 0.6 per cent. Soil application at the rate of 10-12.5 kg/ha has been reported to be effective against *Rhizoctonia solani*.

Hydantoin
  Glycophene
  Rovral, or 1 prodione
  3-(3' 5'-dichlorophenyl)-1-isopropyl carbomyl hydantoin.

The compound has a broader spectrum of activity apart from its effectiveness against *Botrytis cinerea* and other *Monilia* spp., can control a number of other pathogens particularly members of sclerotiniaceae. Rovral was first marketed as a protectant fungicide again *Botrytis* rot of grapes, tomatoes, ornamentals. It has a protective action against a number of soil and foliar pathogens. It is not effective against downy and powdery mildews. The chief merit of this fungicide is its activity against the strains of fungi which have been found to be resistant to benomyl and thiophanate methyl. It has low mammalian toxicity and is quickly degraded in soil into nontoxic compounds. It has a poor systemic effect, but is absorbed slowly by the leaves.

Vinclozolin (Ronilan) [3-3, 5-dichlorophenyl)-5-methyl-5-vinyl-1, 3-oxazolidino-2-4-dione] is an important compound of this group. It has been reported to be specific against *Botrytis cinerea, Monilia* spp. and *Sclerotinia sclerotiorum*.

Morpholines

Two compounds dodemorph and tridemorph are two important fungicidal compounds of this group. Dodemorph is 4-cyclodecyl-2, 6-dimethyl morpholine and tridemorph is N-tridecyl-2, 6-dimethyl morpholine. Both are eradicants but can be easily absorbed through leaf or root, translocated to growing sites and have protective and therapeutic action. Tridemorph is more commercially successful and is marketed as calixiin. Calixiin is used in the dose of 500-750 g/ha. Both these compounds have widespread use for control of powdery mildew of cereals, particularly of barley, likely to be phytotoxic on certain varieties of wheat and many other crops. Calixiin has been found to be effective against blister blight of tea (*Exobasidium vexans*).

Piperazines

Trifarine (Saprol)

1,4-di(2,2,2-trichloro-1-formainido-ethyl) piperazole.

It has a rather broad spectrum of activities and comparatively safe on a number of plants. It is effective against powdery mildew, can control both apple scab and powdery mildew, has been claimed to control leaf spot fungi, besides powdery mildew in barley. It has low aqueous stability, but systemically controls powdery mildews of cereals by soil application, and seed dresser against cereals rusts and some other powdery mildew. The fungicide inhibits ergosterol synthesis, thus induce alteration of mitochondria, cell membrane and cellular ultrastructure.

Metalaxyl (Ridomil)

methyl-dl-N-(2,6-dimethyl phenyl)-N-(2′ methoxyacetyl)-alaninate.

The fungicide is systemic and curative against the downy mildews and pythiaceous diseases. It is available as 25 per cent w.p. 1.5 per cent granule. It is recommended as seed and tuber treatment (0.3-0.5 per cent a.i.) for cotton, maize, sorghum, sugarcane and for potato against late blight as foliar spray against downy mildews, and soil drench (24-100 ppm.) against Phytophthora root rots and downy mildews. It interferes with fungal colonization, affects nucleic acid synthesis and may stimulate phytoalexins, thus promoting resistance.

Prothiocarb

S-ethyl-N-(3-dimethylamino propyl thiocarbamate hydrochloride). It is specifically active against the Peronosporales by soil application (5.6 kg a.i. per ha) alone or when combined with 0.3 per cent a.i. foliar sprays.

Cymoxonil (curzate)

2-cyano-N-(ethylaminocarbonyl)-2-(methoxyimino)acetamide. It is effective against downy mildews and potato late blight by soil and foliar application.

Thiophanate-methyl (Topsin M, Cercobin M)

1,2-bis-(3-methoxycarbonyl-2-thiureido) benzene. It is effective against a wide range of fungal diseases, e.g. scab of apple and pear, sigatoka of banana, rice blast, cercospora leaf spots. The chemical breaks down to carbendazim or its ethyl ester in plants and during prolonged storage. It is available as 50 per cent and 70 per cent w.p. and applied at 30-50 gm/100 litre.

Oxine (Chinosol)

8 hydroxy quinolinol or 8-quinolinol sulphate.

Fungitoxicity of this compound is related to chelation of metal that causes deficiency of the metal concerned. It is also believed to inhibit certain enzymes related to biosynthesis of pyridine a precursor of riboflavin. Though Oxine is an excellent fungicide, it has not been much used because of its high cost.

Pyrazophos (Afungal, Cura mil)

O-O-ethoxycarbonyl-5-methylpyrazolo-[1,5a] pyrimidin-2-yl-O.O-diethyl phosphorothionate.

It is quite stable when pure, but is soon decomposed in the presence of acid or alkali. This is systemic in action but has high LD_{50} (rats, acute oral value 286-632 mg/kg). It is recommended against powdery mildew of cereals and several other crops, specially cucurbits of 0.5-0.7 a.i./ha. It is rapidly absorbed by green tissues, but poorly by roots. It has a limited insecticidal and acaricidal action. Pyrazophos is converted into more toxic pyrazolopyrimidine which strongly inhibits oxygen consumption and growth in susceptible fungi and synthesis of nucleic acids and protein synthesis.

Ditalimfos (Plondrel)

0,0-diethylpithalimidophosphonothioate.

It is known for its curative effect against the apple scab. It has a good compatibility as tank mix.

Aluminium tris (Aliette)

Aluminium ethyl phosphite, has low mammalian toxicity (acute oral LD_{50}-6800 mg/kg). It has both protectant and curative action against oomycetes, particularly *Phytophthora* spp. except *P. infestans*. It can be used as a foliar spray (4 kg a.i./ha) or soil drench (4-16 gm a.i./m²) or granule.

Tributazil (Indar, Dithane R-24, RH 124)

4-N-butyl-1,2,4-triazole. It has been·found to be highly effective for control of brown rust of wheat. It is available as 70 per cent liquid concentrate for foliar spray or as seed treatment.

Triamiphos (Wepsyn)

P-5-amino-3-phenyl-1,2,4-triazol-l-yl-bis- (dimethylamino)-phosphate. It is a systemic fungicide against the powdery mildew. It has good insecticidal and acaricidal properties. It has high phytotoxicity and mammalian toxicity (acute oral LD_{50} for mole rats—20 mg/kg).

Triadimefon (Bayleton)

1-[H-chlorophenoxy]-3,3-dimethyl-1-[1,2,4-triazol-1-yl-butan-

2 one]. It has wide spectrum both prophylactic and therapeutic, effective against powdery mildew, rust, leafspots, can be used as seed dressing for smut of cereals.

(7) *Organophosphorus compounds*

There has been considerable interest recently in the use of certain organophosphorus compounds for control of rice blast in Japan evidently to replace use of phenyl mercury acetate and possibly to control leaf hopper vectors of dwarf virus also. The compounds that have shown promise are hinozan, kitazin etc.

Hinozan (0-ethyl-S, S-diphenyl phosphoroditholate), kitazin (0-0-isopropyl-S-benzyl, phosphorothiolate) and inazin (0-ethyl S-benzylphenyl phosphonothiolate) are equally effective in the control of blast in sprays of 400-500 ppm, above which there is danger of phytotoxicity. These chemicals have both eradicant and protectant action like kasugamycin and are best effective two to three days after infection. Of these kitazin is by far the most effective being most systemic and readily absorbed by the roots and translocated to the leaves. The chemicals can also be applied in the form of granules (10-15 kg/ha) in the water at the time of appearance of blast and the leaves remain resistant to infection for 30 days. Strains resistant to kitazin have been reported.

An account of the promising major therapeutic systemic fungicides is given in Table 6.2.

G. Organo-mercurial fungicides

Mercuric chloride has been known to be a very potent fungicide and bactericide. In view of its extreme toxicity including phytotoxicity, it was not used for control of plant diseases. Hiltner first demonstrated control of Fusarium disease of rye which is internally seed-borne by treatment of seeds with mercuric chloride. In pharmaceutical industries, organomercurial derivatives gradually replaced mercuric chloride. Following the line, chloro-phenol mercury was introduced as a successful chemical for seed treatment in 1914. The compound was to be used as a dip or steep treatment. It contained 18.8 per cent mercury. Later on seed treating chemicals in the form of powder which can be used dry was developed, and mercury content has been gradually reduced to less than one per cent. Though primarily intended for seed treatment, some organomercurials are being used as foliage and fruit protectant. True organomercurial derivatives have the general formula. R. Hg. X, R=hydrocarbon with or without substituent groups, X=represents an acidic radical. Mercury is attached direct to carbon atoms with one or more valency bonds. A large number of compounds are now available in the market with different compositions under different trade names.

TABLE 6.1. List of some organomercurial fungicides

Common name	Name of the compound
Agrosan GN (India, U.K.)	Ethyl mercury chloride and phenyl mercury acetate in equal proportion
Agrox (U.S.A.)	Phenylmercuryurea
Aretan, Agallol, Ceresan wet (India)	Methoxyethyl mercury chloride
Ceresan (Germany)	Methoxyethyl mercury silicate
Ceresan (U.S.A.)	Ethyl mercury chloride
Ceresan M (U.S.A.)	N-(ethylmercuri)P-toluenesulphoanilide
Ceresan dry (India)	Phenyl mercury acetate
New Improved Ceresan (U.S.A.)	Ethyl mercury phosphate
Mergamma (U.S.A.)	Phenyl mercury urea and lindane
Ortho 1M (U.S.A.)	Methyl mercury 8-hydroxyquinolinate
Panogen (U.S.A.)	Methyl mercury dicyandiamide
Puraturf (U.S.A.)	Phenyl mercury triethanol ammonium lactate
Semesan (U.S.A.)	Hydroxymercury chlorophenol
Semesan Bel (U.S.A.)	Hydroxymercury nitrophenol 12 per cent and hydroxymercury chlorophenol 3.2 per cent

Dry seed dressings contain from 0.6 to 1.5 per cent mercury, expressed as metal and liquid dressings 0.6-2.0 per cent mercury. Organomercurials are extremely toxic and precautions have to be used as per specifications. Treated seeds must not be used for food for man and livestock. In recent times organomercurial seed treating chemicals are mixed with gamma-BHC or dieldrin for the control of insects affecting young seedlings.

Mercury is absorbed by the spores or mycelium either as vapour or ion and acts upon SH groups of susceptible enzymes. Mercury compounds may injure seeds in higher concentrations and when stored under conditions of inadequate moisture control.

H. Formaldehyde

Formaldehyde was first employed as a seed disinfectant in the last decade of the nineteenth century in Germany and gradually spread to the U.S.A. and the U.K. as a substitute of copper sulphate, as it was found to be less injurious than the latter that was then in vogue for the treatment of seeds. Formaldehyde treatment was found to be injurious to seeds in many cases and subject to reinfection by other organisms.

Formaldehyde now finds wider use as a soil disinfectant. It is used in solution 37.5-40.5 per cent weight in volume of formaldehyde and up to 14 per cent methyl alcohol is sometimes added to delay polymerisation into paraformaldehyde which is useless for fungicidal purpose. Formalin solution is applied in the dilution of 1 : 50 as drenches and it easily penetrates into soil. Treated soils should not be planted until all smell of the chemical is gone which takes about three weeks in light soils and may be longer in heavier soils. Formaldehyde is extremely phytotoxic. but it leaves no residue in the soil.

1. Adjuvants or Auxiliary spray materials

Spreaders

Auxiliary spray materials are used to improve contact between the pesticide that is being sprayed and the surface on which it is sprayed and promote formation of the liquid/solid interface by reducing the surface tension. The objective is to prevent the spray solution to form large droplets and their roll down the leaves so that the material gets evenly spread out. A spreader has to do wetting before it can spread and wetting-spreading are closely related. Long chain alcohols, caesin, sulphite lye (available as a byproduct of paper industry), soap, gelatin, some oils are examples of spreaders. Spreaders need not have sticking properties.

Stickers

Efficacy of a spray or dust is determined amongst other factors by its retentivity. Substances which are adhesive in nature and improve the retentivity of the spray deposit are termed as stickers. Some spray spreaders namely detergents which reduce surface tension reduce tenacity, whereas gelatin, caesin, etc., which becomes insoluble on drying enhance tenacity. Synthetic resins, polyvinyl acetate, polybutenes, fishoil-resins, gelatin, etc., are good sticking agents.

Deflocculating agents

Materials used to keep the particles separate and prevent their aggregation into lumps are termed as deflocculating agents. They may also be called dispersing agents as they prevent sedimentation of soils in water when the solid particles are kept suspended in water. Sodium carboxymethyl cellulose, methyl cellulose, etc., are good dispersing agents. They act normally by increasing the viscosity of the medium and forming a liquid round the particles with layer of density similar to the medium or liquid.

Emulsifying agents

Materials which stabilise the emulsion of two liquids which are not miscible and tend to separate out or soluble in each other are emulsifying agents. There are a number of sprays in which the active ingredient is dissolved in an organic solvent, but later this solvent had to be dispersed in water for spraying. The emulsifying agents maintain the stability of the concentrate within limits.

QUESTION FOR DISCUSSION

1. How can different fungicides be placed into different categories ? Can you state the categories in which the different fungicides should be placed ?

2. On what criterion does efficiency of sulphur depend ? How does elemental sulphur act on the pathogen ?

3. What are the advantages and disadvantages of elemental sulphur vis-a-vis lime sulphur ?

4. Why is Bordeaux Mixture considered to be an excellent fungicide ? What are the disadvantages in the use of Bordeaux Mixture ?

5. What do you understand by fixed copper compounds ? Which is the most widely used fixed copper compound ?

6. Why have thiocarbamate fungicides gained popularity in recent years ? How are they used ?

7. What are the different groups of thiocarbamate fungicides ? Between metallic dithiocarbamates and ethylene bisdithiocarbamates which would you prefer to use as a protectant fungicide ?

8. What are the modes of action of thiocarbamate fungicides ?

9. Apart from thiocarbamates and copper fungicides, which fungicide is widely used for diverse purposes ?

10. Which organotin compounds have shown promise ?

11. In which disease is petroleum oil used ? What is its mode of action ?

12. Can you mention the antifungal antibiotics which have shown potentialities in disease control ? Why could not some of them make any impact inspite of efficacy ? Can you state some antibiotic preparations which are widely used against specific diseases ? Can you suggest a serious drawback in the use of antibiotics in the control of plant diseases ?

13. Why have systemic fungicides not been effective so far against Phycomycetes ?

14. Which organophosphorus compounds are in wide use against blast disease ? Which compound do you consider to be most effective and why ?

15 Why are systemic fungicides called therapeutic fungi-

TABLE 6.2. Major therapeutic Fungicides (Systemic)

Chemical group	Chemical Name	Trade Name	Manufacturer	Remarks
Organophosphorus	Ditalimphos	Plondrel	Dow	Non-systemic, excellent powdery mildew fungicide used on apples.
	IBP	Kitazin P	Kumiai	Effective against blast of rice, acropetally systemic; disease can be controlled by granular application in water.
	Edifenphos	Hinosan	Bayer	Systemic, effectively used against blast of rice.
	Pyrazophos	Afugan, Curamil	Hoechst	Systemic, excellent against powdery mildews, widely used in a number of crops specially cucurbits.
Benzimidazoles	Carbendazim	Derosal Bavistin	Hoechst BASF	These systemic compounds are known to be converted into carbendazim in plants and this latter compound often known as MBC is fungitoxicant. Benomyl produces another antifungal compound. Wide sprectrum activity except against Phycomycetes. Resistant strains have developed, hence not recommended in many countries.
	Benomyl	Benlate	Dupont	
	Thiabendazole	Tecto	Merck	

TABLE 6.2 (Contd.)

Chemical group	Chemical Name	Trade Name	Manufacturer	Remarks
Thiophanates	Thiophanate methyl	Topsin M Cercobin M	Nippon Soda	
Carboxilic acid anilides	Carboxin	Vitavax	Uniroyal	Usually selective against certain Basidiomycetes. Plantvax however is effective against a number of fungi of group Imperfecti. Carboxin and fenfuram are particularly useful against loose smut, can control externally borne smuts and bunts, but ineffective against other seed-borne fungi. Oxycarboxin, pyrocarbolid and benzodanil are more effective against leaf rusts, but none of them are equally suitable for all such rust fungi ; pyracarbolid is used against coffee rust, while oxycarboxin, benzodanil against leaf rusts of cereals and vegetables.
	Fenfuram	Panoram	Kenograd	
	Oxycarboxin	Plantvax	Uniroyal	
	Pyracarbolid	Sicarol	Hoechst	
	Benodanil	Calirus	BASF	
Morpholines	Tridimorph	Calixin	BASF	Widespread use for control of powdery mildew of cereals, particularly of barley ; likely to be phytotoxic on certain varieties of wheat and many other crops. Calixin has been found to be effective against blister blight of tea (Exobasidium vexans)
	Dodemorph	BAS328F		

TABLE 6.2 (Contd.)

Chemical group	Chemical Name	Trade Name	Manufacturer	Remarks
Pyrimidines	Ethirimol	Milstem	I C I	Success of ethirimol in controlling powdery mildew of barley is its systemic movement in plants following seed treatment and long term protection.
	Dimethtrimol	Milcurb	I C I	Used for control of powdery mildew of cucurbits and ornamentals when used as soil drench.
	Bupirimate	Nimrod	I C I	For powdery mildew control in apples, as spray, because of intralaminar movement in leaves.
Piperazine	Triforine	Saprol	Cela	Has a rather broad spectrum of activities and comparatively safe on a number of plants effective against powdery mildew, can control both apple scab and powdery mildew, has been claimed to control leaf spot fungi, besides powdery mildews in barley.
Triazoles	Fluotrimazole	Persulon	Bayer	More prophylactic, used for powdery mildews as foliar application.

TABLE 6.2 (Contd.)

Chemical group	Chemical Name	Trade Name	Manufacturer	Remarks
	Triadimefon	Bayleton	Bayer	Wide spectrum, both prophylactic and therapeutic, effective against powdery mildew, rusts and leaf spots, but can be used as seed dressing for smut and of cereals.
Oxazolidine	Vinclozolin	Ronilan	BASF	Has been reported to be specific against *Botrytis cinerea*, *Monilia* spp. and *Sclerotinia sclerotiorum*.
Hydantoin	Glycophene	Rovral	Rhone-Poulenc	Broader spectrum of activity apart from its effectiveness against *Botrytis cinerea* and other *Monilia* spp. can control a number of other pathogens; non-systemic although slowly absorbed into plant tissue where it is degraded.

The systemic and therapeutic fungicides so far discovered have not been found to be effective against Phycomycetous fungi. Recently Dupont DPX 3217 (2-cyano-N-[ethylamino]carbonyl)-2-(methoxyimino acetamide) has been claimed to be effective against grape downy mildew (*Plasmopara viticola*), potato late blight (*Phytophthora infestans*) and other Peronosporales, so also Ciba-Geigy 38140 and Prothiocarb.

(Ref :—Pearson, A. J. A. 1977).

cides ? Can you name promising ones of different groups ? Why have always been an interest for systemic fungicides ?

16. Why are organomercurials widely used ?

17. What do you understand by adjuvants ? How are they useful ? Can you mention the broad categories of adjuvants ?

18. What is the difference in the mode of action of spreader and sticker ; deflocculating and emulsifying agents ?

FUNGICIDES

Can you have promising uses of these are known.
have always been an interest for systemic fungicides?
b. Why are organochlorine widely used?
14. What do you understand by adjuvants? How are they
use? Can you mention the broad categories of adjuvants?
15. What is the difference in the mode of action of specific
and actinomycetous and inhibitory fungicides?

CHAPTER 7

HERBICIDES

Application of chemicals for control of weeds is a difficult proposition than the same for pesticides for eliminating diseases or pests or affording protection against them, because weeds are very closely related to the crops with which they remain associated. Naturally application of a chemical intended to kill the weed plants may affect both. Hence weedicides or more precisely, herbicides call for a high degree of selective toxicity which will enable them to save the crops, but exterminate the plants associated with them as weeds. Based on their mode of action herbicides may be classified as selective and non-selective. They may also be divided into contact and translocated. Both categories of chemicals, selective and non-selective herbicides, may act on the areas on which the deposition of the chemical takes place or they may be absorbed by the plant surface and get distributed over the plant body. Besides selective and non-selective herbicides, there may be a third category—soil sterilants, which are used to render the soil unamenable for support of plant growth either for a short period (less than a year) or for long duration more than a year by suppressing all vegetation on the land treated with chemicals.

Herbicides may be applied to the foliage of the weed or to the soil at any of the following stages : (a) before planting of the crop (pre-planting), (b) after planting but before emergence of seedlings (pre-emergence), and (c) after emergence of seedlings (post-emergence).

A contact herbicide in the preplanting stage is used shortly before planting to get rid of annual weeds. These weedicides used are non-persistent and leave no toxic residue, whereas persistent translocated herbicides are used at the preplanting stage to control perennial weeds. They are to be applied in advance before the seeds are intended to be sown so that there is no toxic residue.

Pre-emergent treatment of contact herbicides is given in those cases where weed seeds germinate quicker than the crops and the weed plants come out of the soil sufficiently ahead of the crops. In such cases weed seedlings are killed before seedlings of the crops can emerge and become liable for damage due to exposure to herbicides. Otherwise sowing has to be delayed till the weed seeds have germinated and herbicide is to be applied immediately after sowing of seeds of crop plants. Application of persistent translocable herbicide in the pre-emergent stage is attended with risk, unless the chemical applied is selective in action.

In the post-emergent stage both weeds and crops are exposed to the herbicide. Due to the selective action of the herbicide, only weeds get destroyed, while crop plants remain unaffected.

Classification of herbicides

Selective—*Contact*

dichloral urea, sodium isopropyl xanthate, calcium cyanamide, dinoseb.

Translocated

2-4-D, 2-4-5-T, 2-3-6-TBA, 2,4,5-TES, TCA, MCPA, MCPB, dalapon, CIPC, CDAA, maleic hydrazide, propham etc.

Non-selective—Contact

mineral oils, sulphuric acid, endothal, pentachlorophenol, sodium pentachlorophenolate, sodium arsenite, sodium chlorate, diquat, paraquat.

Translocated

aminotriazole, ammonium sulphamate, borates, urea-based compounds, anilides, phenyl arsenic acid, etc.

Soil sterilants—arsenite, ammonium thiocyanate, sodium trichloroacetate. Besides vapam (sodium-methyl thiocarbamate), methylbromide, chloropicrin which are used as soil sterilants or fumigants for other purposes can also act as herbicides.

Classification of herbicides based on chemical structures :

In 1942, while investigating the action of synthetic growth regulators, it was discovered that certain substituted phenoxyacids while stimulating growth in smaller concentrations can bring about death in plants when applied to them in larger quantities due to overstimulation and abnormal increase in physiological activity. MCPA was first of these compounds, to be discovered to act as a weed killer, later on 2,4-D compounds came into the picture. This discovery ushered in a new era and revolutionised weed control. These compounds have been found to be selective in action—causing damage to the broad leaf weeds and not the cereal crops in the active stage of growth ; cheap in cost ; can be easily applied either as spray or dust and are easily translocated within the plant once the leaf surface receives a small quantity. In comparison to contact weedicides which are used for the destruction of the vegetative parts, rate of application of these compounds is comparatively low. These compounds are marketed as : (a) a powder in the form of dust, (b) concentrated solution in water, or (c) emulsified concentrate. A large number of compounds, which are acids are available as such, or as their metallic salts, amines or esters. Besides derivatives of MCPA and 2,4-D, various other organic compounds have come into the arena of weedicides and they are being profitably utilised for control of weeds. These weedicides are degraded in soils or within the plants in varying periods of time depending on the nature of the chemical, soil and plant. These compounds are generally non-toxic to warm blooded animals and fish.

(1) Derivates of Phenoxyacetic acids

(a) 2, 4-D. 2, 4, dichlorophenoxyacetic acid commonly known as 2, 4-D, introduced in 1942 is a selective translocative herbicide normally applied as a post-emergence treatment. It is one of the earliest compounds to be used as a selective herbicide. It is a white powder, which is normally not very soluble in water, hence metallic salts, esters or amines of 2,4-dichlorophenoxyacetic are marketed. Sodium or potassium salts ; methyl, ethyl, isopropyl or butyl esters, diethylamine, dimethylamine or isopropyl amines are normally available. Esters are recommended in lower doses. Because of greater toxicity, esters are more effective against woody plants.

(b) *2, 4, 5-T*. 2, 4, 5 trichlorophenoxyacetic acid is a translocated selective herbicide applied as post-emergence treatment. It is particularly effective against woody plants besides the weeds which are normally controlled by 2, 4-D. Though it is equally effective as 2, 4-D, nevertheless in view of its higher cost, it is mainly used in control of woody plants, also used for treatment of barks, stumps and injection of trees.

(c) *MCPA. 2-methyl, -4 chlorophenoxyacetic acid*. It has been marketed at the same time as 2, 4-D. Properties of MCPA are similar to 2, 4-D. It is manufactured in European countries owing to easy availability of the raw materials for manufacture. MCPA is claimed to be more selective and less toxic to crop plants, but higher doses are usually recommended for control of weeds.

(d) *2,4-Des-Na(Sesone)*. Sodium 2,4-dichlorophenoxy ethyl sulphate is harmless to the plants when applied on foliage, but in moist unsterilised soil it is converted into active compounds. This compound is known as sesone, while its corresponding benzoate is known as sesine. The active breakdown product is 2,4-D, but the process of breakdown is not fully understood.

(e) *2,4-DEP tris*. (2,4-Dichlorophenoxyethyl) phosphite marketed under the name "Falone" in the U.S.A. was introduced in 1958. It has shown promise as a selective pre-emergence herbicide for control of weeds in maize, groundnut and potato.

(2) *Derivatives of Phenoxybutyric acids*

MCPB, 4-(4-chloro-2-methyl phenoxy) butyric acid and 2,4-DB, 4-(2,4-dichlorophenoxy) butyric acid and their esters have been found to be as successful selective herbicides. They can be used as salts or esters for control of weeds in crops which are normally sensitive to 2,4,-D or MCPA as for example, legumes. These compounds are oxidised to 2,4-D or MCPA by susceptible weeds. In most of the legumes, this oxidation is not affected, hence they are tolerant.

(3) *Derivatives of phenoxypropionic acids*

If in the phenoxyacetic acid series, one hydrogen of methylene group is replaced by methyl, then the corresponding propionic acid derivatives are formed. These propionic acid derivatives have been found to be growth promoters and from them two compounds have been evolved which can be used as weedicides.

(a) 2,4-D propionic acid 2-(2,4-dichlorophenoxy) propionic acid has been introduced in 1956 for control of 2,4-D resistant weeds in cereal crops. Later 2,4,5-T propionic acid 2-(2,4,5-Trichlorophenoxy) propionic acid has been introduced in 1961 for control of 2,4-D resistant weeds and woody plants. The latter has been found promising in control of aquatic weeds. It can be used both as pre-emergence and post-emergence treatment.

Regarding mode of action of herbicides of phenoxy acid origin the following mechanism suggested fits in with research findings, namely (1) the molecule of the herbicide must have an acid group, (2) it must have an aromatic ring or arrangement of atoms equivalent to the ring, (3) the acid groups must be able to assume a position outside the plane of the ring, and (4) there must be a certain hydrophillic and lipophillic balance.

So far action of the herbicides of phenoxy acids is concerned, the most plausible one is that when 2,4-D is sprayed on a susceptible annual weed, the normal growth pattern is remarkably changed. Meristematic cells stop dividing as if the cell division mechanism has been fixed in place. Elongating cells cease to grow longer, but grow wider. Mature functional parts of the root and shoot are also altered. First the cells swell and become meristematic, producing cell growth and root primordia. These changes result in production of a plant with little or no root elongation and abnormal growth in basal stem. Young leaves do not expand properly, but develop an abnormal compact mesophyll with low chlorophyll content. Thus functions of the plant become impaired. Roots lose some of their ability to take up salt and water, photosynthesis is reduced and downward translocation of food through phloem is disrupted. Consequently due to extensive malfunctioning of vital organs of the plant result in death of plants.

Phenoxy acid herbicides after their application on the plant is degraded by enzymes which is the reason for lack of toxicity of these compounds for certain plants. Degradation by soil micro organisms is to a large extent responsible for their persistence, similarly apparently inactive ones may be activated by the action of plant or microbial enzymes.

(4) *Aliphatic acid derivatives*

(a) Trichloroacetic acid (TCA). Several products of TCA are in use of which sodium salt is extensively used. TCA is a selective translocated herbicide which may be used as a pre-emergence treatment for sugar cane, sugar beet, tomato, cabbage, cauliflower for control of annual grasses and certain broadleaved weeds. It is also used in higher doses for control of perennial grasses. TCA is mainly absorbed through the roots, hence mixing of the chemical with the soil plays a very important part in determining its activity. Residual toxicity of high rates of application of TCA for control of perennial grasses may remain for a few weeks to a year or longer depending on rate of application, and soil factors. In high doses (80-100 kg per hectare) it may be used as a soil sterilant. TCA is a direct toxicant and acts by precipitation of protein.

(b) Dalapon, 2,2-dichloropropionic acid possesses properties similar to TCA, but is more effective and can be applied on the foliage through which it can be absorbed and translocated. It can also be applied in the soil. It is more effective than TCA for control of annual and perennial grasses. It is selectively toxic towards monocotyledonous plants. Sodium salt of the acid which is highly soluble in water is used mostly. It is hygroscopic. Sodium salt of dalapon is less toxic than sodium salt of TCA to cattle and human beings. Cultivation after a short time appears to increase the effect of dalapon. In cold weather, it may remain in the soil for a long period and may inhibit germination of seeds. Like TCA, it is also a direct toxicant and acts on protein and protoplasm by precipitating them. It is used at the rate of 3-6 kg per hectare for selective control and in higher doses for exclusive control of grasses.

(5) *Benzoic acid derivatives*

Amiben, 3-amino-2,5 dichlorobenzoic acid in a selective

translocative herbicide to be applied in the pre-emergence stage for control of weeds in soybean, groundnut, vegetable crops, beans, sweet potato, safflower, etc. It is effective against foxtail, crab grasses, many other grasses and wild mustard.

Besides amiben, there are other benzoic acid derivatives namely benzamidooxyacetic acid (TOPCIDE), 2,3,6-trichlorophenylacetic acid (TRI-FEB); 2,3-6 trichlorobenzyl oxypropanol (TRITAC) etc.

(6) Carbamate compounds

Propham. Isopropyl N-Phenyl carbamate (IPC) is sensitive towards monocotyledonous plants, but must dicotyledonous plants are resistant to it. Soil applications have been found to given better results than foliage application but it gets easily decomposed in the soil by the soil microorganism, hence chloroderivatives have been found to give better results. It interferes with mitotic cell division.

(b) CIPC or Chloropham. Isopropyl N-3 (Chlorophenyl) carbamate shown better results than propham, though the mode of action is the same.

Both IPC and CIPC are absorbed through roots and can better be used as a pre-emergent treatment for des*ruction of grasses and other weeds during the process of germination. They are now effectively used for pre-emergent treatment in cotton, soybean, lima bean, spinach, pea, sugar beet and other field and horticultural crops. These compounds are also effective as selective post-emergent application for control of grasses and other broad leaved weeds in lucerne, clover etc. They are generally applied at the rate of 2-8 kg per hectare in the form of spray.

(c) Barban. 4-chloro-2-butynyl N-(3-chlorophenyl) carbamate introduced in 1954 is a general pre- and post-emergent herbicide having little effect on plants beyond the seedling (2-leaf) stage. It has been widely used for control of wild oats in wheat, barley and some other crop at the rate of 1/4—1/2 kg per hectare. It causes cessation of growth with only little and much delayed recovery.

(d) Eptam. S-ethyl-dipropyl thiocarbomate in a selective herbicide applied as a preplanting treatment in the soil at the rate of 2-4 kg/hectare in 800 litres of water with the help of disc harrow or rotary hoe. It is volatile. It is effective in controlling a number of annual weeds and grasses, including some perennials in vegetables and pulses.

(e) *Avadese.* S-2, 3-dichloroal-N, N-diisopropyl-thiocarbamate, is a pre-emergence selective, herbicide, soil incorporation is required for its effectiveness. It is marketed as emulsified concentrate. It is non-corrosive and non-inflammable. It is specific against wild oats.

(f) *Thiobencarb (Benthiocarb) (Saturn. Bolero).* S-4-chlorobenzyl diethylthiocarbamate. The chemical is a pale yellow liquid, soluble in most organic solvents. It is stable under acid and alkaline conditions.

Thiobencarb is a herbicide for use in rice fields. For direct seeded wet land rice the dosage is 3-6 kg a.i/ha to the surface of water 3-5 days or 5-10 days after sowing. For transplanted rice it is applied at 3-6 kg a.i./ha to the water 3-7 days after transplantation. Formulations include e.c. and granules.

(g) *Diallate (Avadex).* S-2, 3-dichloroallyl di-isopropyl thiocarbamate. The chemical is an amber-coloured liquid, miscible with ethanol and other organic solvents.

Diallate in pre-drilling herbicide for particular value of control of wild oats. Being volatile, its immediate incorporation in the soil is necessary at rates of 1.5 kg a.i/ha. The formulations include e.c. and granules.

(h) *Tri-allate (Avadex BW, Fargo).* B-2,3,3-trichloroallyl diisopropyl thiocarbamate. The herbicidal action is similar to Diallate. Carbamate herbicides have been found to have several effects on physiological and biochemical process of the host plants, namely photosynthesis, respiration, growth, disruption of mitosis etc., but the exact processes involved in phytocidal effect is not fully understood. Hydrolysis and conjugation are the major degradative pathways of carbamates.

(7) *Phthalic compounds*

Phthalamates, in general are more phytotoxic than carbanilates, but are used mostly in the pre-emergence stages for control of grasses and broad leaved weeds. They are less effective on the foliage. Normally they are applied to the soil. Exact mode of action is not known, but seedlings treated with these compounds show similar symptoms of interference as in the auxin mechanism namely epinasty, parthenocarpy, loss of geotropic and phototropic response. Important compounds of potentiality are NAP (N-i-naphthyl phthalamic acid), Alanap 2 (N-napthyl-phthalamide).

Endothal, Endothal-Sodium. 7-oxabicyclo-(2,2,1) heptane—2.3—dicarboxylic acid. The chemical is a colourless solid, stable to light, but is converted to anhydride at 90°. It is nonflammable, non-corrosive to metals, of the three isomers *exo-cis*-isomer has the greatest biological activity.

Salts of endothal are recommended for pre- and post-emergence control of weeds in a number of crops at 2-6 kg a.i./ha. It is sometimes used in combination with propham to control aquatic weeds and algae.

(8) *Anilides*

(a) *Propanil* (Stam-F-34) 3,4—dichloropropionanilide is a selective post-emergence herbicide which has become popular for control of weeds, particularly grasses in the rice fields. It is effective selectively against grass, but not against rice which is resistant to the action of Stam F-34. Hence it finds popularity and wide use particularly in upland rice fields where grasses pose a serious problem. Chemical should be used when most of the grass weeds are in one to three leaf stage. Treated fields should be irrigated preferably with flood irrigations one to five days after application of the chemical. Extremely hot and cold weather is not conducive for action of the chemical. It is not compatible with pesticides and liquid fertiliser. It is formulated in the liquid form, the dose being 3-4 litres of active ingredient in 300-400 litres of water per hectare. If applied to plant treated with organo-phosphorus insecticides, severe phytotoxicity may appear.

(b) *Alachlor* (Lasso)—2-chloro-2,6-diethyl-N-(methoxymethyl) acetanilide is used as a pre-emergence soil treatment or may be applied on soil surface. At the rate of 2-4 kg per hectare, this chemical can control most annual grass weeds and many broad leaved weeds.

(c) *Butachlor* (Machete)—N-(butoxymethyl)-2-Chloro-2', 6'-die-thylacetanilide.

Butachlor is a pre-emergence herbicide for control of annual grasses and certain broad leaved weeds in rice both direct seedᴛᴜ and transplanted. Effective rates range from 1-3 kg a.i./ha. Activity is dependent on water availability such as rainfall following treatment, overhead irrigation or standing water in rice fields.

(d) Propachlor (Ramrod). 2-Chloro-N-isopropylacetanilide.

Propachlor is pre-emergence herbicide effective against annual grasses and certain broad-leaved weeds in a number of crops. Dosage is normally 3.5-5 kg a.i./ha. It usually persists in the soil for 28-42 days.

(e) Butralin (Amex, Tamex).
N-Sec-butyl-4-tert-butyl-2,6-dinitroaniline.

The chemical is yellow-orange crystals with a slight aromatic odour, practically insoluble in water, but soluble in organic solvents. It is not corrosive to metals, but will permeate certain categories of plastics or swell certain types of rubber.

Butralin is a pre-emergence herbicide applied to the soil and incorporated into the soil soon after application, at the rate of 1.12-3.4 kg a.i./ha (depending on soil type) for weed control in different leguminous crops, potato, flax, cotton. Mixtures with prometryne are used to control weeds in groundnut and cotton. It is sold as emulsified concentrate.

(f) Trifluraline (Treflan)

a, a, a-trifluoro-2, 6-dinitro-N, N-dipropyl-P-toludine or 2, 6-dinitro-N, N-dipropyl-4-trifluoromethylaniline.

The chemical is a orange crystalline solid, stable, compatible with most pesticides, non-corrosive.

Trifluralin is a pre-emergence herbicide with little post-emergence activity. It is effective for control of annual grasses, broad-leaved weeds when applied in soil in a number of field crops. Trifluralin plus 2, 4-D is used as a post-planting herbicide in transplanted rice.

It is sold in the market as e.c.

(9) Heterocyclic compounds

(a) *Maleic hydrazide.* 1, 2, 3, 6-tetrahydro-3, 6-dioxypyradizine is known to be a growth inhibitor. Its properties are better represented by the structure 6-hydroxy-3 (2H) pyradizinone

growth inhibitor antigibberellin effect in shortening the internodes and anti-auxin effect, by oxidising indole-3-acetic acid.
These properties have been taken advantage in utilising this
compound as a herbicide mainly for inhibition of growth of
grasses. It is used both as pre-emergent and post-emergent
herbicides at the rate of 3-6 kg per hectare in 400-800 litres of
water. The compound finds its application in preventing sprout-
ing and suckering of potato.

(b) *Amitrole*. 3-amino-1, 2, 4-triazole. Possible use of this
compound as a translocative herbicide was first made known
in 1954. This compound is fairly soluble in water and is translocated within the plants from roots and leaves. As herbicide
is non-selective and used in the post-emergence stage at the
rate of 2-10 kg of active ingredient per hectare for control of
perennial broad leaved weeds. It has been found to be of use
in defoliation of cotton in the mature stage for facilitating
harvesting. Its activity is enhanced by addition of ammonium
thiocyanate.

(10) *S-triazine compounds*

(a) *Simazine*. 2-chloro-4, 6-bisethylamino-1, 3, 5-triazine first
introduced in 1955 is of low water solubility and high persistence in soils. It is a non-selective pre-emergence herbicide in
lower concentrations being virtually ineffective on deep rooted
plants. It is applied for control of weeds in maize, sugar cane,
pineapple, grapes, etc. It is also an effective soil sterilant at
higher doses. About 1-4 kg of active ingredient per hectare is
applied for control of weeds and 5-40 kg for its use as soil
sterilant.

This compound is fairly toxic even in low doses to cucurbits,
tomato, tobacco, oats, carrot, spinach, rice, beets, soybean.

(b) *Atrazine*. 2-chloro-4-ethylamino-6-isopropylamino-1, -3,
-5-triazine is similar to simazine in essential properties as a pre-
emergent herbicide, but it is more soluble in water, less depen-
dant on soil moisture. It is absorbed through the foliage, hence
may also be applied as a post-emergent herbicide. It is much
less toxic to maize. It is used for control of weeds in orchards,
ornamentals, sugar cane, etc. About 1-4 kg of active ingredient
per hectare are applied for herbicidal action and 10-40 kg as
soil sterilant.

(c) *Trietazine*. 2-chloro-4-diethylamino-6-ethyl amino-1, 3, 5 triazine which is prepared by the replacement of one of the primary amino group with a secondary amino group is useful in weeding potatoes.

(d) *Ametryne (Gesapax, Evik)*. 2-ethylamino-4-isopropylainino-6-methylthio-1,3,5-triazine.

Ametryne is a selective herbicide used pre- or post-emergence to control broad-leaved and grassy weeds in pineapple and sugarcane at 2-4 kg a.i./ha, in coffee, tea, oil palms at 1-5 kg a.i./ha. It is also used in combination with atrazine.

(e) *Aziprotryne (Mesoranil)*. 2-azido-N-(1-methylethyl)-6-methylthio-1,3,5-triazine.

Introduced in 1967 by Ciba-Geigy under the trade name 'Hesoranil'. It is a pre-emergence herbicide effective against wide range of broad-leaved weeds and some grasses. It is mainly used in transplanted brassica crops (excepting cauliflower) at the rate of 1.5-2.5 kg a.i./ha after emergence of weeds.

(f) *Propazine (Gesamil, Milogard)*. 2-Chloro-4, 6-bis (isopropylamino)—1,3,5-triazine.

Propazine is a pre-emergence herbicide recommended for control of broad-leaved and grass weeds in sorghum and umbelliferous crops at 0.5-3.0 kg a.i./ha.

(g) *Cyanazine (Bladx)*. 2-(4-chloro-6-ethylamino-1,3,5-triazin-2 Ylamino)-2-methyl propionitride.

The technical product is a colourlless crystalline solid, stable to heat, light and to hydrolysis in neutral and slightly acidic or basic media.

Cyanazine is a pre- or post-emergence herbicide of short persistence. It is valuable as a general weed control.

(h) *Prometryne (Gesagard, Caparol)*. 2, 4-bis (isopropylamino)-6-methylthio-1,3,5 triazine.

It was first introduced in 1962 by Ciba Geigy with Trade Mark Gesarard, and Caparol.

The chemical is a colourless crystalline solid, sparingly soluble in water, but readily soluble in organic solvents.

Prometryne is either pre- or post-emergent selective herbicide for weeds control in cotton, peas, carrots sunflowers, onions etc. The recommended dose is 1.0-1.5 kg a.i./ha.

It is marketed as wettable powder.

(i) *Desmatryne* (*Semoron*). 2-isopropylamino-4-methylamino-6-methylthio-1,3,5-triazine.

The chemical is a crystalline solid, readily soluble in organic solvents, stable in neutral, slightly acidic or alkaline media. Desmatryne is a selective post-emergent herbicide of brief persistence in soil. It is effective for control of *Chenopodium album* and other broad-leaved and grassy weeds in brassica crops, at the rate of 500 g a.i./ha.

It is sold as wettable powder.

(j) *Dimethametryn* 2-(1,2-dimethylpropylamino)-4-ethylamino-6-methyl thio-1,3,5 triazine.

The chemical is a solid soluble in polar organic solvents.

Dimethametryn is a selective herbicide active against broad-leaved weeds in rice. Under the trade mark "Avirosan", it is marketed in combination with piperophos for the control of mono and dicotyledonous weeds. Rate of 1-2 kg total a.i./ha are used for transplanted rice and 2-3 kg a.i./ha for direct seeded rice.

Effect of triazines on the plant is reduction of starch accumulation in the leaves due to inhibition of Hill reaction. Plants are able to metabolise triazine compounds.

(11) *Substituted urea group*

(a) *Monouron*. N-p-chlorophenyl-N'-N8-dimethyl urea is the first substituted phenyl urea discovered in 1951 as possessing herbicidal properties. In controlled low doses, it can act as a selective herbicide for pre-emergent treatment. It is mainly used to render soil barren of plants. As a selective weed killer 1/2 to 5 kg of active ingredient per hectare is sufficient. It is effective against a number of annual grasses, wild mustard and many other crops. It may be used in orchards, sugar cane, pine apple, cotton and such other deep rooted plants. It is recommended for areas of average rainfall.

In substituted urea compounds of phenyl urea group water solubility and resistance to microbial degradation can be achieved in the desired level or regulated in the desired manner by alteration of substituents. Greater stability in soils can be obtained by increase in chlorination and resistance to leaching by both increase in chlorination and larger alkyl group.

(b) *Fenuron*. N-phenyl-N'-N'-dimethyl urea is highly water soluble and used for control of deep rooted perennial woody plants, dosage being 10-60 kg of active ingredient per hectare. It may used as a pre-emergent treatment in many cases.

(c) *Diuron*. N-(3-4-dichlorophenyl)-N'-N'-dimethyl urea has a very low solubility in water and is recommended for use in high rainfall areas. It has the same properties as monouron otherwise it can be used both as a pre-emergent herbicide (1/2 to 5 kg of active ingredient per hectare) and sterilant (10-40 kg of active ingredient per hectare).

(d) *Neburon*. N-(3-4-dichlorophenyl)-N'-butyl-N'-methyl urea is sparingly soluble and recommended for use where long persistence in the soild is desired. Usually 12-40 kg of active ingredient per hectare is recommended for control of different weeds. At a very low dose, it may be somewhat selective.

(e) *Fluometuron (Cotoran):* 1,1-dimethyl-3-(oc. oc, oc-trifluoro-m-toluyl) urea is a selective pre- and post-emergence herbicide. It is marketed as wettable powder. Beets, cole crops, cucurbits and egg plants are sensitive to this chemical.

(f) *Metoxuron (Dosanex)*. N8-(3-chloro-4-emthoxyphenyl)-N, dimethyl urea is used in cereals as a pre- or post-emergence herbicide. It is marketed as a wettable powder.

(g) *Buturon (Eptapur)*. 3-(4-chlorophenyl)-1-methyl-1-(1-1 methylprop-2 ynyl) urea. It is a pre- and post-emergence herbicide absorbed mainly by roots. Recommended for use in cereals and maize for control of shallow-germinating grasses and broad-leaved weeds at 0.5—1.5 kg a.i./ha.

(h) *Isoproturon (Arelon, Graminin)* 3-(4-isopropylphenyl)-1, 1-dimethyl urea.

Isoproturon controls annual grasses and broad-leaved weeds in wheat, and barley at 1.0-1.5 kg a.i./ha.

(i) *Methabenzthiauron (Tribunil)* 1-(benzothiazol-2-y1)-1, 3-dimethyl urea. It is a selective herbicide recommended for control of different grasses and annual weeds either pre- or post-emergence in wheat, oat, barley and other winter cereals. Dosage vary from 2.p to 3.5 kg a.i./ha. The soil should be moist at the time of treatment, as the chemical is absorbed mainly through the roots. It is broken down in the soil and has no after-effects on subsequent crops.

These substituted urea compounds are easily translocated in the transpiration stream when absorbed through roots. They accumulate in the leaves causing collapse of parenchyma cells in the leaves and of tissue at the ground level in maize. Like triazine compounds, these compounds also inhibit Hill reaction and reduce photosynthetic activity. Physical poison nature disrupting the cell surfaces may also be responsible for phytotoxicity.

12. *Quarternary ammonium group*
The related compounds namely diquat-(1-1'-ethylene-2-2'-bipyridillium ion) and paraquat (1-1'-dimethyl-4,4-bipyridillium ion) available as di-(methylsulphate) have become increasingly popular since its introduction in 1958. They can kill the aerial parts of plants very quickly, but in soil activity is lost rapidly due to its decomposition. Hence the areas treated by these compounds (dosage: 2-4 kg of active ingredient per hectare) can be planted within a very short time of application of herbicide. Cultivation with no or minimal tillage is possible after application of these compounds. These herbicides effectively kill the annual and perennial weeds. They are quickly defoliated with loss of tops, but they may regenerate again.

Paraquat is more effective than diquat, and is used for control of grassy weeds in orchards. It can be successfully employed in tea gardens, pineapple orchards, etc., for control of under-growth of grasses.

Both compounds exert their phytotoxic action through reduction into a free radical during photosynthesis, the free radical again oxidised to original quaternary salt. This process interferes with normal redox potential and oxidation-reduction in the plant.

(13) *Other compounds*
Pentachlorophenol. (PCP) and its sodium salt are used as contact and residual pre-emergent treatments on a variety of crops. It is applied as a fine power fortified with oils for general contact pre-emergent weed control. These compounds are very irritating to the skin, throat and nose, and are poisonous when taken orally by warm blooded animals including man. So safety precautions are to be taken in its application. Usually

dosage is 4-20 kg per hectare mixed with 500 litres of oil and then 1000 litres of water forming an emulsified preparation.

Trifluralin (Treflan). oc, oc, oc-trifluoro-2, 6-dinitro N, N-dipropyl-p-toludine, is a selective pre-emergence herbicide. It should be incorporated at 5-10 cm depth. Within four hours of application, it is absorbed on clay colloids and organic matter and does not have much downward movement. This compound does not stay active in soil more than four to six months. Maize, sorghum, beet and spinach are sensitive to this chemical. It is marketed as emulsified concentrate and granules.

Planavin (Nitralin). 4-(methylsulphonyl)-2, 6-dinitro-N, N-dipropylaniline is a selective pre-emergence herbicide. It kills weeds as they are germinating. There is residue problem when applied at recommended doses. It is marketed as wettable powder-dispersible liquid.

MSMA. Monosodium methanearsonate (Ansar-529) is a contact post-emergence herbicide. It is completely inactivated in soil by absorption or ion exchange. It is used as a directed spray. The chemical has some fungicidal action.

Nitrofen (Tok-E-25). 2,4-dichlorophenyl-4-nitrophenyl ether is a pre-or post-emergence herbicide. Other pesticides and fertilisers cannot be mixed with this herbicide. It is inflammable and smoking is prohibited while spraying. It is promising for cole crops as pre-emergence treatment.

Sinbar (Terbacil). 3-terbutyl-5-chloro-6-methyluracil needs moisture for activation. It is nonvolatile. Incorporated treatment gives good control of nutsedge. It is not to be used in sandy, loamy sand, and gravelly soils nor on soils low in organic matter (less than 1 per cent) as crop injury may result. It is marketed as a wettable powder.

Glyphosate (Roundup)

N-(phosphonomethyl) glycine. It was introduced in 1971 by Monsanto Co. The principal formulation trade mark 'Round up' is the mono (isopropylamine) salt.

Glyphosate is a non-selective, non-residual post-emergence herbicide, very effective on deep-rooted perennial species and annual and biennial species of grasses, and broad-leaved weeds. Excellent control of most of the species is obtained at rates 0.7-5.6 kg a.i./ha, annual species requiring the lower rates. Better control of most weeds is obtained if applications are made at the later stages of plant maturity. It is sold as a water-based solution.

Benzoyl prop-ethyl (Suffix)—Ethyl N-benzoyl-N-(3,4-dichlorophenyl)-DL-alaninate. The chemical was introduced in 1969 by Shell Research Ltd. under the trade name 'Suffix'.

The technical material is an off-white crystalline powder, photochemically stable and hydrolytically stable at intermediate pH.

Benzyl-prop-ethyl is applied to wheat at the rate of 1.0-1.5 kg a.i./ha between the end of tillering and second node formation stage for control of wild oat (*Avena fatua, A. barbata, A. sterilis, A. ludoviciana*)

It is sold as e.c.

Kavainetone-4-methoxy-3, 3'-dimethyl benzophenone

The chemical is fine colourless crystals, soluble is most organic solvents, stable under acid and alkaline conditions but slowly decomposed in sunlight.

It is a selective pre-emergence herbicide effective against grasses and broad-leaved weeds in rice and vegetable crops at 3-5 kg a.i./ha, also used in rice pre-emergence in combination with bensulide at 30-50 kg granules/ha. It is bio-degradable and leaves no residue. It is toxic to fish.

Chlorfenprop-methyl (Bidisin)—methyl 2-chloro-3(4-chlorophenyl) propionate.

The herbicide was introduced in 1968 by Bayer AG under the trade name 'Bidisin'. The pure compound is a colourless liquid with a fennel like odour, soluble in fatty oils and organic solvents.

Chlorfenpropmethyl is specific herbicide, used for control of wild oats. It is used only after the emergence of wild oats which are most susceptible between leaf stage and tillering. The recommended dose is 4 kg a.i./ha.

It is marketed as emulsified concentrate.

Dicamba (Banyel, Mediben). 3-6-dichloro-o-anisic acid.

The chemical is colourless solid, readily soluble in ethanol and ketones.

Dicamba is a translocable post-emergence herbicide for weed control in cereals. It is usually formulated with one or more phenoxyalkonic acids or with herbicides of other class. It is rapidly degraded in oil. It is metabolised in plants to herbicidally inactive 2,5-dichloro-3-hydroxy-6-methoxybenzoic acid.

Quinonamid (*Alginex*)—2,2-dichloro-N-(3-chloronaphthequinon-2-yl) acetanids.

The chemical is yellow needles, soluble in most organic solvents. It is decomposed in the presence of acid or alkali.

Quinonamid is effective against algae in the open, as well as algae and mosses in glass houses. It can be used as seed dressing or spray for control of algae in rice fields.

The formulations available are wettable powder or granules.

Bifenox (*Modown*) methyl 5-(2,4-dichlorophenoxy)-2- nitrobenzoate.

Bifenox is used for pre-emergence and directed post-emergence treatment to control important broad-leaved weeds and some grasses in rice, maize, sorghum, and small grains. Pre-emergence applications of 1.68-2.24 kg a.i./ha are recommended to provide effective weed control over a range of soil types and diverse climatic conditions. Post-emergence directed sprays are recommended at 1.12-1.68 kg a.i./ha. There is no problem of carry over or residue.

Oxadiazon (*Ronstar*) 5-*tert*-butyl-3-(2,4-dichloro-5-isopropoxyphe-nyl)-1,3,4-oxadizol-2-one.

The chemical was introduced in 1969 by Rhone-Poulene Phytosanitaire. It is colourless white crystals almost insoluble in water, but soluble in many organic solvents. It is stable under normal storage conditions and non-corrosive.

Oxadiazon is a selective herbicide effective against mono- and dicotyledonous weeds in rice at 1 kg a.i./ha. The product is marketed as e.c. or granules.

Piperophos (*Rilof*, *Avirosan*,). S-2-methylpiperidinocarbonyl methyl O, O-dipropyl phosphorodithicate.

The chemical is an oil at room temperature, miscible with most organic solvents.

Piperophos is a selective herbicide active against annual grasses and sedges in rice. It is marketed in combination with dimethametryn under the trade name 'Avirosan' to control both mono- and dicot young plants through roots, coleoptiles and leaves. In tropical countries piperophos is to be applied with hormone herbicide.

Chlormethoxynil-2,4-dichlorophenyl 3-methoxy-4-nitrophenyl ether. It is a herbicide used mainly for paddy field and upland rice at 1.5-2.5 kg a.i./ha.

(14) *Dinitro herbicides*

DNOC. Dinoseb. 3,5-dinitro-O-cresol was introduced in 1932 as a selective herbicide and became popular in Europe and the U.S.A. Later on, it was found that from compounds formed by substitution of the methyl group by higher alkyl groups 2-4-dinitro-O-sec-butylphenol (2-(1-methyl-n-propyl)-4,6-dinitrophenol) commonly known as Dinoseb is more effective than DNOC on many weeds. It has greater solubility in oil. Both DNOC and dinoseb are efficient herbicides on a wide range of annual weeds, and do not affect cereals. Selectivity is due to differential wetting and consequent absorption of the chemical by plants. It is difficult to wetten narrow upright linear leaves of monocotyledonous plants. In broad leaved plants, leaves can be wetted easily. Dinoseb herbicides are usually formulated as ammonium salts or amine (1-2 kg per hectare in 800 gallons of water).

Both these compounds are yellow staining and highly toxic to men and animals, whether taken orally or by inhalation, or by contact through skin. So safety precautions are to be taken in use.

Other herbicides in use are calcium cyanamide, sodium arsenite, ammonium sulphamate, borates etc.

METHODS OF WEED CONTROL

Different methods of weed control may be classified as follows: (1) Mechanical and cultural methods; (2) Chemical methods; and (3) Biological methods.

Any measure for control of weeds should take into account their life cycles, methods of propagation, localities in which they thrive best and their distribution in relation to space and time.

Mechanical and cultural methods aim at control and eradication of weeds by manual and cultivation practices including use of implements—hand or power operated. These include removal by pulling out by hand, or hand hoeing, tillage operations which may include ploughing, harrowing, discing, or mowing. Normally these measures may completely eradicate the weeds or only top portions. Apart from removal by hand drilling, summer ploughing may be helpful in destruction of underground competition. Burning of residue of plant parts and mulching with straw or paper may also be helpful in control of weeds.

Selective herbicides should be carefully selected and applied so that damage to susceptible crop should be avoided. Symptoms of the herbicide damage include distortion or twisting of stem branches and foilage. To avoid damage due to herbicides drift, certain precautions are to be taken, namely, herbicide spraying should be done on calm days, only to avoid drift even under conditions of high volume of spraying. Under hot summer conditions, less volatile formulations should be used. Instead of esters, amine salts may be preferred under such conditions. Suitable droplet size under appropriate pressure conditions is necessary to avoid harmful effect of drift.

Herbicides for weed control may be applied with any of the spraying or dusting equipments. It is known that 2,4-D and similar compounds are equally effective when applied in dilute or concentrated from provided the same quantity of active ingredient is applied per acre, but the application should be uniform over the area. Hence low-pressure, low gallonage sprays are generally used. Low-pressure pumps also eliminate mist sprays that may float away and cause damage to adjacent crops by drift. Through these spray pumps which may operate at low pressure of 2 kg/cm^2 with proper nozzles and speed (6 km per hour) as small a quantity as 15 gallons of concentrated solution can be applied per hectare. Nozzles producing a fan-type with ice (solid CO_2) to provide low pressure may also be employed in chemical weed control. The sprayers may be hitched with or mounted on moving or other suitable machines.

QUESTIONS FOR DISCUSSIONS

1. Why is the application of chemicals for the control of weeds a more difficult proposition than the use of pesticides?

2. What do you understand by selective toxicity of herbicides?

3. What do you understand by 'contact' and 'translocated' herbicide?
4. What are pre-emergence and post-emergence applications? Why is selective toxicity of great importance in post-emergence applications?
5. In which cases is the pre-emergent treatment of contact herbicides effected or recommended?
6. Why are 2,4-D compounds advocated as selective herbicides? What are their advantages? Can you state some 2-4-D compounds which have successfully been used as herbicides?
7. Why are anilides advocated for control of weeds in rice fields?
8. How can water solubility and degradation in soils in substituted urea compounds be regulated?
9. Why have bipyridilium compounds become increasingly popular as weedicides? What are their advantage over other weedicides? Under what conditions can they be profitably used?
10. What are the other methods of weed control besides the use of chemicals?
11. What steps should be taken to avoid damages due to the use of herbicides? What are the symptoms of herbicide damage?
12. What type of spraying should be advocated for herbicides and why?

CHAPTER 8

STORAGE OF PESTICIDES

Ideally a store for pesticide should be located on a carefully selected site and used solely for storing pesticides. In reality, there is little opportunity to design and construct a new store. Normally a building or a portion of the building designed for some other purpose, is adopted for storing pesticides.

The essential features of a pesticide store should be as follows:

Site

a) pesticide store should be isolated from stores for other commodities, especially food and drink, animal feed, seed and fertiliser to avoid contamination;

b) there should be no danger of flooding;

c) the store should be away from water sources, wells and other sources of domestic water which may be contaminated as a result of spills, leaks or a major emergency like fire;

d) the store should be shady to keep store temperatures down, as high temperatures may destabilise some pesticides; and

e) the site should have good access not only for delivery vehicles, but also for fire fighting equipment.

Building

a) the store should have direct access to the outside, not through some other store or building;

b) office accommodation should be outside the store;

c) provision should be made for washing facilities and separate storage of protective clothing;

d) if the pesticides are to be repacked or bulk handled, then a ventilated working area should be kept;

e) if herbicides are to be stored, then they should be stored completely separately from pesticides to avoid the risk of contamination;

f) the building should be fire proof and free from rodents;

g) floor should be smooth concrete or other impervious material so that spills of pesticides can easily be removed;

h) there should be adequate arrangement of ventilation to prevent build up of toxic vapours which would endanger the health of those working in the store, inflammable vapour and to keep the store as cool as possible, as high temperature can destabilise many pesticides. Hence there should be arrangement of ventilators as well as windows. Windows should be shaded to prevent entry of strong sunlight;

i) electrical fittings should be well insulated and fittings should be flameproof and dust proof; and

j) there should be adequate arrangement of water supply and drainage.

Storage system

i) storage system should be so arranged as to minimise the need for handling pesticide containers as this often leads to mechanical damage to the containers and subsequent leaks,

ii) floor space should be kept clear and uncluttered,

iii) stock should be so arranged that the oldest stock is used first,

iv) containers of pesticides should never be placed on the floor,

v) dusts, granule and wettable powder formulations can cake if subjected to pressure. Where these formulations are supplied packed in polythene bags in cardboard or fibre cartons, they should be properly stored.

Staking Heights

Package	Single package No. of package
Large drums (200 litre)	
Steel	3
Plastic	2
Small drums (20 litre)	
Steel	6 to 8
Fibre	4 to 6
Plastic	3 to 4
Sacks (25 kg)	
Paper	15
Polythene	10-15
Fibre board cases containing soft packages	
(plastic bottles, sachets)	8 to 10
Wood cases	15

Shelf life

The period of time over which pesticides may safely be stored before deterioration is sufficiently serious to affect their use is known as Shelf Life. Almost all pesticides have a limited shelf life. Most pesticides have a shelf life for at least two years from the date of manufacture.

Pesticides in sealed containers may change with time with the result that the active ingradient may change as a result of chemical reactions to give breakdown products which may or may not have pesticidal properties and the concentration of original active ingradients may fall. Formulation of the pesticide may breakdown making it impossible to mix the insecticide or to spray with it.

Where only a few kilogrammes or litres are involved, the best method of disposal is by burial. This method may be practical, but it can lead to problems of public health and environmental contamination. The main problem is to keep the buried pesticide in place to prevent the contamination of surface run-off water and ground water through leaching. There may be further problems in areas subject to flooding.

The best method of disposing of large quantities of pesticides is incineration (at 1600°). However incineration requires special equipment with provision for "scrubbing" the combustion products and is outside the scope of most pesticide store keepers.

General procedures for chemical disposal of pesticides on a small scale

A trench measuring 60-90 cm. long, 45 cm. in depth and 15 cm. wide should be suitable for disposal or 5 litres of liquid pesticides or 5 kg. of solid pesticide. The soil of the trench should be of a clay type so that the liquid component of the pesticide is not drained quickly. If the formulation of the pesticide is very concentrated, say 70 per cent or more, the mixture of pesticide and solid diluent should be further diluted with an equal volume of water. The pesticide should be mixed with an approximately equal volume of solid diluent such as sand or soil in the trench dug in the ground. The decomposing agent should be slowly added to the pesticide layer and again mixed with the stick. The stick should be left in the trench and filled with tightly packed top soil.

Chemical agents and amounts

Ideally, the amount of agents for decomposition should be calculated for the particular combination of the agent involved, pesticide and formulation, so that there is sufficient chemical agent to ensure complete destruction, but not so much excess that it becomes a hazard in itself. The following provides a rough guide for the purpose.

a) alkalis (for organophosphorus esters and N-methyl carbamate)

1 part by weight of active ingradient of pesticide to 0.5 parts of sodium hydroxide or 1 part of slaked lime.

b) Acids (for thiocarbamate fungicides, some esters):

1 part by weight of active ingradient of pesticide to 0.2 parts of hydrychloric or sulphuric acids.

c) Oxidizing agents (for some organophosphorus esters, cyanides):

1 part by weight of pesticide active ingradient to 1 part of sodium hypochlorite or bleaching powder.

d) Reducing agents (for chlorate herbicides).

1 part by weight of pesticide active ingradient to 1 part of sodium thiosulphate and 1 part of hydrochloric acid.

The amounts suggested refer to the chemical agent itself and the corresponding quantities of diluter solutions, which are the only ones available or which are prepared for safety reasons, must be calculated. To cite an example, 0.2 parts of sulphuric acid may be supplied as battery acid which has a strength of 34% w/v and the equivalent amount of this would be about 0.6 parts. The 0.5 parts of sodium hydroxide could be dissolved in 5 parts of water to make a 10 per cent w/v solution before mixing with the pesticide.

Precaution

(a) As pesticides and chemical agents are hazardous, protective clothing, gloves and safety gloves should be worn.

(b) It is desirable to use stronger chemical agents in diluted form, as the same is less hazardous.

(c) Concentrated pesticides and chemical agents should never be mixed in containers such as buckets or drums. A rapid uncontrolled rise in temperature or evolution of gas may eject the decomposing mixture with consequent danger to the operator.

(d) In case the quantities of the pesticides to be decomposed are larger than 5 litres of liquid or 5 kg of solid, they should be dealt with as several small batches.

(e) The efficiency of decomposition depends on the pesticide formulation and the decomposition agent used. Destruction may not be completed in all cases under usual conditions, and specific instructions for each pesticide formulation are needed.

Decontamination of traces of pesticides after accidental spillage

As much as possible of the spilled pesticides should be gathered up for disposal. In case of spills of liquid formulations should be soaked up with absorbent material, e.g. sawdust, sand, earth, then swept up and placed in marked containers for disposal. Solid formulations, e.f. dusts, w.d.p. granules should be swept after adding damp sand or sawdust and placed in marked containers for eventual disposal. After sweeping, the area affected by the spill should be scrubbed with water, strong soap, or detergent or bleaching powder (Sodium hypochlorite solution).

QUESTIONS FOR DISCUSSION

1. What are the essential features of a store for pesticide?
2. What should be the storage system of pesticides?
3. What should be the staking heights of pesticide packages in store?
4. What are the general procedures for chemical disposal of pesticides?
5. What are the procedures of decontamination of pesticides after accidental spillage?

CHAPTER 9

APPLICATION OF PESTICIDES

Chemical compounds used for control of pests, diseases or weeds need to be applied in such a manner that they come into contact with pests in case of direct action or eradication or in case of prophylaxis, distributed evenly over the surface of the plants to form an uniformly distributed persistent deposit to secure a protective covering. The active ingredient of a pesticide is rarely in a form to be applied as such. They are mixed with a diluent which may be a solid dust or liquid. The small quantity has to be used to cover a large area. Hence effective means of their dispersion have to be evolved and appropriate diluting material (diluent) or base has to be used. Depending on the nature of the diluent, liquid, solid or gas, the methods of application are classified as spraying, dusting and fumigation, respectively.

Spraying

Many pesticides are applied in liquid form, the diluent being water or, in some cases, oil. Pesticides may be used as : (a) aqueous solution, if the material is soluble, (b) suspension of particles, or (c) emulsion, in cases where the material is insoluble. When an active ingredient is easily dissolved in water, the material is available in the form of a concentrated solution, e.g., nicotine (95 per cent solution), or nicotine sulphate (40 per cent solution). By addition of requisite amount of water at the time of spraying, the desired concentration is achieved. If necessary, wetting and spreading agents are incorporated in the dissolved material to secure an uniformly spread out layer of pesticide over the sprayed surface. These solutions from the point of even distribution of the pesticide are satisfactory as the active ingredient is dissolved and the solution is homogeneous.

Many pesticides do not keep well in solution in storage, moreover it may be inconvenient and irksome to transport a bulk material containing water. Hence it is convenient to have a formulation in powdered form containing the active material finely ground to increase the rate at which it will get dissolved quickly. Adjuvants like spreaders and stickers may have to be added to increase the efficiency. Zineb and Maneb are examples of water soluble fungicides, which are, marketed in powdered form.

Large number of pesticides are, however, sparingly soluble in water. They are formulated as wettable or water dispersible powders to make them amenable for spray application. Active ingredients are finely ground and mixed with appropriate quantities of wetting, dispersing, and spreading, if necessary, deflocculating agents. A distinction has to be made between wettable and water dispersible powders. A wettable powder does not resist the penetration of water and can thus be wetted with the help of appropriate chemicals. Some examples of wetting agents are long chain alcohols, flour, esters of many fatty acids. One disadvantage of wetting agents is that in their presence, spray residues are more easily wetted by water and leached out and deposits have a shorter duration of effectiveness. A water dispersible powder which when wetted with water or mixed with water will remain as individual particles in suspension for a considerable period of time. Both these types of formulations are unsuitable for low volume spraying.

Wettable powders may have to be agitated continuously to maintain an even dispersion to have an uniform deposit of active ingredient. Hence they should be applied by sprayers provided with an agitator or some auxiliary mechanism of agitation.

Water dispersible powders may be applied through most types of sprayers as the particles of the pesticides will remain suspended for a long time even without agitation. Sometimes preparations of colloidal solution are used. In colloidal solution, water dispersible particles are so finely divided that they will never sediment out. Some copper fungicides are marketed in colloidal form.

There are a number of chemical compounds which are not

soluble in water, but soluble in many organic solvents. These solvents are usually highly volatile, and highly expensive, besides when applied in large quantities they are phytotoxic. Hence a large quantity of active ingredient has to be dissolved in a small quantity of organic solvent and the solution is mixed with water which acts as a carrier of the solution and an emulsion is prepared. To prevent coalescence of dispersed particles and to impart stability to emulsion, emulsifying agents are used.

Generally two types of emulsions are possible : (a) oil in water (O/W) in which oil may be dispersed as fine droplets suspended in water which is a continuous phase, or (b) water may be the dispersed in oil (W/O) emulsion. The principal function of the emulsifier is to modify the properties of interphase between the dispersed and continuous phase. Generally oil in water emulsion is used for spraying. Many pesticides and herbicides are marketed as emulsified concentrates. Prior to application, the desired amount of the concentrate has to be poured into the requisite quantity of water as indicated by the manufacturer. If more water is added, then the emulsion may break and the suspended particles are likely to separate out.

The most important property of the pesticidal emulsion is its rate of breaking or separation into immiscible constituents. This factor can be controlled by agitation and quantity of emulsifier employed. Breaking of an emulsion is the usual method by which the toxic dispersed phase comes into play ; breaking usually occurs immediately after application of the spray or evaporation of the greater part of continuous phase, i.e., water. Quick breaking emulsions are preferred in agricultural sprays because they produce heavy deposits, but they may be deficient in wetting and spreading properties

Creaming of emulsion may take place due to differences in specific gravity between the continuous and the dispersed phase, rate of creaming being dependent on the differences in densities of the two phases and size of droplets of the dispersed phase. Agitation normally restores the homogenous state of emulsion and creaming can be prevented by closely matching specific gravities of the two phases of emulsion.

10

Phase reversal of emulsion may take place in some cases, changing O/W type to W/O type or vice versa. Such occurrences can be rectified by the use of suitable emulsifier, e.g., alkaline soaps—sodium and potassium esters of long chain fatty acids ; organic amines which form amino soaps with fatty acids —triethanolamine (both incompatible in hard water) ; sulphonates of long-chain alcohols, stable in hard water acids and alkalies. Sulphonated, aliphatic esters and amides—dioctyl sodium sulphosuccinate, sodium sulphoethyl-m-ethyl-dioctyamide, mixed aliphatic aromatic sulphonates, e.g., sodium decylbenzene sulphonates ; non-ionic types—ethers, alcohols, esters of polyhydric alcohols ; polyoxyalkene derivatives, polyalkene ether alcohols ; and natural agents lipids, saponins (glucosides) ; albumen ; caesin, gelatin, lecithin, etc., are excellent wetting, spreading or dispersing agents. Soap is a good emulsifying agent when the active ingredient is dissolved in kerosene or diesel oil.

For effective coverage, spraying should be done in such a manner that a known concentration of the toxicant is effectively spread over the area under the crops to be treated in an uniform and controlled manner. Liquid has to be distributed in the form of droplets which will be of convenient size to have good coverage with minimum wastage of the liquid and minimum expenditure in terms of active ingredient and cost involved. The general principle adopted in spraying is that the liquid which is kept in a closed container is passed under pressure through special orifices called nozzles. Passage through nozzles results in the formation of desired drops which are thrown into the distance by the pressure of air through the liquid. This is known as high volume spraying. This system of high volume spraying has certain limitations : it is expensive in terms of cost of machinery ; transportation of the machinery, which is often bulky or weighty, may be difficult particularly in wet soils or terrains ; availability of water and its cost of transportation ; and the difficulty of passage through the crop for spraying. In ground crops, inconveniences may be much less as the nozzles can be held close to the crop. In the spray of systemic toxicants, the problem is much less as the plant itself, once it has the supply of toxicant, carries the task of distribution within itself. Special nozzles are used for spraying of herbicides. They can be used in conventional sprayers with certain formulations of concentrates to reduce the time and cost of spraying.

An alternative method has been taken recourse to for the coverage of large areas in lesser time and cost, in which the liquid is broken down to small droplets and dispersed by the action of a large volume of air directed against the stream of the liquid. In this mechanism energy required for the purpose is transmitted through the air stream and not through the liquid as in conventional spraying. For spraying from the ground, this stream of air is produced by fan, propeller or turbines (axial flow plan). In aerial spraying, the speed of the aeroplane alone is sufficient for the formation of small droplets and in helicopters this job may be done by the rotor. Under this system very fine droplets are produced and smaller volumes of spray per unit area is required. Use of organic solvents carrying a larger concentration of the active ingredient is economical. When aqueous solution is used, concentration of the active ingredient may be increased in this type of low volume spraying.

Smoke generation

For spraying in inaccessible areas, say in forests, active ingredient may be dispersed in the form of smoke or gas. Pyrotechniques are followed in generating the smoke or gas. Artificially produced air streams may be used for the dispersal of the gas over a large area. In the smoke generation process, low combustion temperatures are required to prevent the destruction of the material during the process of volatilisation due to heat. Pyrotechnic compounds are so chosen that combustion products are not phytotoxic.

Dusting

Dusts in which the active ingredient is mixed or diluted with suitable powdered carriers are more convenient for use. Properties of the carriers or diluents are largely determined by the quality of finished products. Selection of a diluent may be made on the basis of compatibility with desired insecticide (including pH, moisture content and stability), particle size, abrasiveness, absorption capacity, specificity and cost. Organic flour, soybean flour or wood bark may be used as diluents. Minerals such as sulphur, silicon dioxide, diatomite, tripolite, CaO-lime, gypsum, silicate, bentonites, kaolinites, pyrophyllites, talc, volcanic ash may be used as diluents in dust formulations.

Diluents like montmorillonite clays and silicic acid, form constituents of sorptive dusts in which these materials absorb the lipid protective layer of insect cuticle and insect dies by dessication. These are effective as carriers of fluorides and organophosphorus toxicants.

Dusting appliances are much lighter and easier to operate. Dust formulations are ready-mix and can be used as such. The irksome factor of water supply is not existent. Above all, the process of dusting is much less time-consuming and expensive. It is said that the efficiency of dusting is ten times more than that of spray. As the dusting machines are much lighter consisting essentially of a fan or blower, dusting can be taken recourse to in such areas where spraying machines cannot be transported.

In spite of all the advantages stated, dusting has its own disadvantages. First of all, dust formulations should be such that uniform content of active ingredient is maintained throughout its particle size and distribution. Particle size and weight of both the components namely active ingredient and carrier should be so regulated that they will be easily carried by a slight breeze to the desired distance. Dusting cannot be done in breezy weather due to the effect of drifting. The carrier may have hygroscopic properties, so dusts may form lumps or balls and may be discharged from the machines in lumps or aggregates rather than as a free flowing powder. Dust particles may be retained in the duct due to electrostatic charge generated in the process of blowing due to friction. Various carriers have different capacities to be dusted off or dusting tendencies. Besides dusts do not have the same sticking properties or tenacity and retentivity on the surface as spray materials have. It is commonly said that an average job of spraying is equivalent to very good dusting. Dusting is associated to some extent with health hazards.

In recent times active ingredients are applied in the form of small granules (particles of 30-60 mesh size) through specially designed simple machineries. They may be applied by hand in some cases with suitable precautions. Granular formulations have some advantages that they may be conveniently applied on ground crops and advantageous in aerial spraying in forest canopy. Hazards of health due to spraying and dusting of toxic materials are negligible. Granular formulaticns, though they dispense with costly machineries are more expensive and have other technical limitations. In the use of granular formulations, drift hazards and contamination of border areas are minimised.

Fumigation

Toxicity of a gas is proportional to its concentration c and time of exposure t of the said concentration to the organism and is governed by the equation $ct = k$, k being constant. In measurement of activity of vapour phase k is usually 5 mg/l/hour or above. For effective fumigant atcion in the open k should be at least 1 mg/l/hour. Unless the insecticide is of exceptional potency, fumigation is successful only in the closed space, hence it is practised in glasshouse, or covered godowns or under temporary covers.

Fumigation for control of insects may be taken recourse to in some special cases. Fumigation of stored plant products has to be done inside a closed space. Because of the poisonous nature ot the chemicals, the chamber or the enclosed spa^e should be air-tight to prevent escape of gases which may make treatment ineffective and cause health hazards. Fumigants may be applied to the soil in case of pest infestation in the soil.

Chemicals to be applied as fumigants should have low boiling point when they are intended particularly for treatmeni inside an enclosed space. For soil fumigation, it is desirable to have a higher boiling point so that the toxicant is not lost too quickly. The choice of a fumigant is made on its cost in relation to the benefit derived, easy application, toxicity hazards, and effect on viability in case seeds are to be treated. Fumigants are used under different conditions against a variety of insects infesting living plants and plant and animal products.

For fumigation purposes certain general directions are to be followed, (a) safety to operators who should take adequate measures including use of gas masks, (b) dosage and duration have to be carefully determined, (c) fumigation should preferably be done to the living plants in darkness since sunlight during or within an hour before or after is likely to cause injury, (d) windy and cold weather is to be avoided, (e) in case of seeds, moisture content should not exceed 14 per cent, otherwise viability may be affected, and (f) phytoxicity to living plants has to be considered.

Success of fumigation is dependent on attainment of a critical value often termed or expressed as cf (product of dosage per unit volume and period of exposure) which will kill 99 per cent insects. Lower the concentration longer will be the period of exposure. Since critical value or cf is considered to be a physico-chemical process, Vant's Hoff's law $Q_{10} = 2$ is followed in a particular direction. Between 10°C (optimum temperature) and 32°C, for every 10°C rise in temperature, rate of reaction is reduced by 50 per cent. Sorption of the fumigant by the objects to be fumigated, is dependent on the nature of the substances and moisture percentage.

Tent fumigation

Citrus or walnut trees are fumigated under gas proof tents with HCN to control scale insects. Liquid HCN is volatalised by vapourisation and the gas is introduced through a hose. For fumigation of packed comodities of stored products, plastic sheets (of 300 gauze) should be rolled and firmly clamped at the edges and sealed at the ends with gunny tubings filled with sand. Polyethylene and polyvinyl chloride sheets of 0.01 cm to 0.015 cm thickness are effective in retaining methyl bromide.

Vacuum fumigation

Delay in penetration of fumigants under ordinary atmospheric pressure may be overcome and the time required for effective fumigation may be reduced if partial vacuum can be created, particularly with HCN, CS_2, ethylene dioxide, ethylene dibromide or methyl formate. In fumigation for articles for industrial purposes, vacuum to 1/3rd of atmospheric pressure needs to be created. The fumigant is then introduced after heating to 49°C till the atmospheric pressure is reached. After the fumigation, the gas is pumped out and air is introduced and circulated till all traces of fumigants are removed. Cotton, dried fruits, coffee, cereals, potato, seeds to be used for commercial purposes are treated by fumigation in large quantities for quarantine and other purposes. Special chamber-fumigators of different sizes are available for the purpose.

Fumigation may be taken recourse to in field crops if the method of spraying is not likely to give the desired results (as in groundnut, strawberries, etc., where bushy habit prevents successful spraying), and such measures are sufficiently remunerative to compensate for the cost of fumigation. It may be accomplished by the adoption of pyrotechnic methods already discussed, or by use of an electric heater on which the active ingredient can be placed for vapourisation.

Aerosol

In the aerosol method, the toxicant is dissolved in a suitable solvent, e.g., methyl chloride, "Freon" (dichlorodiflouromethane) which is normally gaseous but can be liquified under pressure at ordinary temperature. On release of pressure the liquid containing the toxicant vapourises leaving a residue over the entire surface.

Baits

Toxicants are mixed with dry or wet food sufficiently attractive to the organisms which are intended to be exterminated by allowing them to consume the food mixed with toxicants. Normally rice or wheat bran, molasses are used along with toxicants for control of locusts and other insects, toxicants used being gamma-BHC, DDT or copper aresnate. Zinc phosphide or aluminium phosphide is used as toxicants for baits for rodents. Molluscas are also controlled by poison baits.

Paste

A few pesticides are marketed in the form of a paste which contains a low percentage of the active ingredient, together with a filler—some form of liquid medium and possibly some sticking agent. These pastes are normally used for banding of trees for control of some specific diseases and pests. Sometimes paste formulations are marketed to keep the fine state of division which cannot be maintained as wettable or water dispersible powder.

Spraying : Volume

Toxicant is normally available in the concentrated form in different dilutions. A specific amount of active ingredient which is usually small has to be distributed over a large area and the volume used will have to ensure that an effective coverage has been achieved. When a large volume is used, a complete coverage is usually ensured. When the same amount of pesticide is applied in a smaller amount of water, there will be an increase in concentration of the active ingredient. Spray in such cases needs to be accomplished in the form of fine droplet spectrum or mist. Similarly, the amount of spray per hectare will depend on the crop to be treated. In trees and shrubs a larger amount of spray is required than in the case of ground crops. Depending on the volume that needs to be used for coverage, sprays are categorised into high volume, medium volume, low volume, very low volume, etc.

Normally the requirements of volumes of spray liquid needed for coverage under different conditions are approximately as below :

	Litres per hectare on trees and shrubs	Litres per hectare on ground crops
Very low volume	Under 250	Under 60
Low volume	250—600	60—250
Medium volume	600—1200	250—700
High volume	Over 1200	Over 700

In ultra low volume spray, where the carrier or diluent is much less volatile only a few litres per hectare are sufficient for the protection of the crop.

Normally the concentration designates amount of active ingredients of spray materials used in 100 litres of water or amount of active ingredient per hectare. The volume of the spray has to be adjusted in consideration of the concentration of the liquid that has to be used. Doses also designate or imply the concentration of the chemical used. In case of protective spraying, a complete coverage has to be ensured. In case of compounds which are systemic in nature, absorbed and translocated within the plants such complete coverage may not be needed, nevertheless the material has to be carefully applied.

Sprayer

Depending on the source of energy for operation of sprayers, they may be broadly divided into two categories : (a) manually operated or hand sprayers, (b) power operated sprayers which are commonly termed as power sprayers. Whatever may be the source of energy, sprayers consist of the following essential parts :

(a) *Tank*—depending on the source of energy and its use, the size of the tank may vary from a litre in a hand atomizing sprayer to many hundred litres in heavier power driven orchard sprayers. Tanks are usually made of galvanised iron or steel coated with anti-corrosion material particularly on the inside. Tanks from cold rolled brass sheets or copper lined tanks are in use in spray of copper fungicides. Tanks with resinbound glass fibres or lined with inert plastic materials have also appeared in the market. Aluminium or light metal alloys may be used particularly in smaller machineries, but they are usually expensive. Depending on the job, the capacity of tank should be suitably devised so that time is not wasted on repeated fillings of the tank. Volume of the spray to be used/volume of the tank will give an approximate idea of the number of fillings needed. The liquid in the tank should be sufficient for operation for at least 15-20 minutes spraying. Besides, in manually operated sprayers, the tank should be so designed that the operator does not feel any discomfort.

The tank is provided with a large filler hole fitted with strainer not only to have an easy pouring of liquid into the tank, but also for cleaning. It is desirable to have rounded corners and bottom and there should be a drain plug at the bottom of the tank for better cleaning facilities. There should be airtight devices during the spraying operation.

(b) *Pump*. This is the most expensive and essential component of the sprayer which is needed for atomisation of the liquid through air. While selecting a pump for a sprayer, two factors need to be considered : nature of the spray liquid—its formulation as concentrate, viscosity, corrosiveness ; and delivery—the time and the pressure at which the liquid has to be delivered ultimately through nozzle.

Depending on the mechanism of action, the source of power, etc., pumps in use in the sprayer may be broadly put into three categories : (i) pneumatic ; (ii) displacement pumps, plunger, piston, rotary and diaphram ; and (iii) centrifugal or impeller.

Air or pneumatic pumps : —Used mostly in hand compression or pneumatic sprayers. It is used to force air into the airtight tank up to a certain pressure. The compressed air forces the liquid or release of valve, exerts pressure on the liquid and forces it through the nozzle. It is really a force pump.

Positive Displacement pumps : Positive displacement pumps are those which take in a definite volume of liquid from inlet and without possibility of escape transfer it to the outlet.

Plunger or Piston Pump : It is used in power sprayer and can generate high pressure up to 70 kg/cm². It essentially consists of a piston which operates inside a cylinder. In the suction stroke it sucks in the liquid through an inlet valve. This valve closes on the pressure stroke and forces liquid through the outlet valve into the delivery circuit. Two or three cylinders are commonly used and an air vessel is introduced to even out the pressure. These pumps are stoutly made and consist of a number of component parts, hence are expensive. Pistons are fitted with plunger caps (washers) which are of moulded rubber and other materials and they must fit in well with or be seated against, the inside of the cyclinder. The inside of the cylinder is usually lined with enamel. When muddy water with fine sand and clay is used then there is a rapid wear of the plunger cap with possible corrosion and unevenness of the lining. The capacity of the pump depends on the number of cylinders, their diameter, number of strokes per unit time and length of the stroke. The rate of movement of water is slow, hence these types of pumps are unsuitable for viscous liquids. They are suitable for high volme spraying.

Rotary pumps : These have the advantage that they can be directly coupled with the driving shaft of the power source without involvment of the crank shaft needed in a reciprocal pump. Common types used are : (a) gear types and (b) roller vane type.

(a) In *gear types,* pumps are fitted with a continuous delivery system which is maintained by movement of a pair of gears running together in mesh in a casing. The liquid enters between the teeth as it comes out of the mesh and is carried round between casing and teeth to be discharged at a later point before the teeth enmesh once more. The pump is made of brass, bronze or high carbon steel and can be operated smoothly only when clean water is available. The use of formulations of wettable powder will cause heavy wear in such pumps, hence such pumps are not recommended for use with wettable powder. They are used for low volume spraying as the pressure generated is low, 4.2 kg/cm^2.

(b) *Roller vane pump* is operated by a single rotor which is eccentrically mounted inside a casing. The casing is divided into a section of vanes into which rollers are fitted and thrown against the casing by centrifugal force. The liquid enters through the inlet part (opening for intake of liquid) and is trapped between the rollers and then ejected through the outlet subjected to pressure being trapped between the rollers. These pumps are simple, but more expensive than gear pumps, but they develop higher pressure than gear pumps (8.4 kg/cm^2). Most medium output sprayers are fitted with pumps of this type.

Diaphragm pumps : These operate on the same principle as plunger pumps. The pressure is generated on the rear of the diaphragm by a reciprocating plate which acts like a plunger. The diaphragm is moved up and down a short distance by means of a rod. With the downward movement of the diaphragm a vacuum is created above it and the liquid enters through an inlet valve. The diaphragm is then pushed up and the liquid is ejected under pressure through the outlet valve. Liquid is sealed from moving parts, hence there is little damage due to abrasion, and wear and tear are much less. But these pumps are limited in capacity conditioned by strength and movement of diaphragm. High pressure as obtained in plunger pump cannot be generated in these sprayers fitted with these types of pumps. Diaphragm pumps are not in common use.

Centrifugal or impeller type of pumps : These take in the liquid at its axis and throw it to the periphery by centrifugal force, where it is delivered. The liquid moves out quickly because of high speed. Due to the absence of reciprocating action liquid discharge is even. Pressure generated reaches up to 7 kg/cm^2, hence these pumps are not suitable for high pressure sprayers. They are workable with viscous liquids and muddy water. They are more expensive than gear and vane types but easy to maintain.

Agitators : For maintenance of an uniform dispersion of suspensions and emulsions, there is necessity of agitation which is usually provided in the power sprayers by a return flow pipe as well as separate mechanical agitators.

Nozzles : Nozzles are specially designed apertures to break up the liquid coming out of the spray tank into fine droplets. Many different types of nozzles have been devised to suit different types of work : (a) hydraulic energy, (b) gaseous energy, (c) centrifugal energy, (d) kinetic energy, and (e) thermal energy. Hydraulic nozzles consist of three essential parts : (1) disc or jet bearing aperture through which the liquid has to pass to get into the air, (2) swirl plate which directs flow of the liquid, and (3) swirl chamber between swirl plate and disc. Besides these three parts, a filter may be present. Hydraulic nozzles used on sprayers may be of two types : fan and cone. Disc is composed usually of tungsten steel or ceramic plate to counteract the wear and corrosion. For fan shaped distributions, the ceramic disc has a V-shaped orifice through which the spray liquid is emitted under pressure to give the desired pattern of droplets. In cone or swirl nozzles, the disc which is a hard ceramic plate has a central aperture. The spray, before being emitted through this aperture, is given a swirling motion in a chamber which it enters by angular slots in the swirl plate. The pattern of disc and swirl plate is changed to suit different patterns of spray. A required output can be obtained from a larger aperture at low pressure or a smaller aperture through

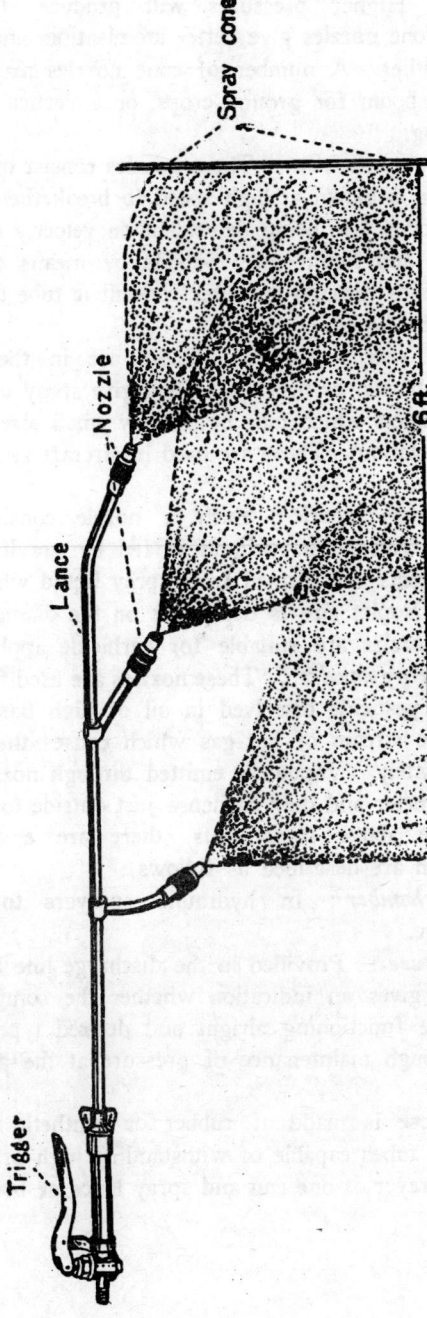

FIG. 16. Diagrammatic representation of discharge of liquid through nozzle (courtesy American Spring & Pressing Works (Private) Ltd.)

high pressure. Higher pressures will produce droplets of smaller size. Cone nozzles give better atomisation, and are more efficient but costlier. A number of cone nozzles may be fitted on a horizontal boom for ground crops, or a vertical boom for orchard spraying.

Gaseous energy nozzles : These nozzles consist of an orifice across which air is blown at high speed to break the liquid into small droplets, drop size being dependent on velocity of air. In some cases air and liquid meet outside by means of annular nozzle. Air passes through an outer concentric tube circling the tube carrying the liquid.

Centrifugal energy nozzles : These are in the form of spinning discs, drums, or brushes. They give spray in the form of fine mist droplets which may be of very small size at a very high speed of rotation. These are used in aircraft and ultra low volume spraying.

Kinetic energy nozzle : Such a nozzle consists of an oscillating tube with holes. Liquid is fed by gravity into the nozzle and filaments are formed from spray liquid which breaks into droplets. Droplet size is dependent on the diameter of the holes. These nozzles are suitable for herbicide application.

Thermal energy nozzles : These nozzles are used for producing fogs. The pesticide dissolved in oil of high flash point is injected into the stream of hot gas which causes the liquid to vaporise immediately. Vapour is emitted through nozzles which are of large diameter and they condense just outside to form fog.

Besides the above components, there are a few more accessories which are described as follows :

Pressure chamber : In hydraulic sprayers to maintain uniform pressure.

Pressure gauge : Provided in the discharge line to note the pressure which gives an indication whether the components of the machine are functioning alright and desired type of spray is achieved through maintenance of pressure at the appropriate level.

Hose : Hose is made of rubber or synthetic rubber or nylon or plastic tubes capable of withstanding high pressure. It is fitted with sprayer at one end and spray lance or boom on the other.

Spray lance : It is a brass tube of variable length with a minimum diameter (6 mm) and a minimum thickness to withstand pressure. One end of spray lance is fitted with a hose and other with a nozzle. The nozzle end is usually bent. When more than one nozzle is fitted either horizontally or vertically, it is usually called a spray boom.

Different types of sprayers

Hydraulic energy sprayers

In this category of sprayers, hydraulic pressure is thrust upon the liquid by the hand operated pumps. As a result, the liquid is forced through the nozzle in the form of a spray of droplets, which are mostly 300-400 mμ in diameter. Sprayers of this type are high volume, high pressure, and suitable for complete coverage of both ground and field crops. Different types of hydraulic energy sprayers commonly used are enlisted as follows :

(a) *Syringe* : Syringe consists of a cylinder into which the spray liquid is drawn through the nozzle aperture on the return or suction stroke of the plunger and thrown out on the compression stroke. The liquid is kept in a separate container from which the syringe has to draw in each stroke or operation. It is very tiresome to operate. It is difficult to control the rate of application by this sprayer. It can be used for drenching.

(b) *Bucket pump sprayer* : It consists of a hand operated pump with a lance and nozzle outlet. Suction hose is placed in the bucket. One hand operates plunger, while another hand keeps the pump in stable position. Plunger rod is hollow and serves as the compression chamber. Liquid is discharged in both suction and delivery strokes, hence a continuous application can be made. It is very tiring and rate of application cannot be controlled.

(c) *Stirrup pump sprayer* : The sprayer consists of a double action pump which can be suspended in a bucket container with the help of suitable clamps and supported by a foot stirrup reaching the ground. A long flexible outlet pipe with a lance and desired type of nozzle is provided for delivery. Two operators are necessary, one for pumping and agitating the suspension if necessary and another for spraying. By continuous pumping, steady uninterrupted spraying can be made. Normally a pressure of 4 kg/cm² is achieved. These sprayers give satisfactory performance.

When bigger areas are to be covered, a bigger sized double action pump can be fitted permanently in a large container, which may be mounted on wheels. The entire assembly may be pushed through the area to be sprayed in the manner of wheel barrow. The pump is usually operated by a lever mounted on it. Provision of a long delivery system enables a larger area to be covered. Such sprayers are often known as wheel barrow type of sprayer.

(d) *Knapsac sprayer* : The sprayer consists of a flat or slightly curved rectangular assemblage which can be carried on the back with the help of straps. It is provided with a double action lever operated pump which may be either inside or outside the sprayer. The operator with his one hand, usually left hand, operates the lever which is extended along the left hand side of

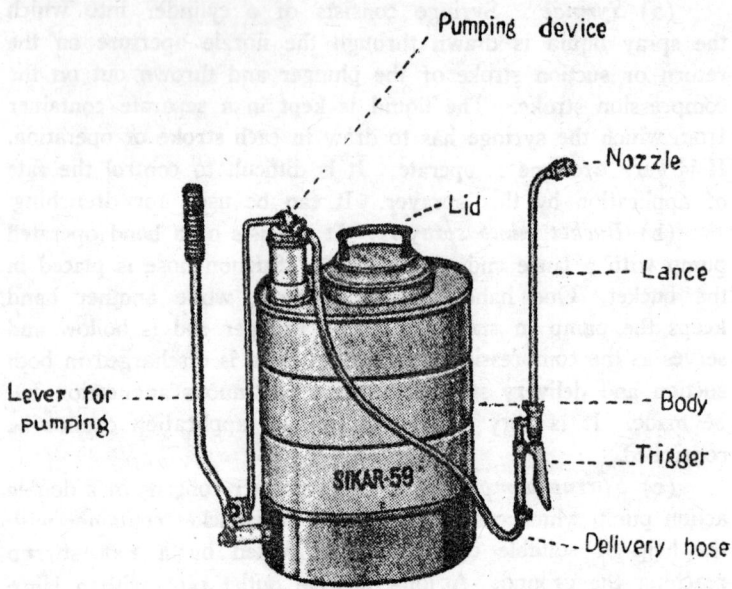

FIG. 17. Knapsac sprayer (Courtesy American Spring and Pressing Works (Private) Ltd.)

the operator. Spray liquid is delivered through the delivery system, consisting of lance and nozzle, which is connected with the pump by a flexible hose. Spraying is done by right hand. Coarse nozzles are normally used to undertake spraying of any type of material. At present these sprayers may be fitted with low volume nozzles to achieve low volume spraying. These sprayers are useful because of their simplicity in opeartion, durability and for diverse use including spraying bushes of tea and coffee. Different types of nozzles and tailboom may be fitted to suit desired conditions.

(e) *Rocking sprayer* : This type of sprayer consists of a lever operated pump assembly which rests on a wooden platform.

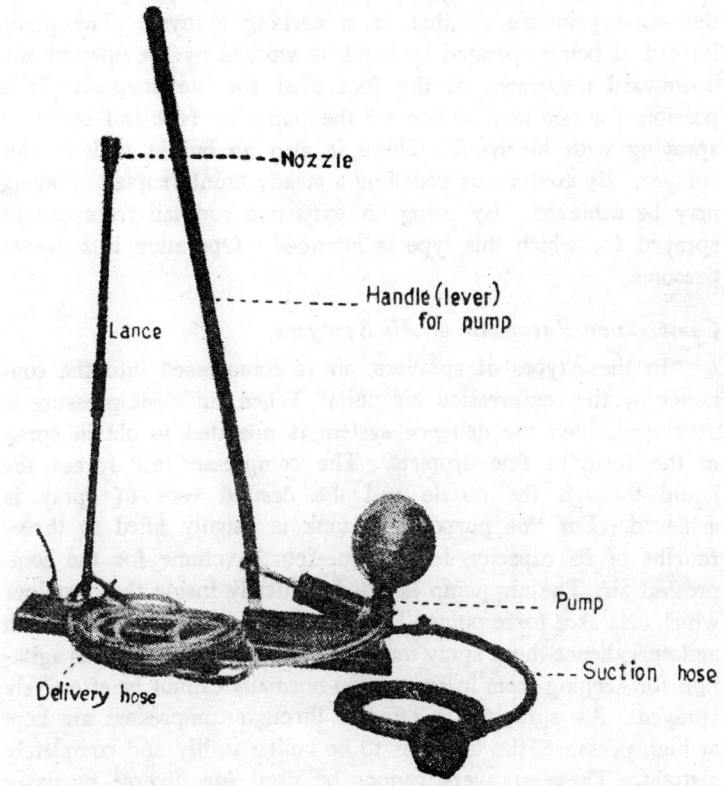

Fig. 18. Rocking tree sprayer (courtesy Shaw Wallace Co. Ltd.)

Suction hose with a strainer is immersed in a separate container containing the spray liquid. Delivery system consists of a separate pressure chamber, a flexible hose, spray lance, and a spray nozzle. The lever attached to the pump is operated by the rocking—forward and backward movement—of the handle. Pressure is developed in the pressure chamber, which may attain pressure of 14-18 kg/cm². Such sprayers are used for spraying tall plants like coconut and arecanut trees, and sugar cane plants. Uniform spraying can be done if sufficient pressure is maintained in the pressure chamber. It needs two persons to operate the sprayer, one for operating the pumping system and another for the application of spray liquid.

(f) *Foot or pedal pump* : Foot or pedal pump works on the same principle as that of a rocking sprayer. The pump instead of being operated by hand, is worked by the upward and downward movement of the foot used for the purpose. It is possible for one man to operate the pump by foot and carry on spraying with his hand. There is also no builtin tank in this sprayer. By continuous pedalling a steady uninterrupted spraying may be achieved. By using an extension rod tall trees can be sprayed for which this type is intended. Operation is however tiresome.

Compression Pneumatic or Air Sprayers

In these types of sprayers, air is compressed into the container by the compression air pump. When sufficient pressure is developed, then the delivery system is operated to obtain spray in the form of fine droplets. The compressed air forces the liquid through the nozzle and the desired type of spray is achieved. For this purpose the tank is usually filled to three-fourths of its capacity, leaving one-fourth volume for the compressed air. The air pump is fitted vertically inside the container which acts as a force pump. These sprayers are not provided with agitators, hence those spray materials which need continuous agitation for keeping them in suspension normally cannot be effectively sprayed. As spraying is effected through compressed air kept at high pressure, the tank has to be built sturdily and completely airtight. These sprayers cannot be used for diverse purposes like hydraulic types nevertheless various types are available.

Depending on the size, these sprayers may be compression hand sprayers or compression knapsack sprayers.

Pneumatic or compression Hand sprayer : This type of sprayer consists of a tank of small capacity varying from 0.5 to 3.5 litre with a pump inside the tank. The tank itself acts as pressure chamber. The outlet pipe is suspended in the liquid in the container, the end running into the bottom, the other end or the outlet terminates in a nozzle. Before spraying, air is forced into the tank by action of the pump till sufficient pressure is built. Release may be made either through a stopcock or trigger. A continuous fine spray is obtained till the liquid is emptied out. Due to the provision of fine nozzles in these sprayers, solutions and emulsions can be effectively applied. Suspensions tend to clog the nozzle system to prevent which an easily removable strainer is provided.

Pneumatic Knapsack sprayer : This sprayer works on the same principle as the pneumatic hand sprayer. The capacity of the tank, which is cylindrical, varies between 10 and 20 litres. It has to be fixed on the back, with suitable adjustable straps. A curved backrest is provided for easy carrying of the sprayer. Air is charged into the container or tank by action of the pump and a pressure gauge is provided to indicate the pressure. Air capacity and pressure are usually adequate to discharge the liquid contents out in the form of fine spray without repumping. For discharge of the spray at constant pressure a pressure regulating valve may be provided. Instead of pumping air, compressed air may be provided artificially. Pressure of 4-5 kg/cm² is considered sufficient for the purpose.

In conducting spray of plantation crops, a large number of sprayers are used. In such cases a battery system in which 6 or 12 sprayers can be charged with air from outside can be adopted. Liquid is also pumped in from outside. After discharge of the liquid, these sprayers are again recharged with the liquid, as air under pressure is retained in the sprayer due to provision of a ball valve system. This system proves to be economical only when a large number of sprayers are in operation. Different nozzle systems can be fitted on the lance to have the desired type of spray. In this sprayer, the operator can concentrate his attention on spraying, since charging of the air is done beforehand.

Gaseous energy sprayers

Atomisers : These sprayers consist of a small cylinder or chamber usually with capacity 0.5 to 1 litre. A plunger pump is fitted which is moved forward and backward to produce the compressed air. Valves are not usually present. The leather piston allows air to pass into the compression side of the chamber in the return stroke. Outlet is modified into a fine nozzle and fixed at a 90° angle to a narrow liquid feed tube suspensed inside the container. In the compression stroke the vaccum created draws in the liquid which is broken into fine droplets as it comes out by the force of the air delivered out of the pump. This sprayer is suitable for experimental work on individual plants. It is tedious to continue spraying for a long time with atomisers.

Power sprayers

Power operated sprayers have been put into use wherever conditions justify their use : intensity of pest infestation, suitability of such machines being conveniently operated and cost involved. These machines evidently can cover much larger area, and do the job efficiently. They are of particular value when there is an outbreak of pest or disease and also economical from the operational point of view. There are also some snags, namely many power sprayers cannot be conveniently taken into many fields, particularly wet rice fields. It is only justifiable to use such sprayers when large areas are to be treated or protected. Besides, skill on the part of the operator is needed for its operation, care and maintenance. In recent years power sprayers have been put into use for timely and effective pest control. A wide range of power sprayers, beginning from the knapsack type which can be carried on the back, portable to tractor drawn ones are now in the market. Different mechanisms have been adopted for the economy of spray fluid, quickness of operation and effective coverage.

As in the case of manually operated sprayers, power sprayers may be broadly divided into different categories, depending on the mechanism of action : (i) hydraulic energy, (ii) air pump, (iii) gaseous energy or blow applicators, and (iv) centrifugal energy sprayers.

Hydraulic energy sprayers : These sprayers essentially consist of a double action reciprocal pump driven by an aircooled or watercooled engine, with a pressure chamber, pressure regulator and a delivery system consisting of a variable length of hose terminating in a hand operated spray gun or fitted with booms and fixed nozzles. In smaller portable ones they are without any tank, and in traction type tractor drawn or truck mounted large ones, tanks of variable size from 160 litres to more than 2000 litres are fitted with the sprayer. These are essentially high volume sprayers and operate under high pressure which may be up to 40 to 50 kg/cm².

Small portable ones may be either stretcher type or wheel barrow type. In small portable ones normally two delivery hoses can be fitted with hand adjustable spray guns to regulate the type of spray. They are usually run by aircooled engines of 1-3 hp. In stretcher type there is no built-in tank, and a

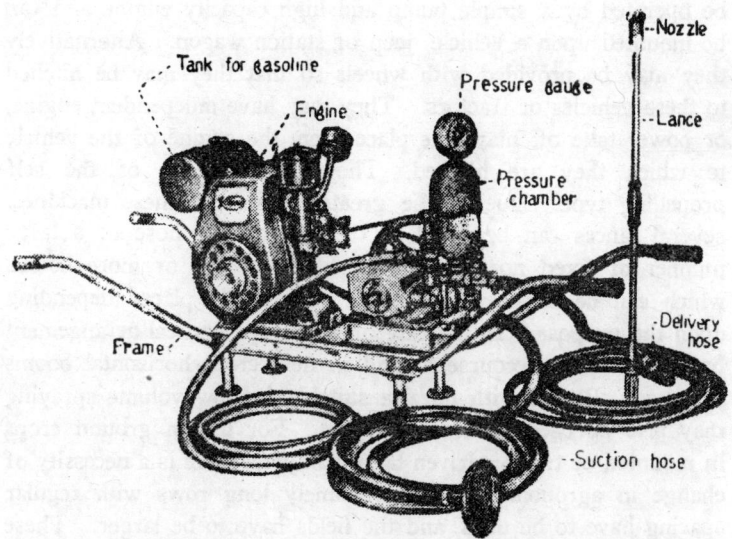

FIG. 19. Hydraulic energy power sprayer (courtesy American Spring and Pressing Works (Private) Ltd.)

separate container has to be used into which a suction hose fitted with a strainer is put. Stretcher type sprayers are light (usually 20 kg in weight) and fitted with contrivances for easy transportation for which two persons are needed. They can deliver up to 23-27 litres/minute at pressures of up to 10 kg/cm².

Wheel barrow types are heavier in weight and are fitted with one, two or four wheels depending upon the weight and structure of the entire assemblage and easy transportation in the area for which it is intended. These sprayers may or may not have a built-in tank attached to them. If a tank is provided, its capacity lies between 50 and 80 litres. It can deliver 7-14 litres/minute at a pressure of up to 14 kg/cm².

Both these types are suitable for spraying both ground crops and orchard or plantation crops. They are of special value in smaller areas where the plot size is variable and cultivation with regulated spacing in lines cannot be practised.

Large sprayers : They have a tank capacity ranging from 200 to 2000 litres or more. Some of them are capable of high delivery at a very high pressure of up to 28 kg/cm². They may be operated by a simple pump and high capacity engine and can be mounted upon a vehicle, jeep or station wagon. Alternatively they may be provided with wheels so that they may be hitched to these vehicles or tractors. They may have independent engine, or power take off may take place from the engine of the vehicle to which they are hitched. They may also be of the self propelling type. Due to the greater output of these machines, several lances can be fitted at the end of the hose or a large number of fixed nozzles may be fitted on one or more booms which can be either vertically or horizontally placed depending upon the purpose. In the case of tree crops vertical arrangement has to be taken recourse to and in field crops horizontal booms are used. Booms with nozzles suitable for low volume spraying may also be used in these sprayers. For use in ground crops in mounted or tractor driven large sprayers, there is a necessity of change in agronomic practices, namely long rows with regular spacing have to be used, and the fields have to be larger. These sprayers cannot be conveniently taken inside for the spraying of tall field crops like jute, sugar cane at a later stage nor inside the wet rice fields. They may be used for spraying of orchards and plantation crops.

Compression sprayers : In these sprayers, no pump is involved. Engine power is employed to create a layer of compressed air over the spray liquid in the tank which is very heavily built to withstand pressure of up to 14 kg/cm^2. The size and capacity of the sprayer varies. They may be easily portable ones to large mounted types. Since no pumping action is involved, these sprayers can be used for the spraying of corrosive materials provided the inner surface of the tank has an anticorrosive lining.

Portable and small mounted sprayers : These sprayers may range in size and weight to the extent that small one can be carried by one man, to the larger capacity one which can be carried by two men or mounted on chassis to be either pulled or pushed. Compressed air from a single cylinder compressor driven by a V-belt from an air cooled engine is forced into the liquid in the tank which may hold up to 50 litres of spray liquid. The delivery system contains a hose from the tank of the sprayer which may bear lances ending in nozzles or boom with a number of nozzles. There is no system of agitation in the tank in these sprayers, hence it is not suitable for spraying of materials which sediment quickly.

Larger mounted sprayers : These sprayers are essentially similar to the smaller portable ones in their working mechanism. They have a larger tank capacity which may be of up to 2000 litres. Smaller machines can be mounted upon vehicles and larger ones hitched to them, the power take off in both the cases may be from the vehicle or tractor as the case may be. In the case of those sprayers which are fitted with independent engines, the engines are more powerful. The delivery or spraying system may be elaborate, with a number of lances or booms fitted with a number of nozzles as in the case of large sized hydraulic operated sprayers. Since these sprayers have arrangements for agitation of the spray liquid, all types of spray formulations can be used. These sprayers are expensive and can only be used on level lands. They cannot be employed in small undulating areas.

From the preceding descriptions of both the manually

operated and power sprayers, it is evident that though they are efficient, they have their limitations for use. The idea of spraying concentrates instead of high volume of comparatively lower concentrations which necessitates a large supply of liquid, has gradually gained momentum in the past. The present trend is to use concentrates for application with smaller sprayers, smaller pump, lower pressure, smaller weight, more uniform air velocity, faster rate of travel and convenient control. Such machines are cheaper because of their lower cost of manufacture, their portability, and the cost of labour needed for spraying being lower.

Gaseous energy sprayers : Gaseous energy power sprayers which are low volume, low pressure, work on the blowing of a strong current of air towards the liquid coming out of the sprayer which atomises the liquid into fine droplets. These essentially consist of a source of power, an air cooled engine of 1.2 to 3 hp, a fan or blower to throw air at high velocity, a pesticide container and a delivery system terminating in fine nozzles. These sprayers are suited to spray concentrates but suffer from one defect, namely, the difficulty in regulation of uniformity of spray deposits at different distances. They are commonly known as blow applicators. Some machines have been so designed that both dusting and spraying can be done with them.

Droplets produced in this fine spray vary in size from 100-400 mμVMD (volume median diameter).

Motorised knapsack sprayer is a blow applicator. It is gradually becoming popular, particularly in this country where large-sized conventional machineries are too bulky and expensive for use and ordinary hand sprayers are inadequate. As against 0.4 hectare which may be covered by a very efficient hand sprayer, more than 3 hectares can be treated very conveniently by a motorised knapsack sprayer by one person.

Hopper made of high density polythene has a capacity of 7-12 litres. Besides the hopper there is a small tank of 0.75 to 2.25 litre capacity for fuel. These sprayers are operated by an air cooled engine 1.2-3 hp. These sprayers weigh 7-15 kg when empty. The machine is put on a suitable frame which is provided with a shock-proof cushion so that the operator does not feel any inconvenience when the machine is fitted on his back. Spray liquid is blown by an air current produced in the machine. The location of engine throttle on the delivery line enables the operator to control the air velocity. The nozzle is connected with the container through a flexible hose and control of discharge is effected through manipulation of a series of discs or restrictors with different bores. Normal discharge of air is 2.7 to 9.1 m^3/min at a velocity of 175-320 km/h. Discharge rate of the liquid varies from 0.5-5 litre/min and fuel consumption 0.6-1.86 litre/h. Normally the tank should not be full, but a small space should be left for the air cushion to facilitate the uniform discharge of spray liquid. Motorised knapsack sprayers may be used for dusting and ultra low volume application.

Motorised knapsack sprayers may be enlarged with a larger tank capacity with greater delivery rate by providing more powerful engines. Such machines which are heavy cannot be taken on the back of the operator and have to be mounted on stretchers or wheel barrows or may be vehicle mounted. Stretcher or wheel barrow types are used where greater acreage has to be covered. Vehicle mounted ones are used for the spraying of orchards and bushes.

Centrifugal-energy sprayers : These sprayers are designed to produce fine drops (50-100 mμ) in the form of mists, hence are often known as mist sprayers or mist blowers. A fan or blower of the axial flow or centrifugal type supplies air at high velocity and throws the small droplets to the target. The sprayers which may be of hand mist type or electrically operated, are lightweight and meant for indoor use in glasshouses or in the treatment of livestock in sheds. Thin metal discs are mounted on a shaft in the fan outlet. Spray liquid gets collected to these discs in revolution and they are thrown out as fine droplets by the axial flow of air at high velocity.

Knapsac mist blowers are also in common use. They operate like motorized knapsac sprayers. They are high concentration, low volume, low pressure sprayers. Atomisation of spray liquid is accomplished by a rotary atomiser of cylindrical wire gauze fitted to aerofoil. Spray liquid is fed to the motor which moves at a predetermined speed at low pressure and is broken into fine droplets by the action of a rotor. Centrifugal force of r which can be regulated by its size and speed determines the of the droplets. A constant pressure usually of 0.7 kg/cm² is maintained. A container is placed at the top of the framework of the engine. Mist sprayers may be used for blast spraying in which the droplets are deposited on the target by the blast of the air from the machine, or for drift spraying in which the droplets are deposited being carried over by the wind in its direction effectively due to the very small size of the droplets. In drift spraying concentrate formulations based on oil are used evidently for low evaporation. By adjustment of the outlet, swarth can be increased.

Large sized mist blowers, either stretcher or wheel barrow type or vehicle mounted, are also available.

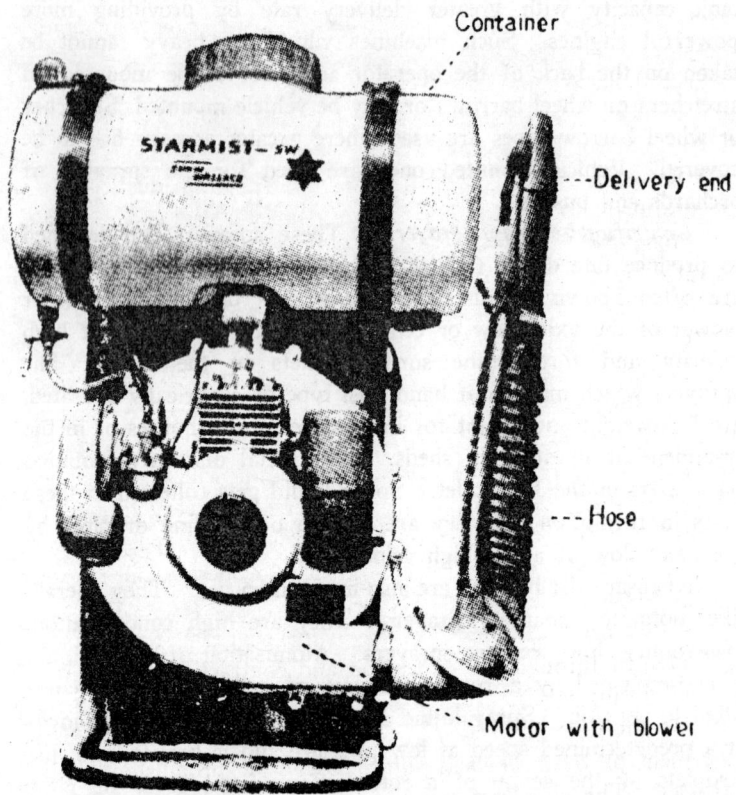

FIG. 20. Mist blower (courtesy Shaw Wallace Co. Ltd.)

Dusting machineries

Pesticides may be applied in dust formulations with specially designed machineries called dusters. All dusters essentially consist of a container which is termed as a hopper, a system of keeping the dust particles agitated and in motion and to be fed into a current of air so that the particles are ultimately discharged in the form of fine dusts dispersed both horizontally and laterally over a large area in the form of a cloud. The desired current of air may be produced by bellows, fan or venturi. The power for operation may be obtained by hand, or from an engine, drive from the wheels, or take off from tractors or suitable vehicles.

Hand dusters

Package or container dusters : These are very primitive types in which the dust is kept inside a plastic, rubber or leather container. By squeezing or through a crude valve system of cardboard diaphragm, dust is discharged in the small stream of air produced. Dusts may be kept in plastic bottles with a small outlet. On squeezing dusts may come out. Small muslin packets may be made which on gentle shaking will emit dusts. These are useful in kitchen gardens or for domestic use.

Hand pumps : These types of duster are usually small and operated by a plunger pump. The container which is normally air-tight, and cylindrical in shape contains an inlet of detachable lid for filling in the dust and an outlet, the aperture of which may be suitably controlled. Dust is thrown out through the outlet by the compression stroke of the plunger pump. A separate dust chamber may be kept at the bottom of the container. Provision of a double action plunger pump will enable a continuous dusting operation. These dusters have very limited use in kitchen gardens, houses, or on animals where spot treatment is needed. They are very cheap and easy to operate, but tiresome over a considerable period of time.

Bellow-type dusters : As the name suggests, these dusters are operated through the expansion and contraction of a pair of bellows during which process dust is sucked in and then thrown out into the delivery system. A current of air produced by the movement of the bellows is sufficient for the propulsion of dust particles. The simpler types, consist of a pair of bellows with the hopper placed between the air inlet and the bellows. When the bellows are stretched then dust is taken in and during contraction of the bellows, the dust is thrown out. A simple lever type of agitator from the scissors of the bellows helps in feeding dusts into the incoming air. These are very tiresome to operate.

A more convenient type of bellow duster is designed in the knapsac fashion. The hopper is of larger capacity and may contain 4-8 kg of dusting material and is mounted on a frame with straps for being fitted on the back of the operator. A pair of bellows are fixed on the top or back of the hopper and are operated by means of a lever or rod through an upward and downward movement extended on one side of the operator parallel to his body. A rotable blade agitator also fixed inside coupled with this lever system helps in feeding the dust into the mixing chamber. The delivery system consists of flexible hose from the hopper which is adjusted or controlled by one hand of the operator. Leakage in bellows may make the machines inoperative. Leather bellows are liable to deterioration in moist hot climates due to action of moulds. They may be satisfactorily replaced by plastics specially developed for the purpose. The weight of the assemblage may be reduced by the use of light alloy metals, though it may increase the cost of the machine.

Rotary hand dusters : These dusters are operated by the current of air produced by a fan which is driven by the rotary motion of a handle through a reduction gear. The fan system may be either on one side or below the hopper. Dust is fed into the fan housing either direct from the chamber or into the stream of air that is sucked in from the inlet vent in the fan. The feeding mechanism is carried by a rotating brush inside. An agitator inside the hopper helps to move the dust and prevent caking or the formation of lumps inside the hopper.

Delivery or outlet is direct from the fan. It may be of flexible hosepipe or of rigid tubular structure. There may be two outlets instead of one to dust two rows at a time. Nozzles may be spoon type or fan tailed for effective dispersal. Improvement in nozzle types has been effected.

These dusters may be carried on the shoulder with the help of straps, or on the belly or on the back like knapsacs. The shoulder and belly types are more convenient to use. The weight of the machine is usually 5-6 kg and the hopper can contain

5 kg of material. For proper operation, the hopper should not be completely full, but an empty space should be kept for the movement of air. Normally 0.5 to 150 gm/minute may be discharged at 35 rpm. The belly types are preferred to the shoulder types, provided the toxicant or formulation does not emit any offensive odour. Rotary dusters are quite popular for use in different crops and in different situations.

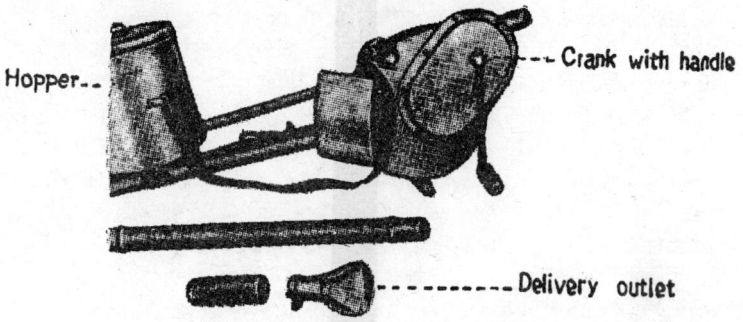

Hopper

Crank with handle

Delivery outlet

FIG. 21. Rotary hand duster (courtesy American Spring and Pressing Works (Private) Ltd.)

Power dusters : The basic components of a power duster are essentially the same as in a hand duster, except that in the former the power for operating the machine is generated from the wheels, engine or vehicles on which they are mounted.

Traction dusters : These are used in advanced countries. They are mounted on wheeled structures. No separate power source is needed. The movement of the wheel in which the duster is mounted generates sufficient power to drive the fan to carry on dusting. The capacity of the hopper varies from 20-45 kg. As operation of the machine is dependent on the speed of the wheel, uniform discharge may not be maintained, in case movement of the vehicle is slowed down or impeded.

Engine-operated power dusters : Engine operated power dusters are in wide use. Normally 1-3 hp. aircooled engines are used and the hopper capacity is usually 10-20 kg. Small engine operated dusters may be stretcher or wheel barrow type or may also be shoulder mounted. In shoulder mounted or knapsac

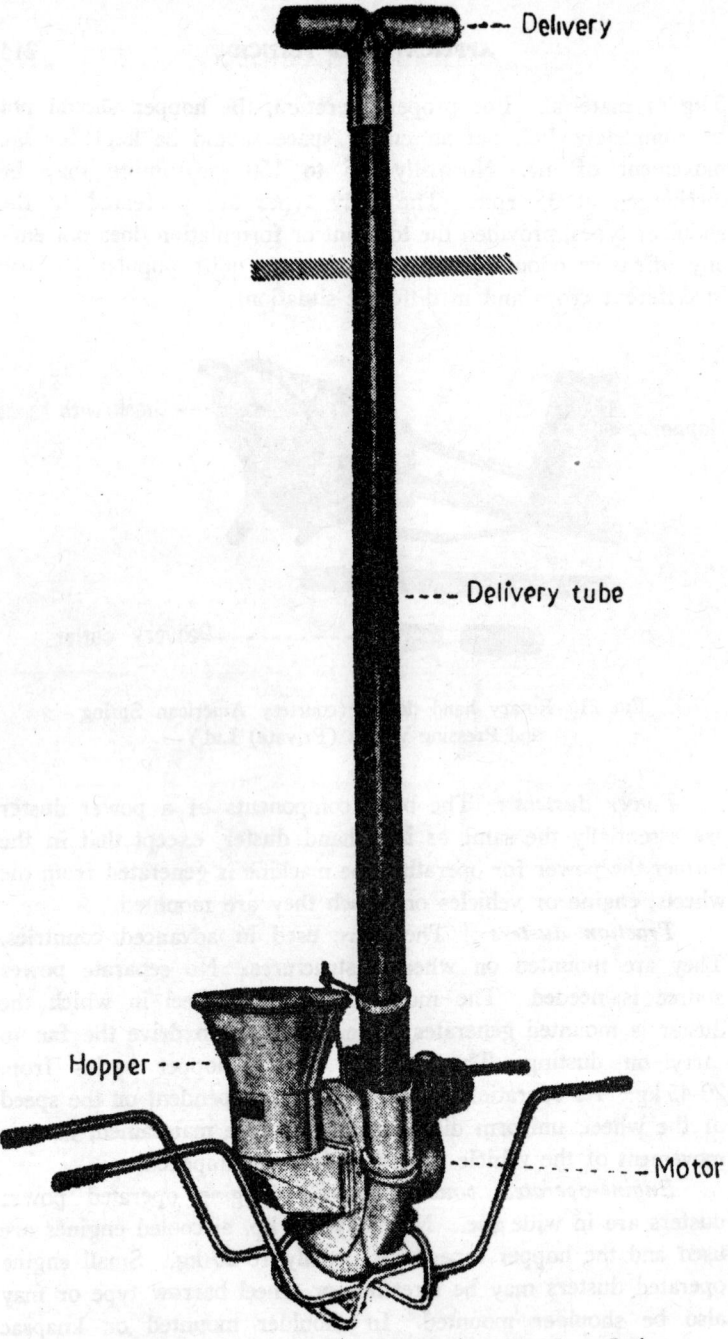

FIG. 22. Motorised power duster (Courtesy American Spring and Pressing Works (Private) Ltd.)

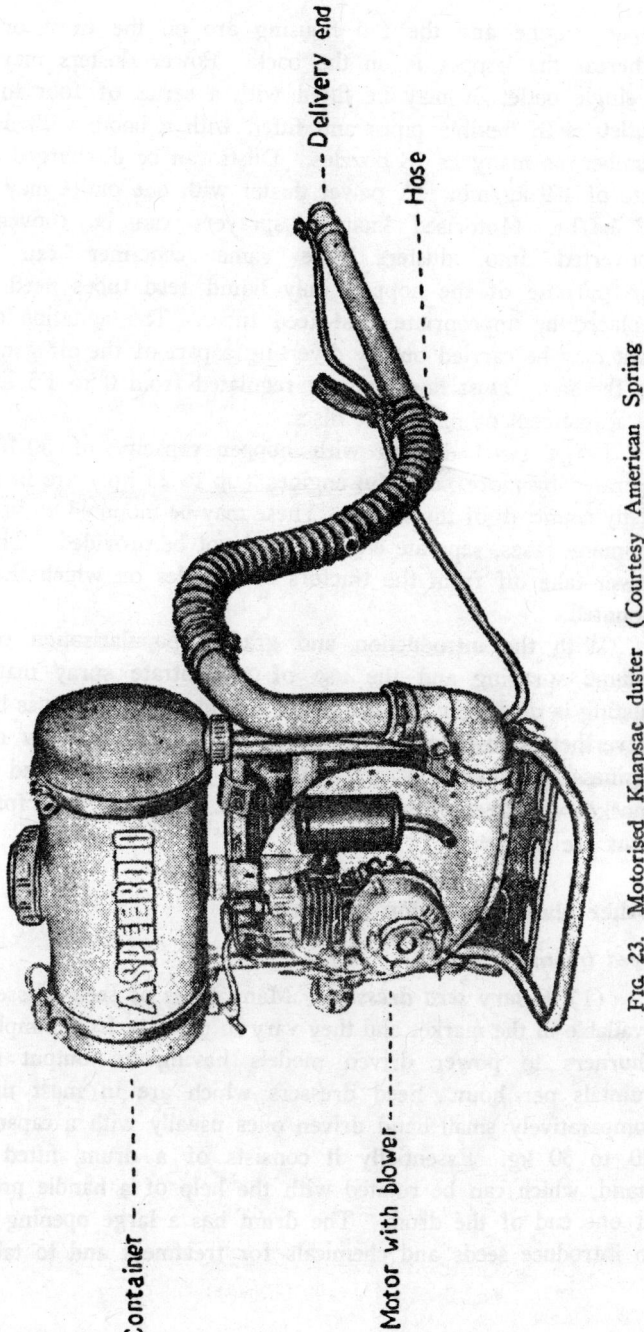

Fig. 23. Motorised Knapsac duster (Courtesy American Spring and Pressing Works (Private) Ltd.)

types, engine and the fan housing are on the chest or belly whereas the hopper is on the back. Power dusters may have a single outlet or may be fitted with a series of four to eight outlets with flexible pipes and fitted with a boom with a large number, as many as 18 nozzles. Dusts can be discharged at the rate of 1-9 kg/min. A power duster with one outlet may cover 12 ha/hr. Motorised knapsac sprayers can be conveniently converted into dusters. The same container can serve the purpose of the hopper, only liquid feed tubes need to be replaced by appropriate dust feed tubes. The agitation of the dust may be carried out by diverting a part of the air generated by the fan. Dust flow may be regulated from 0 to 1.5 kg/min by adjustment of multi-hole discs.

Large sized dusters with hopper capacity of 50-100 kg operated by more powerful engines (up to 25 hp.) are in use in many countries of the world. These may be mounted on vehicles. In many cases, separate engines need not be provided. There is power take off from the tractors or vehicles on which they are mounted.

With the introduction and gradual popularisation of low volume spraying and the use of concentrate spray materials, dusting is no longer popular to the extent to which it was before. Nevertheless dusting is still being carried out in view of the easiness of operation and simplicity of machine and quick knockdown effect of many insecticides. Besides, dust formulations are cheaper.

Other plant protection machineries

Seed treating machines

(1) *Rotary seed dresser* : Many types of seed dressers are available in the market and they vary in capacity from simple milk churners to power driven models having an output of 50 quintals per hour. Seed dressers which are in most use are comparatively small hand driven ones usually with a capacity of 20 to 30 kg. Essentially it consists of a drum, fitted on a stand, which can be rotated with the help of a handle provided at one end of the drum. The drum has a large opening or lid to introduce seeds and chemicals for treatment and to take out

seeds after treatment. Iron baffles are fitted inside the drum at right angles to the drum. The drum should be rotated slowly 30-40 times during which process seeds will be thoroughly coated with the chemical. The drum should be filled up up to two-third of its capacity to enable proper mixing. Power driven models are used in large seed farm or centralised distribution agencies.

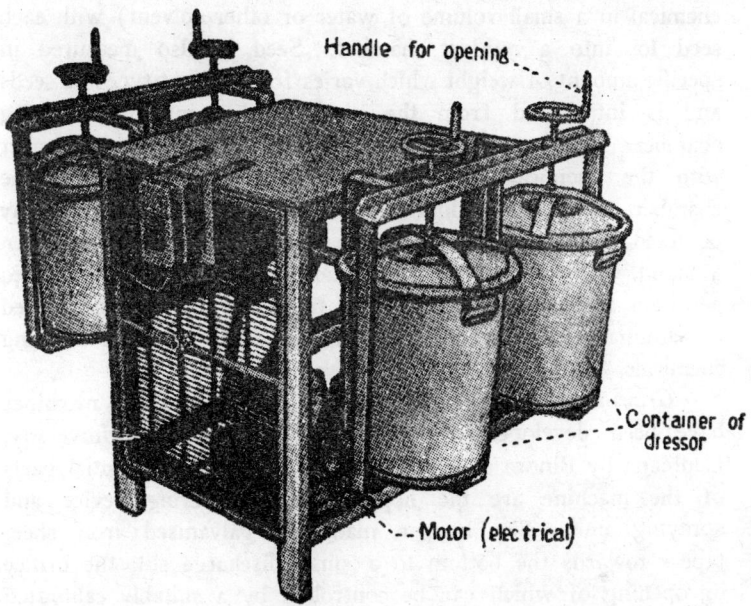

FIG. 24. Motorised seeddresser (Courtesy American Spring and Pressing Works (Private) Ltd)

(2) *Cascade or Gravity seed dresser* : The machine consists of a suitable steel drum fitted with cones and baffles which are perforated. A lid is provided at the top through which the seed and chemical are introduced. Delivery of the treated seeds is taken from a small opening provided at the bottom. Seeds while coming downwards slowly due to gravity get mixed with the chemical. This type of dresser is suitable only for seed treating chemicals which are comparatively volatile in nature. Treatment takes more time, but does not involve any manual labour. Doubts have been expressed whether this treatment is as efficient as that with the rotary seed dresser.

12

(3) *Slurry seed treaters* : These machines are driven by 3-4 hp. electric motor or oil engine. A unique feature of this machine is that the measured quantities of seed and the chemical (slurry) are automatically fed into the treatment chamber in a synchronised manner. The slurry cup introduces a given amount of slurry (a thick suspension of the seed treating chemical in a small volume of water or other solvent) with each seed lot into a mixing chamber. Seed is also measured in specific amount of weight which varies for different types of seeds and is introduced from the overhead bin into the treating chamber. The seed is thoroughly mixed in the chamber with the chemical. Slurry paddles are provided inside the chamber to prevent sedimentation. The smaller machines may be fixed on a platform, while the larger ones may be fitted on a stand. Approximately 30 quintals of seeds may be treated per hour in this machine. Slurry treatment has been advocated to eliminate hazards of objectionable dusts of seed treating chemicals.

Grain-treating machines. Simple grain treating machines have been developed in the Punjab Agricultural University, Ludhiana by Bindra and his co-workers (1977). Essential parts of the machine are the hopper, grain metering device and spraying unit. The hopper made of galvanised iron sheet tapers towards the bottom to a small discharge slit, the orifice or opening of which can be controlled by a suitably calibrated knob serving the purpose of a grain measuring device. The spraying unit consists of a hand compression sprayer usually of 12 litre capacity with a spray lance fitted with pressure gauze and pressure regulating mechanism. Through flat fan low volume nozzle, the spray liquid is discharged at the rate of 290 ml/min at a pressure of 2.1 kg/cm². Automatic spraying is arranged through a bigger control mechanism. As grains which are fed on the top of the hopper fall slowly downwards due to gravity, they are treated with the chemicals. Different calibration scales can be worked out for different types of grains and concentration of chemicals can also be worked out. These machines can treat up to 100 quintals of grains in eight hours. Two persons are needed to operate the machine.

Another manually operated grain treating machine (for

pulses) has been devised in which the pulse grains from the hopper are moved through corrugated rollers into the collecting container through a trough. While the grains are being collected in the trough they are treated with the chemicals which is applied from a hand compression sprayer at the rate of 500 ml of the spray liquid per 100 kg of grain at a pressure of 0.7 to 0.84 kg/cm². Feeding from the hopper which is normally of 50 kg capacity into the treating rollers is done by the rotation of crankshaft at a speed of 360 rotations per minute. About 100 kg of grain are fed in a period of 8-10 minutes. Three persons are needed for the treatment of grain in this machine.

Sack-impregnating machine : Central Food Technological Research Institute, India has developed a simple machine for the treatment of gunny or cotton sacks meant for storage of food grains against the possible infestation of insects. In this machine sacks move through an endless conveyer belt (supported horizontally by rollers powered by 0.25 hp. engine) slowly during which process they are impregnated with the chemical which is applied from a compressor in an 18 litre tank maintained at a constant pressure and powered by 0.25 hp engine. The liquid is applied through a low volume nozzle so that drying is not necessary. Only one side is treated at a time. It has been claimed that this treatment gives an effective protection against the infestation of insects. Approximately 3000 sacks can be treated per day.

Soil Injectors : For the treatment of soil with fumigants which are mostly available in liquid form, a very convenient equipment is a soil injector gun. It essentially consists of a pump, a container for the liquid to be injected and a hollow injection needle. The nozzle can be driven up to a depth of 25-30 cm in the same manner as that of a garden fork. The pump is then pushed down by the hand to force the liquid into the soil through the injection aperture. The containers are of the capacity of 2.2 to 3.5 litres. The process is laborious. Approximately 0.3 hectares may be covered in a day. This is useful for treatment in kitchen gardens or in experimental areas.

In the treatment of large areas, tractor mounted field applicators have to be used. Field applicator consists of a container from which large tines come out. They are drawn in the same manner as harrow. The liquid is fed into the tines and to the soil, as the tines move through the soil.

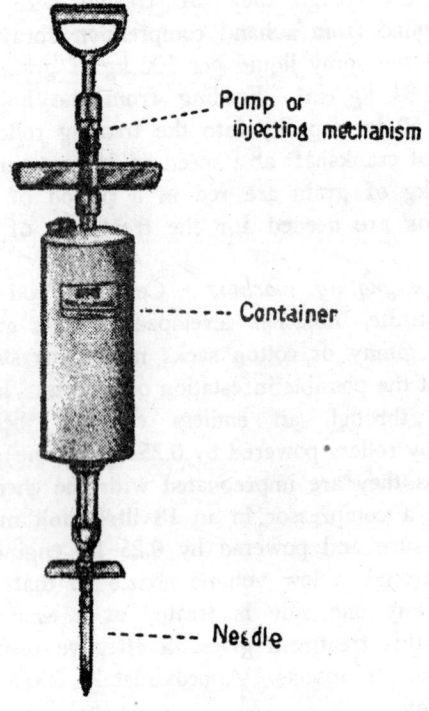

Pump or injecting mechanism

Container

Needle

FIG. 25.   Soil injector (Courtesy American Spring and Pressing Works (Private) Ltd.)

Granule applicator : Insecticides in the form of granules have been introduced in recent years, particularly in the application of systemic insecticides cn crops with broad leaves and leafsheaths or leafwhorls, so that the insecticides can be lodged in the leaves and are slowly released for a period of time. For this purpose, special machines, granule applicators, have been devised. The essential components of a granule applicator are the hopper in which the granules are kept, a long flexible discharge tube with a nozzle at the distal end and a finger controlled mechanism for the regulation of constant flow of granules. At each end of a stroke from the mechanism which is at the base or middle a specific quantity of granules is released into the tube through the exit hole of the hopper. These granules are discharged from the tube at the other end of the stroke. The granule applicators which are usually made of plastic (except the calibration unit) are light and have a granule capacity of 1.0-1.3 kg. Comparatively large sized granule applicators with hopper capacity of 10 kg, of knaspac type are also available.

In the advanced countries row application of granules is carried out by fitting the granule applicator to planters, high clearance tractors or other farm machines. Granules can be applied to the soil along with interculture operations.

Dust applicator for rat burrows (cyanomag foot pump). These machines are used for applying poisons for rats in the burrows, or other insecticides, for the extermination of ants, etc.,

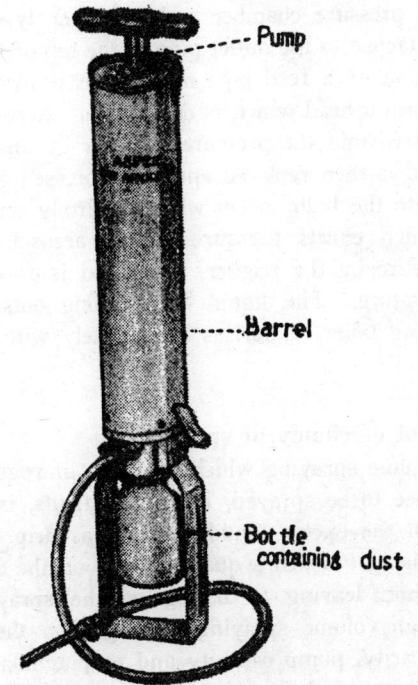

FIG. 26. Cyanomag foot pump (Courtesy American Spring and Pressing Works (Private) Ltd.)

in the soil. It is essentially a plunger type of duster with a dust chamber of 450 gm capacity either of metal or glass air pump, discharge pipe and a delivery tube fitted with a metallic tube at the end. A foothold is provided to hold the pump in position. Hence it is sometimes known as a foot pump. By the action of the pump through a diaphragm type of valve at the lower end of the pump, an airblast is produced which enters the dust chamber. This airblast agitates the powder and forces the same to be discharged through the delivery tube.

Aerosol Projectors : These are useful in large greenhouses or in enclosed spaces as in temporary or permanent godowns or store houses. These projectors consist of a pressure chamber with a feed pipe running from the bottom of the chamber and terminating in a fine nozzle. An inlet tube from a bulb socket enters into the pressure chamber and a trigger type of release mechanism is attached to the outlet pipe at the top of the chamber. There is provision of a feed pipe nozzle release mechanism and a bulb socket form a head which is detachable. Aerosol formulation is first poured into the pressure chamber by unscrewing the head. The head is then replaced and a compressed gas cartridge is introduced into the bulb socket which is firmly screwed. Gas is liberated which exerts pressure on the aerosol liquid. On release of pressure by the trigger, the liquid is forced through the nozzle aperature. The liquid on reaching outside, because of its low boiling point, vaporises immediately with the formation of a fine mist.

Measurement of efficiency of spraying

In high volume spraying which has been in vogue for many years, the surface to be sprayed, foliage or fruits, is brought to a state in which the excess liquid runs off as drip. An excess of liquid is applied of which a quantity falls on the object below or on the ground leaving a coating on the sprayed surface. Efficiency in high volume spraying is judged on the one hand, by the tank capacity, pump capacity and output, nozzle pressure and output, distance of throw of the spray, type of spray swath or cone, and on the other, the time taken to wet or spray the surface and the deposit of residue. Efficiency of spray which also includes economy of spray liquid depends upon the droplet size or atomisation. Hence recently emphasis has been laid on droplet size, number of droplets per unit area, their diameter and amount of chemical present in the droplets, apart from hydraulic and mechanical aspects of spraying which have been looked into hitherto. The droplet density and distribution can be measured in a number of ways in the field by direct visual observations with the help of direct prints, leaf prints or flourescent deposits. In such cases droplets containing certain specific reacting materials, e.g., aluminium chloride sprayed on haematoxylin paper ; or metallic ions, or tracers on leaf surface later transferred to filter paper and developed in reagents or flourescent dye tracer technique in case of application on leaf surface. Direct measurement of the droplets may be made after collection of the same on slides coated with soft grease and covered with paraffin. Various techniques of indirect measurement, are adopted namely, formation of water on slides coated with black deposits of carbon of magnesium ribbon, or use of stains along with sprays collected on glossy papers, etc. Then droplets may be measured direct or indirect as the case may be and data may be analysed.

To determine the rate of deposit of a given method, deposition of the chemical on the surface treated has to be found out. This can be done by washing off the deposit from a test surface of a given area, using a given amount of solvent and then determining the concentration of the chemical in solution. Concentration of the chemical in this washing may be determined by chemical analysis, if possible or determinable ; or by colorimetric method in which case a given amount of strong dye, e.g., nigrosin has to be added to the spray liquid to make it more easily detectable. Accuracy of this method can be greatly improved upon by the addition of soluble flourescent dye and flourescence analysis. From these methods, the amount of pesticide deposited per unit area can be determined from the measured amount of dye or flourescent material on the assumption that dye or the flourscent substance and the pesticide was homogeneously distributed in a given ratio as in the spray liquid.

Another possibility of deposit analysis is the direct evaluation of the substances deposited through determination of the wetted area. Drop sizes can be determined in a number of ways and spray deposit can be found from the set medium concentration in the spray liquid. Determination of individual drop size is a laborious process and should be used for microanalysis and adhesion study of spray fluid. Analysis can also be made by chromatography or bioassay. For the choice of the optimum droplet size of a particular concentrate which involves the deposition of a given dose of chemical on an unit surface area, the smallest droplet size possible with maximum number of droplets per unit area would be desirable, as this will ensure the best distribution of the active material uniformly over the sprayed surface. To achieve this objective as far as practicable, various factors have to be taken into account. Terminal velocity has to be sufficiently low to allow the droplets to remain in suspension in air while being capable of acting as a toxicant, terminal velocity being directly proportional to density. Wind speed, airflow, convection and inversion air currents, density of air and evaporation are some of the pertinent factors to be considered for the purpose. It has been assumed that droplets on falling on the surface will be attached to the surface. A number of factors including the chemical nature of the spray materials determine attachment, as in lime sulphur droplets above 80 μ of surface mean diameter are phytotoxic.

In fungous diseases, where protective spraying is aimed at, the surface should be thoroughly covered leaving no gaps, otherwise spores falling on the gaps may germinate and bring about the infection. Hence coverage and redistribution play an important part in such cases. When a large number of nozzles are used on a boom, in coarse or high volume spraying transverse distribution over the entire boom is a function of each individual nozzle whereas in fine low volume spraying it is a function of overlap of the individual distributions of all the nozzles. Besides flow rate from the nozzle is considerably influenced by wear and tear due to use. Whereas in brass nozzples increased flow may be 50....60 per cent, in plastic and stainless steel nozzles it is virtually absent. Overlapping may occur also in the longitudinal direction. Cone spray nozzles give a better distribution and overlapping system than jet fan spray nozzles, in both transverse and longitudinal direction. In view of the possible roughness of the surface, hollow cone nozzles are considered to be the most efficient.

In recent years there has been a shift in the nature of spraying from high volume, high pressure to low volume, low pressure with the use of active ingredient in more concentrated form. Efficiency of high volume spraying is judged from the factors governing wettability of the sprayed surface, whereas the droplet size and spreading power of the droplets determine the same in low volume spraying. Hence more emphasis has been laid on the droplet size because with control of the droplet size it may be possible to cover a large acreage per unit time even with reduced consumption of pesticide. This depends on distribution accuracy. Inadequate distribution can cause damage to plants, residue problems etc. It would also be pointless to adopt sophisticated and expensive application techniques if low level of toxicity of the chemicals, and economic factors would permit high volume distribution with coarser particle or larger droplet size. In this context the formulation of the chemical or active ingredient plays an important part. Adequate information is not yet available on pesticide formulation in relation to concentrate spraying. Wettable powders are not suitable, but water dispersible powders may be used provided a stable suspension is obtained. Agitation may cause frothing with the result that there may be an increase in the concentration of wettable powders. Emulsified concentrates may serve better in such cases for their stability. Other factors, namely, miscibility of the emulsion in oil or water, nonphytotoxicity of carrier oil, tenacity of the spray deposit and evaporation rate are also to be considered.

Choosing, operating and maintaining equipment

Planning of Pesticide application

In planning pesticide application it is necessary to know what pests and diseases are likely to appear and need to be controlled, the area to be covered, number of applications that may be required, time of application and quantity of pesticides—both in terms of formulations and active ingredient. Records of past results are helpful in planning for the future, hence pertinent informations regarding infestation or infection and control achieved should be kept. Timings of spray schedules, to a great extent will be determined by the weather conditions favouring infection or infestation and the build up of inoculum potential in case of diseases or economic threshold of pest population. The scale of operation has to be decided on the basis of the magnitude of the problem. In the case of possible severe infestations spreading rapidly, or the outbreak of epiphytotics, largescale operations will be necessary and resources will have to be mobilised accordingly. Short-term forecasting of diseases and pests is helpful in this context.

Choice of pesticide

The pesticides to be used are to be chosen carefully partioularly in relation to the disease or pest and economic returns involved. Main requirements of a satisfactory pesticide should be : (1) high level of performance in the field which will be determined by inherent toxicity, ready availability of the active principle, adequate coverage, initial retention of the pesticide applied, and tenacity of the residue ; (2) low phytotoxicity ; (3) low level of toxic hazards to animals and men including operators ; and (4) stability in storage as well as after dilution to the desired strength. Survey conducted by National Council of Applied Economic Research (1967) on the use of pesticides in India has shown that either wrong pesticides or wrong formulations have been responsible in making the growers sceptical of the efficacy of pesticides. Apart from inherent toxicity to pests, the choice of pesticides should depend to some extent on hazards involved on the operator, and surroundings due to drift of spray, dust or fumigant. It is particularly necessary in the case of application of herbicides where such drift may cause injury to neighbouring crops.

In case a mixture of fungicides and insecticides is to be used, compatibility of the mixture should be carefully checked. A compatibility chart is given in the annexed Table.

In storing pesticides for use in the season, shelf life should be carefully checked. Many chlorinated hydrocarbons can be stored for a long period, while many organophosphorus pesticides have a shorter life.

Pesticide application equipment

Equipments for application of pesticides should be procured

in advance. They should be carefully chosen for the job and kept in proper condition for use.

On the choice of equipment for pesticide application first of all it has to be decided what type of equipment is needed—duster or sprayer and if sprayer, high or low volume. Nature of damage, pest or disease involved, area to be covered, and crop or crops involved will determine the type to be procured. Apart from the above considerations, attention should also be paid to conditions of the field where these machines are to be used, availability of water and its quality in case of spraying and operators who will actually use the machines. Weight, portability, manoeuvarability of the delivery system in case of sprayers, and the ability to withstand rough handling and durability are also to be considered in the choice of the right type of machine. While large sized power equipments may be able to give quicker results, they are complicated and involve more expenditure in maintenance. Normally an application equipment is expected to give good coverage with minimum expenditure of labour, power and low level of depreciation. Besides a machine should be easily operated, and convenient to use.

Maintenance

Instructions given by the manufacturers for cleaning, lubrication and adjustment should be carefully learnt. An operator should familiarise himself with these instructions and working of the machines. Since a lot of deterioration occurs when a machine remains stored away, not being used, it must be cleaned, washed and dried prior to storage and should be stored under such conditions where it will keep dry. A thin coating of oil may be placed on the outer surface of the sprayer. Hoses should be cleaned and properly stored in darkness. Wearing parts and bearings should be checked at the end of each operating season and replacements should be made well in advance. Deterioration of equipments is faster under hot and humid conditions, hence special care has to be taken in our country particularly in view of the capital cost involved.

Before putting the machine to use, lubrications should be done according to instructions. The right type of fuel and fuel mixture (in case of 2-stroke engines) should be used. Calibration

TABLE 9.1. Compatibility Chart of Plant Protection Chemicals

	D.D.T.	BHC, Lindane	Endrin, Aldrin, Hepta-chlor, Chlordane	Parathion, Methyl Parathion	Systox, Metasystox, Schradan (OMPA)	Malathion	Diazinon	Phosphamidon	Dimethoate	Carbaryl	Bordeaux Mixture	Fixed Coppers	Zineb, Thiram, Ziram, Maneb	Organomercurials	Captan	Wettable Sulphur	Lime Sulphur
D.D.T.	‡	‡	‡	‡	‡	‡	‡	‡	‡	‡	‡	‡	‡[5]	‡	‡[5]	‡[5]	‡[2]
BHC, Lindane	‡	‡	‡	‡	‡	‡	‡	‡	‡	‡	‡[3]	‡	‡	‡	‡[5]	‡	‡[2]
Aldrin, Endrin, Heptachlor, Chlordane	‡	‡	‡	‡	‡	‡	‡	‡	‡	‡	‡	‡	‡	‡	‡[5]	‡	‡
Parathion, Methyl-Parathion	‡	‡	‡	‡	‡	‡	‡	‡	‡	‡	–[4]	‡	‡[1]	‡	‡[5]	‡	‡
Systox, Metasystox, Schradan (OMPA)	‡	‡	‡	‡	‡	‡	‡	‡	‡	‡	–[4]	‡	‡[1]	‡	‡	‡	–
Malathion	‡	‡	‡	‡	‡	‡	‡	‡	‡	‡[6]	–	‡	‡[1]	‡[1]	‡[5]	‡	–

K E Y

1. Not usually mixed together or compatability not known.
2. When mixed with water decomposes after standing. With a ziram mixture, adding 500 gm skim milk to 500 litres of spray may prevent decomposition.
3. Not recommended, except as direc-

Diazinon	+[1]	++	++	++	++	++	++	++	++	++	++	++
Phosphamidon	−	++	++	++	++	++	++	+[2]	+[2]	−	+[6]	++
Dimethoate	−	++	++	++	++	++[6]	++[6]	++	++	+[6]	+	++
Carbaryl	+	++	++	++	++	++	++	++	++	++	++	++[5]
Bordeaux Mixture	−[3]	−[1]	−[2]	−[1]	−[6]	−[2]	+[1]	+[2]	+[1]	++	++	++
Fixed coppers	+[1]	+[1]	++	++	++	++	+[5]	+[3]	++	++	++	++[5]
Zineb, Thiram, Maneb, Ziram	++[5]	++[5]	++[5]	++[5]	+[5]	+[5]	++	++	++	++	++	++[5]
Organomercurials	−	−	−	−[2]	+[1]	+[1]	−	++	++	++	++	++
Captan	+[5]	+[5]	+[5]	+[5]	+[5]	++	++	++	++	++	++	++[5]
Wettable sulphur	++	++	++	++	++	++	++	++	++	++	++	++
Lime sulphur	−[3]	−[3]	−[3]	−[3]	−[1]	−	−	−	−	−	−	−[5]

++ = Safe   + = Caution   − = Incompatible

of sprayer and duster, in relation to the formulation of pesticide should be made so that the correct dose is applied. Only those formulations for which the machine is designed should be used. In case of dust formulations, the powder should be dry. In preparation of spray liquid, clean water should be used as far as practicable and in cases of suspensions, care should be taken to prevent precipitation of suspended materials by agitation. This is particularly necessary in those cases where the spray liquid is being drawn from an outside container and the equipment does not have any built-in tank, or no agitator even if there be a tank attached to the machine.

Considerations during the use of the equipment

Nozzles should be set properly and adjusted to suit the crop and prevent drift. Attention should be paid to the uniformity of spraying, which will also necessitate watching on the pressure gauge apart from nozzles. Reasons for fall in pressure should be carefully checked and defects rectified. Both in case of dusting and spraying, prior to change from one chemical to another, a machine should be thoroughly cleaned before a new toxicant or pesticide is put into use. Engines must not run continuously, nor should the machines be put into heavy use.

Delivery should be carefully planned to prevent wastage of time, material and labour. Application should be made in a logical sequence. In case of hand operations, or where automation is not taken recourse to the operator should move at a constant speed, and in case of vehicle mounted ones, vehicles must move at a constant speed. In the case of non-systemic insecticides and protective application, care should be taken to ensure complete coverage.

On each day after the spraying operation, the machine should be washed, filters taken out and cleaned, tank flushed out, water pumped through the delivery hose and nozzles. The machine which has been used for the application of a herbicide should not be used for an insecticide and pesticide unless it has been thoroughly decontaminated as per recommendation. In case of duster, the hopper should be emptied out at the end of the day. All ducts should be tapped to dislodge and remove caked dust.

Since many pesticides may be toxic to fish, pesticides must not be drained into tanks, ponds, or water courses, nor the machineries be washed in tanks, ponds, water courses or washings from them should fall into them. Care should be taken to see that pesticides and washing from sprayers or clearings from dusters, seed treaters, etc., are not thrown in the open which may cause hazards to livestock.

It is also necessary that operators should know the correct method of applying a pesticide and the safety precautions that must be observed including the use of protective clothing if necessary. An operator must wash himself well before he takes any food, drink or smoke. Medical advice should immediately be sought if an operator shows symptoms of illness during or after the application of pesticides. Protective clothings should be stored safely and cleaned frequently and should be replaced when necessary. Under the Insecticide Act of the Government of India, 1968, it is the duty of manufacturers, formulators of pesticides and operators to dispose packages or surplus materials and washings in a safe manner to prevent environmental or water pollution. The used packages are forbidden to be left outside to prevent their reuse and the packages should be broken and buried away.

Drift of pesticides during their application into the areas not intended or desired to be covered may cause not only undue loss of pesticidal coverage in the crop to be protected, but may also have undesirable effects. The drift of pesticides which have a high mammalian toxicity, into the food crops in neighbouring areas or pasture may cause serious toxic hazards to men and animals. The drift of herbicides to sensitive crops may result in injury to the crops into which the chemical has drifted. It is, however, impossible to eliminate entirely, the drift of pesticides either as dust or as liquid or mist. Nevertheless, the drift may be minimised by adjustment of distance between the crop and the mist in the case of low volume spraying or mist blowing; use of hoods particularly in low volume sprayings; use of high volume spraying which will give rise to large droplets that will settle quickly or by adjustment of pressure and nozzle, incorporation of additive hygroscopic material like glycerine or glycol; anti-evaporants, particulating agents.

Injuries due to drift may be more marked in dusting where the particles form a cloud and move to a great distance.

Control over drift has been sought in the electrification of the particles. It has long been known that electrostatic charges are developed both during spraying and dusting. Charge is dependent on the particles, particles over 40 microns produce more charge, while those below 2 microns, low charge. At higher relative humidity of over 50 per cent, electric charge cannot be detected. The chemical nature of the particle determines the nature of charge, negative charges being acquired by materials exposing acidic group on fracture and positive charge by alkaline materials. The performance of a dust is often improved by the addition of another material. The intensity of charge is influenced by the electrical resistivity of the dust which varies with the nature of the dust, size of the particles, temperature and relative humidity. For charging of dust particles, electro-static dusters are available in some developed countries. Electrostatic charging may be profitably used when relative humidity, wind and turbulence are conducive for electrostatic effects. Charging of the dust particles results in greater initial deposit and coverage of lower surface and areas not accessible to normal dust operations.

Spraying in Crops

Methods employed in the field generally fall into two categories namely large/high volume and small/low volume. In temperate or advanced countries, spraying methods are largely based on tractor-operated hydraulic sprayers, and mist blowers; or an aircraft. These are automatic in the sense that apart from the driving the machine or piloting the aircraft, virtually no human effort is involved in actual spraying.

In tropical and developing countries, though large areas are to be covered, hand operated equipments, are in use such as knapsac sprayers, shoulder mounted mist blowers, or motor or hand operated sprayers with hand lances supplied through hoses or in these methods, the spray is hand directed and under human control throughout the spraying operation. Hence efficiency of these methods depends upon organization and training of operators who will carry out the spraying operations. It has been shown by experimental results that the most effective system for tackling complex pest situation under wide variety of conditions is the high volume application by hand directed lances with dilute spray liquids with wetting properties. This method, if properly done, has a very high biological efficiency, as it ensures contact of the active chemical with pests and disease producing organisms.

In many situations, this method or technique inspite of its very high level of biological efficiency is expensive in terms of labour and may not be adquate to cover large area as may be necessary under epiphytotic conditions. This situation has resulted in the use of tractordrawn automatic equipments where the labour requirement can be reduced to one quarter and the operating speed three, to four times, but it requires that the plots are large and properly laid out for carrying on such operations.

In automatic spraying nozzles are rigidly fixed in position during the spraying, the speed and liquid throughout are constant, irrespective of variations in size, density of the plants and canopy of the foliage, in as much as there may be spraying without much effective contribution to the control of target pests and disease producing organisms.

In comparison, hand spraying if properly done is much more flexible, enables to direct most of the spray fluids against the target pests and an efficient job can be done keeping in view size, shape, density of the crops including trees. Major disadvantages of hand spraying are its slow speed and high labour requirement. Hence in many situations, particularly in large plantations, this application technique though biologically efficient is often considered economically impracticable being too expensive in time and labour. Naturally this situation has developed into automatic spray machines operated on tractors. But hand spraying cannot be equated with automatic power spraying, results obtained in the former cannot be comparable with the latter because of lack of flexibility, which is particularly important where the crop has very dense foliage and needs a protective coating.

Mist blowing techniques, using the airstream from a fan in which much smaller quantities of liquid are needed, has been developed to reduce costs. Steps have been taken to increase their efficiency to the level of hand spraying. Aircraft spraying is now widely used employing either nozzle booms or spinning cages. Though aircraft spraying is economical in time, labour and cost, yet it has its own problems which are operational in nature.. Besides spraying by this method results in deposition of much of the spray liquid on the upper surfaces of the leaves and tops of trees.

Wide use of systemic and trans-laminar insecticides and newer fungicides and closer planting of the crop together with dwarf habit in many cases have now focussed attention on reexamination of spraying methods for effectiveness as well as reduction in costs.

Successful application by aircraft of very small volumes (1-2 litre/ha) of concentrated insecticide on cotton and rice has created interest in the application of much reduced amount of spray liquids to the crops by ground equipment for this purpose, small droplets are to be developed to prevent too rapid evaporation under tropical conditions.

QUESTIONS FOR DISCUSSION

1. Why is the method of application of pesticides considered to be nearly as important as their toxicity ?
2. How can sparingly soluble pesticides be used in spraying ? Why is the formulation in powdered form of spray pesticides favoured ?
3. In which cases are emulsions used ? What precautions need to be taken in emulsifying preparations ?
4. What steps should be taken for securing effective spraying ?
5. In which cases are smoke generators used ?
6. How are dust formulations made ? What are the merits and demerits of dusting vis-a-vis spraying ?
7. Why are granular applications being favoured recently ?
8. Under what conditions are fumigations normally made ? What prerequisites and directions are necessary for good fumigation ?
9. What is aerosol ?
10. What constitutes a bait ? Against what pests are baits used ?
11. How would you categorise different types of prays depend-

ing on the volume ? in

12. What do you understand by concentration in a spray liquid ? What is meant by dosage ?

13. What are the essential parts or components of a sprayer ?

14. What are the requirements of a tank in a sprayer ?

15. What are the broad categories of pumps used in sprayers ? What are the comparative advantages and disadvantages in each case ?

16. What are the different types of nozzles normally used ? What is the objective of use of the different types of nozzles ? What relation has aperture to the type of spray ?

17. What is a knapsac sprayer ? How does it differ from a compression air sprayer ? What is the essential mechanism of a sprayer ?

18. What are different categories of power sprayers ? How does a hydraulic energy sprayer differ from a centrifugal energy sprayer ?

19. What are mist blowers ? What are their characteristics ?

20. What are the essential components of a duster ? What is the main principle of working of a duster ?

21. How does a bellow hand duster differ from a rotary hand duster ? Which type is more advantageous for working and why ?

22. Do you think that dusting is gradually losing its popularity ?

23. What is the main function of a seed dresser ? What are the different types or categories of seed treating machines ?

24. What are soil injectors ? What are the essential components of a soil injector ?

25. What are the essential parts of a granule applicator ? How is a granule applicator operated ?

26. Why is it necessary to judge efficiency of spraying ? How is efficiency in high volume spraying judged ?

27. Why are droplet size and distribution considered important in spraying ?

28. How can droplet size and density be measured ? What should be the optimum droplet size ?

29. Why is low volume and low pressure spraying advocated now ? What determines efficiency of high volume vis-a-vis low volume spraying ?

30. What considerations should be made in the planning of pesticide application ?
31. What points should be taken into account in the choice of equipment for pesticide application ?
32. What considerations should be made in the use of the equipment ?
33. What steps should be taken for the disposal of empty containers and why ?
34. What hazards are attended with drifts ?   How can such hazards be minimised ?
35. What care should be taken in the choice of pesticide ?

CHAPTER 10

SEED TREATMENT

Many pests during the absence or dormancy of the host perpetuate in the seed or propagative organs till a new host is found. Hence the treatment of seeds to get rid of infection and to secure healthy plant materials constitutes one of the major measures of crop husbandry.

The application of fungicides to seeds before planting has two fold purposes : (a) control of diseases caused by seed borne infection, (b) protection of germinating seeds or seedlings from the attack of soil-borne pathogens. Appropriate treatment of seeds can get rid of the seed-borne pathogens and can control, to a large extent diseases that would otherwise result. Besides, incorporation of a protective chemical on the surface of the seed can reduce the chances of infection, consequently harmful effect of many soil-borne pathogens which are capable of causing decay of seeds, pre- or post-emergence damping off or infections that may persist as chronic diseases of plants. International shipment of seed for planting requires that seeds should be treated in the country of origin before they are exported to prevent possibility of the introduction of new seed-borne pathogens or pathogenic races into the importing country and probable damage to the seeds that may take place due to exposure to new soil-borne pathogens in the importing country. 'Seed' in the phytopathological sense means not only true seeds, but also fruits, tubers, bulbs, corms, rhizomes or other vegetative propagative stocks in which the pests may remain in the dormant stage, and are prepetuated and transmitted. In this process of transmission of pest, seeds may be externally infested with the propagules, as in the case of covered smut and bunt of cereals or tendu disease of wheat, or they may be infected internally by the pathogen as in loose smut of wheat and barley, seed-borne infections incited by *Helminthosporium* spp.

The introduction of systemic fungicides for seed treatment has added further possibilities : (a) control of pathogens located deep inside the seeds, which are inaccessible to other seed treating chemicals as *Ustilago nuda* in wheat, barley, and (b) control of air-borne infection at a later stage of growth of the crop, the toxicant being systemically translocated to aerial parts.

There is no single method or material which can be universally recommended for treatment of seeds. These methods have to be chosen in accordance with the modes of perpetuation of pests. Broadly speaking, they may be divided into three categories : (a) mechanical, (b) chemical and (c) physical.

(a) Mechanical Methods

In many cases due to infection there may be an alteration in size, shape and weight of seeds by which it is possible to detect the infected seeds and separate them from the healthy ones. In ergot diseases, infecting sclerotia are large in size and lighter than grains. They may be separated out by sieving or flotation. In tendu disease of wheat incited by *Anguina tritici,* galls due to infection in grains can be separated by flotation. In many seed-borne infections, infected seeds are usually smaller in size and lighter in weight. They may be separated out by gravity grading, flotation or sifting through sieves, as may be convenient. Such mechanical separation eliminates infected materials to a large extent and in some cases, e.g., ergot and in tendu diseases, this is the only method.

(b) Chemical Methods

Disinfection of seeds by chemicals was in vogue for a long time. The use of wine and brine has been reported in the seventeenth century. In 1809, Prevost carried out his classical work on the use of copper sulphate for the treatment of cereal seeds infested with spores of bunt fungi. Formaldehyde treatment was practised in the nineteenth century. Sulphur dust for the treatment of smuts in grain crops has been recommended in the earlier part of the twentieth century. With the discovery that merɑuric chloride can get rid of seed-borne infection in the case of *Fusarium* disease or rye and later the introduction of organomercuric compounds for disinfection of seeds a new chapter opened in seed treatment in the case of fungal infections. Later on non-merɑurial organic chemicals also appeared in the market and were successfully used for seed treatment. Such treatments besides controlling infestation and or infection, as the case may be, also afford protection to the seeds and young seedlings in the early stages of growth from soil-borne fungi by sterilising the small amount of soil around the seed and keeping it free from organisms as far as practicable during germination and early stages of establishment of seedlings.

Very recently attempts have been made to control deep seated infections by the use of systemic fungicides for the control of : (a) pathogens situated within the seed and previously inaccessible to chemicals, and (b) air-borne diseases using the dressing as a reservoir of fungicide during the growth of the crop at least in the early stages. Success of the use of systemic fungicides as seed dressings is now clearly established and eoxathiin and carboxin are widely used for the control of deep seated infections of loose smut caused by *Ustilago nuda*. By use of pyrimidine ethirimol as seed treatment control of powdery mildew has been effected.

Chemicals used in seed treatment against pathogenic fungi may be classified in Table 10.1

Treatment of seeds by chemicals may be effected by : (a) steeping in liquid, (b) dry seed treatment, or (c) slurry treatment. Steeping in liquid may be done in buckets. Dry seed treatment is usually carried out in rotary or gravity-fed seed dressers. In dry seed dressings, powder which is applied in very fine form adheres to the surface of seeds. Slurry treatment in which the chemical is applied in the form of a thick soup (active material dissolved in a small quantity of water or in any harmless solvent), so that during the process of treatment slurry gets deposited on the surface of seeds in the form of a thin paste which dries up. Seeds treated with dry dusts may be stored for a long period so also slurry treated seeds. But seeds treated by steeping in liquid cannot be stored.

TABLE 10.1

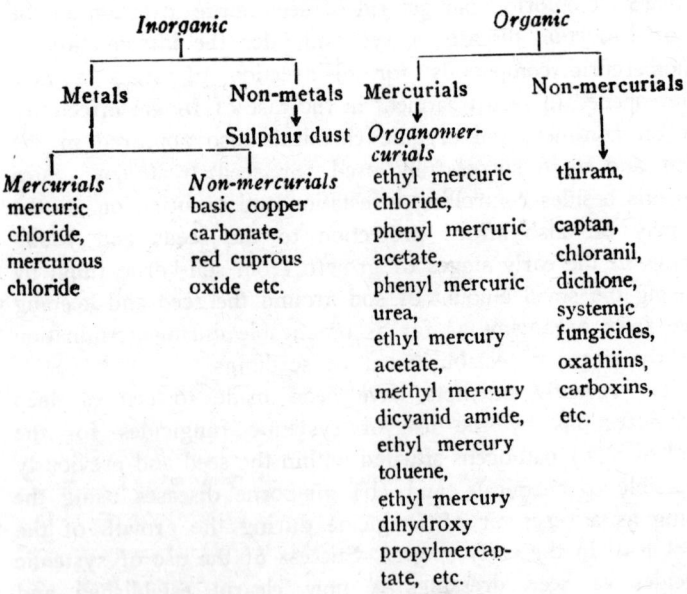

	Inorganic		Organic	
Metals	Non-metals	Mercurials	Non-mercurials	
	Sulphur dust	Organomer-curials		
Mercurials	Non-mercurials	ethyl mercuric chloride,	thiram,	
mercuric chloride,	basic copper carbonate,	phenyl mercuric acetate,	captan, chloranil,	
mercurous chloride	red cuprous oxide etc.	phenyl mercuric urea,	dichlone, systemic	
		ethyl mercury acetate,	fungicides, oxathiins,	
		methyl mercury dicyanid amide,	carboxins, etc.	
		ethyl mercury toluene,		
		ethyl mercury dihydroxy propylmercaptate, etc.		

Seeds treated with chemicals must be stored dry and treatment should at least be done one week before sowing. In case of treatment with liquids, as seeds cannot be stored the treatment has to be done immediately before sowing. The time required for the purpose, as recommended, should be strictly adhered to. Liquid treatment is usually taken recourse to for treating vegetative propagative stocks, e.g., cuttings, tubers, corms, bulbs, etc., which are not amenable to dry or slurry treatment. Storing under damp conditions after treatment has been reported to damage the viability of seeds. Hence proper storage of treated seeds has to be ensured.

With the exceptions of loose smut diseases of wheat and barley, organomercurial seed dressings have been able to control a large number of diseases of food and commercial crops. Organomercuric chemicals have shown effectiveness against a number of diseases in a number of different crops and may be called broad spectrum seed treating fungicides. Major important diseases which can be controlled by organomercurial seed treatment are given in Table 10.1

TABLE 10.2

Crop	Disease	Pathogen
Wheat	Bunt	*Tilletia caries*
	Bunt	*Tilletia foetida*
	Snow mould	*Fusarium nivale*
	Seedling blight	*Fusarium* spp.
	Seedling blight	*Helminthosporium sativum*
	Seedling blight	*Septoria nodorum*
Barley	Covered smut	*Ustilago hordei*
	Leaf stripe	*Helminthosporium gramineum*
	Net blotch	*Helminthosporium teres*
	Seedling blight	*Fusarium* spp.
Rye	Snow mould	*Fusarium nivale*
	Stripe smut	*Urocystis occulata*
	Seedling blight	*Fusarium* spp.
Maize	Leaf spot	*Helminthosporium* spp.
Rice	Blast	*Pyricularia oryzae*
	Brown spot	*Helminthosporium oryzae*
	Stem rot	*Helminthosporium sigmoideum*
Oat	Loose smut	*Ustilago avenae*
	Covered smut	*Ustilago levis*
	Leaf spot	*Helminthosporium avenae*
	Seedling blight	*Fusarium* spp.
Sorghum	Grain smut	*Sphacelotheca sorghi*
Linseed	Seed borne diseases	*Fusarium* spp.
	Pre emergence rots	*Colletotrichum* spp. and others
Groundnut	Crownrot	*Aspergillus niger*
Cotton	Anthracnose	*Glomerella gossypii*
	Blackarm	*Xanthomonas malvacearum*
Jute	Stem rot	*Macrophomina phaseolina*
Seed potato	Black scurf	*Corticium solani*
	Dry rot	*Fusarium caeruleum*
	Gangrene	*Phoma* spp.

In view of the development of systemic fungicides which are capable of controlling loose smut diseases of wheat and barley and the need for the replacement of mercury compounds with safer and less toxic chemicals has led to the identification of other seed treating fungicides. Alternative materials include chlorobenzene, benzimidazole compounds, dithiocarbamates and miscellaneous compounds like captan, carboxin. Effectiveness and limitations of these compounds in relation to mercury compounds are given in Table 10.3

TABLE 10.3

Chemical	Effectiveness and spectrum of activity	Agricultural limitations
Mercury compounds alkyl mercury alkonyalkyl mercury aryl mercury	highly effective, wide spectrum	
Chlorobenzene compounds hexachlorobenzene	effective for wheat bunt	Resistant races developed in some areas
quintozene	effective for wheat bunt, barley covered smut ; *Rhizoctonia* on several crops	Resistant races of bunt in some areas.
Benzimidazole compounds benomyl	effective for wheat bunt, wheat loose smut, rye stripe smut, potato black scurf	Ineffective for *Helminthosporium* infection, high cost may limit use to foundation stocks
thiabendazole	effective on seed and soil-borne wheat bunt	High cost
dithiocarbomates maneb	effective for barley leaf smut and wheat bunt, broad spectrum	

TABLE 10.3 (*Contd.*)

Chemical	Effectiveness and spectrum of activity	Agricultural limitations
mancozeb	similar to maneb	
thiram	mild seed protectant, broad spectrum	Limited effectiveness with high seed inoculum
Miscellaneous fungicides		
captan	Limited effectiveness on seed-borne diseases; protectant against soilborne seedling pathogens	
carboxin	highly effective against seed and soil-borne wheat bunt and wheat and barley loose smut; reported effective against *Helminthosporium* infection in some areas.	Expensive; restricted to seed increase in some areas.

* Source—Use of mercury and alternative compounds as seed dressings. FAO/WHO Publication *FAO. Agricultural studies No. 95*; *World Health Organisation Technical Report* Series No. 555.

Many chemicals used for the treatment of seeds are poisonous or toxic to men and animals and they should be used with caution. Even though the chemicals may not be poisonous nevertheless in the dust stage they may be irritating to the skin, eyes and nose. Chemically treated seeds should never be used for consumption by men and animals even after a lapse of time after treatment and repeated washings. Sacks, bags, or other containers used for storing treated seeds should not be used for other purposes before thorough cleaning. Inhalation of fumes or dusts during the process of treatment must be avoided and precautions have to be taken for the purpose. Extreme care should be taken to avoid the skin coming into contact with mercury compounds which may cause serious injury. Seed treatment should always be carried out in the open. Instructions given by the manufacturers, regarding safety, dosage, handling should be strictly followed. Cereal grains treated with a seed treating chemical should be immediately and permanently distinguishable from undressed seed. All dressed seeds should be distinctly coloured and should be labelled with appropriate danger symbols and other safety measures as prescribed under the Insecticides Act or Rules. Chemicals in overdose may cause injury and in underdose may not give the desired results. In treatment with toxic materials, care has to be exercised to find out the margin of safety, dose at which infection will be cured and the dosage which the material to be treated can tolerate so that viability of seeds is not affected.

There is an apprehension whether legume seeds should be treated with seed treating chemicals, as such treatment may adversely affect the size and number of nodulation due to its effect on *Rhizobium* particularly when seeds are inoculated with *Rhizobium* culture. It has been observed that non-mercurial seed treating fungicides like captan and thiram do not affect nodulation. Organomercurial fungicides may prevent nodulation unless the seeds are sown immediately after inoculation.

(c) Physical Methods

Seeds and planting materials can be subjected to heat treatment in order to eliminate certain internally borne pests including nematodes. Temperature and period of treatment will vary with the infection or pest concerned. They must be closely controlled to within half a degree Centigrade so that the viability of the plants is not adversely affected, while the organisms inside the seed or planting material are killed. The use of heat as an agent for disinfection of seeds was first made by Jensen who attempted to control the internal infection of diseased tubers of potato affected with late blight (*Phytophthora infestans*). It was noted that internal mycelium could be killed by a four-hour treatment at 40°C. This method however did not gain popularity.

Jensen's method of hot water treatment evolved in 1887 is more widely known and accepted for the control of loose smut of wheat, barley and oats. For the hot water treatment seeds are to be presoaked for four hours at 20-30°C, during which period dormant mycelium develops activity and becomes more vulnerable to exposure in hot water at 50-52°C for a few minutes. Seeds after treatment are to be dried very carefully before they can be used for sowing. This method is cumbersome. Small quantities can be treated at a time. Besides, temperature has to be exactly maintained during the period of treatment. Nevertheless it has been widely practised and is still in vogue, because before the discovery of the use of systemic fungicides for the control of loose smut, this has been the only known method.

In India, solar energy has been utilised as a means of disinfection of seeds attacked with loose smut. In this process, which is popularly known as Luthra's method of treatment, after the name of the discoverer seeds are to be presoaked for four to five hours in a shallow vat containing water, during night. During this period the seeds remain immersed in water. Then they should be taken out, the excess water drained off and the seeds spread to dry in the sun, on a clean floor. At the end of the day, by sunset, the seeds are expected to be dry. They may be stored in clean airtight containers. This process is simple and suitable for bulk treatment, but can only be adopted in summer in areas which have uninterrupted sunshine for nearly 12 hours or more, maximum temperature of approximately 40°C and relative humidity low to ensure the complete drying of the seeds in one day. At present, the use of systemic fungicides, e.g., oxathiin, carboxin gives good control of loose smut for which the hot water treatment used to be recommended.

Hot water treatment has been found to be effective for the control of nematodes. Immersion of narcissus bulbs for three to four hours at 43-43.5°C can kill the eelworms (*Anguina dipsaci*) while the viability of bulbs is not affected. A three hour treatment at 43-43.5°C is sufficient for bulbs of iris to be cured of the same nematode. *Aphelenchoides* spp. (chrysanthemum and strawberry) may be killed in chrysanthemum stools and strawberry runners by hot water treatment for 20-30 minutes at 43.5°C.

The practicability of use of hot air treatment for the control of virus in the propagating stocks was first suggested by Kunkel in peach yellows. It was claimed by Kunkel that bud sticks of peach can be cured of yellows by warm water treatment. It was also noted that plants or planting materials kept at 35°C for a fortnight can be cured of infection. Attempts were made to control virus diseases of sugar cane and strawberry runners by hot air treatment, but due to the ill effects of heat treatment on the planting stocks, method has not been adopted on a wide scale.

In the 1950s, hot air treatment was used by Dr. Kassanis in Rothamsted Experimental Station to eliminate potato leaf roll virus. Exposure of potato tubers to 37-39°C in "hot boxes" under high humid conditions together with meristem culture is used for the production of virus-free nucleus stocks. Sugar cane setts can be cured of infection mosaic disease by treatment with hot water at 50°C for two hours.

Seed treatment particularly by chemicals has become popular. It is advocated to have disease-free seeds and planting materials. The cost of treatment per hectare is very low in terms of the potential benefit, and methods of treatment are simple. At present, treatment of seeds is adopted as routine measure by the organisations dealing with seeds.

QUESTIONS FOR DISCUSSION

1. In which way is seed treatment useful for the cultivation of crops ?
2. Why are seeds meant for export treated ?
3. What precautions are to be taken in the case of treatment of seeds with chemicals and why ?
4. Why has it become necessary to replace mercuric chemicals with non-mercurials ?
5. What is a wide spectrum seed treating chemical ? Why is a wide spectrum chemical preferred ?
6. In what form can heat be used for the disinfection of seeds and planting materials ? Why is emphasis ·ing laid now on non-chemical methods ?
7. In which groups of diseases, is heat being mostly used for the treatment of planting material ?
8. How can solar energy be used for the treatment of seeds ? In which disease has it been used ? Is it more convenient than chemical treatment ? What are the limitations of this treatment ?
9. In which ways, can chemical be applied to seeds or planting materials ?

CHAPTER 11

SOIL TREATMENT

Soil treatment includes those methods which are used for the possible destruction of pests in the soil—harmful fungi, bacteria and nematodes, insects and weed seeds. It also aims to cure the soil of a general malady often termed "sickness". Methods may be classified broadly as physical and chemical. Besides, certain other methods which may be termed as agronomic practices or mechanical, are also adopted for the purpose. Biological methods may also be applied. Many harmful fungi, bacteria, nematodes and insects live in the soil as permanent members of soil microflora or they may colonise in the soil for longer or shorter durations depending on the availability of appropriate sources of nutrients in the soil. These soil-borne organisms are responsible for causing maladies in plants and are often very difficult to be got rid of. Control of them poses a serious problem and in many instances may be a limiting factor in the successful growth of the crop in glasshouses, backyards, nurseries as well as in the field.

Besides, as stated earlier, a condition can prevail in the soil which may be termed generally as "soil sickness". Such soils which are usually under intensive mono-cropping do not show lack of nutrients, nevertheless fail to give a sustained good yield. It is generally assumed that under such conditions, pests affecting the crop are encouraged resulting in the general decline in health of the plants. In some cases deterioration may be due to possible altered physical or other conditions in the soil, while in others, such conditions have been attributed to great preponderance of some definite harmful micro-organisms in the soil, namely, nematodes, very often *Meliodogyne* spp., *Anguillulina* spp., fungi—*Fusarium* spp., *Rhizoctonia* spp., *Pythium* spp., *Sclerotium* spp.,*Phytophthora* spp. or even the existence of bacteriophages attacking *Rhizobium* as in lucerne. In such cases there is no other course left but to resort to soil treatment for the improvement of soil condition.

Soil contains both harmful and beneficial organisms. The fertility of soil and availability of nutrients to a large extent is dependent on the biological activity of the beneficial microorganisms. Hence treatment of soil by any method either physical or chemical should aim at the destruction of only harmful organisms, while the beneficial ones will not be affected. If all micro-organisms are killed, then more harm than good will result. Soil treatment for the control of diseases and pests has its differential destruction of phytopathogenic fungi, bacteria and nematodes, as well as insect pests without disturbing as far as practicable beneficial micro-organisms and natural microbial balance in the soil. Hence "partial soil sterilisation" is aimed at.

Various methods of soil treatment may be classified as follows :

(1) *Physical methods* : (a) Steam sterilisation ; (b) Hot-air sterilisation ; and (c) Electrical sterilisation.

(2) *Chemical methods* : (a) Volatile chemicals ; and (b) Non-volatile chemicals ;

(3) *Mechanical methods* :

Physical Methods

Physical methods of soil sterilisation aim in using heat in dry or wet form to kill the destructive organisms. In tropical countries, exposure of the soil to the heat of the sun in summer gives at least partial control of soil-borne organisms. Partial sterilisation of soil in larger areas in the field is not practicable in view of the cost involved and operational difficulties. Soil sterilisation with heat has its own advantages, namely, it does not leave any toxic residue and soil can be used immediately as soon as it cools, whereas in chemical treatment, a "rest" or gap of one to three weeks after treatment has to be given to allow the toxic chemicals to escape into the atmosphere or to be degraded into harmless compounds. Heat treatment does not offer any hazards in operation whereas in chemical treatment, some chemicals which are toxic to mammals may require proper precautions by the operators. Amongst some of the disadvantages of physical treatments are the inconvenience, cost of heating installations in many cases and the possibility of accumulation of soluble salts that may be injurious to plant growth.

Steam is considered to be the efficient source of heat. Most of the heat resistant fungi, bacteria and viruses capable of inciting plant diseases are inactivated by a 30-minute exposure to temperature of 60° C. Most weed seeds are inactivated at 82° C and nematodes at 49° C. The majority of the beneficial micro-organisms can survive the temperature that inactivates the pathogenic ones. To secure the beneficial effects of soil sterilisation, the soil should be maintained at a temperature of 80-90° C for one hour. Various methods are available for heating the soil to the required temperature. Sterilisation by steam is effected by the passage of the steam, at high or low pressure generated from a boiler into the soil until the required temperature is attained and maintained during the period of treatment. A mixture of steam and air has also been recommended for the reduction of cost without affecting quality of sterilisation. Small quantities of soil in glasshouses can be sterilised by heating in the flat pan, and caution should be exercised to attain and maintain the correct temperature during the period of treatment. However, it involves labour and cost of movement of the soil. Methods of heating the soil with electric heating coils are also available and practised in some countries, e.g., in cotton fields in the U.S.S.R.

Chemical Methods

Application of chemicals into the soil for the control of harmful fungi, bacteria and nematodes is a comparatively simple phenomenon, particularly when the soil is fallow and the chemical is volatile and disappears quickly either by voltilasation or decomposition. The chemical besides its toxicity to soil-borne fungi, bacteria and nematodes should be cheap ; have power of good penetration into the soil ; harmless to the operator ; and non-injurious to the plants either directly or due to its action on properties of soil. Soil-treating chemicals should also be non-injurious to the plants in the soil adjacent to the area where treatment has been carried out because there may be standing crop in adjacent fields. Unfortunately most of the chemicals used for the sterilisation of soil do not possess all the desired properties.

The method of application of pesticides for the treatment of soil depends on the characteristics of the chemical and machinery to be used for the purpose. Gases like methyl bromide are introduced into the soil by injecting them at set points through injectors. Similarly point treatments are made with liquids like carbon disulphide. Formaldehyde may be applied at set points or as a drench, as may be convenient. Suspensions or liquids are applied as overall drenches. Solid particles in the form of dust are incorporated into the soil through ploughing.

Most of the pesticides have to be applied before planting or sowing as they act as general sterilants. Some fungicides like maneb, zineb, ziram or copper compounds are however not phytotoxic except in high concentrations and they may be applied as drenches in the standing crop.

Systemic fungicides being highly selective in nature can be applied *in situ* for the control of soil borne organisms. Benomyl and thiabendazole compounds are being used as drenches for control of *Verticillium* wilt. Soil applications are made of thiophanate methyl, benomyl and ethirimol against air-borne infection of mildews.

Kitazin, is applied in the transplanted rice fields in Japan for the control of blast as granules in irrigation water. These granules, being of limited solubility in water remain effective for a period of three weeks or more. Efficiency of the chemicals—general or systemic toxicants—is determined to a certain degree by other factors, namely, adsorptive capacity of the soils, biological and chemical degradation in the soil. Similarly effect of these biocides on the microbiological and chemical properties of the soil affecting fertility may also be a deciding factor in the acceptability of the same.

Volatile soil treating chemicals

These chemicals are often known as fumigants. They are as follows :—

   Chloropicrin :   *Nitrochloroform*   or   *trichloronitromethane*

(CCl_3NO_2) : is a colourless, non-inflammable, poisonous liquid. It is effective for the control of a wide range of soil-borne pests including fungi responsible for damping off, root rots and wilts nematodes, insects, weed and grass seeds. It is to be applied as a preplanting treatment, as it is extremely toxic to living plants and care should be taken so that no chemical is spilt near any living plant. It is marketed as a liquid or as an aerosol formulation.

It is injected into the soil with special injection equipment. About 3 ml of the liquid has to be injected at a depth of 13 cm on 26 cm centre grids throughout the bed. Immediately after treatment, the soil should be wet with water and the entire area should be covered preferably with plastic sheets for 24 hours to allow the chemical to stay. Large areas can be treated with chloropicrin, using field application equipment, at the rate of 400 litres per hectare. Soil in small quantities may be treated in bins by injecting at the rate of 10 ml of liquid in 0.028 cum of the soil. Treated soils should not be used for planting or sowing at least for 10-14 days after treatment. Longer periods may be needed in cold rainy weather.

Chloropicrin is extremely poisonous and should be handled with care. It has been used as a "tear gas" by enforcement officials. In liquid form it can cause damage to the skin and eyes even at short exposures. It may also be absorbed by the skin and body temperature is sufficient to volatalise it inside the body and cause toxic effect including temporary blindness. In the vapour stage, it is irritating to the skin, eyes and nose.

Methyl bromide $(CH_3 Br)$: Methyl bromide does not posses many undesirable properties of chloropicrin. It is not injurious to the plants and has a greater penetration rate. Soil treated with methyl bromide will be ready for use for planting in a few days. It is odourless, but very dangerous to the operaters, and all precautions that need to be taken in the use of chloropicrin have to be observed in this case as well.

Methyl bromide is a gas which is marketed as an aerosol or in cylinders. It may also be available in liquid form containing various percentages of active ingredient. It should be applied in the same manner as chloropicrin, at the rate of 1 kg of compressed gas per 10 sq m. Soil treated with methyl bromide should be covered for a period of 24-48 hours and a seven-day aeration period should be allowed before planting. Treatment with methyl bromide may reduce germination of some seeds particularly ornamentals. Methyl bromide treated soils should not be planted with garlic, onion, carnation or salvia, as these plants are susceptible even to traces of methyl bromide in the soil.

Vapam (properties have been described earlier) ($C_2H_4NS_2Na, 2H_2O$) : It can be applied to the soil mixed with water in a watering can or with the help of a power sprayer at the rate of 500 cc per 10 sq m (active ingredient being 40 per cent). After treatment, the soil should be watered and kept for two to three weeks before use for planting. Vapam should not be used within three metres of adjacent living plants. Vapam and its decomposition products may be irritating to the eyes, but it is one of the safest soil fumigants as it leaves no toxic residue in the soil. It is also very effective against not only pests, but also weed seeds and nematodes.

Mylone : 3, 5-dimethyltetrahydro 1, 3, 5, 2H-thiadiazine-2-thione ($C_5H_{10}S_2N_2$) : It is applied at the rate of 150 kg (active ingredient) per hectare either mixed with sand and applied with fertiliser spreader or may be applied uniformly with a sprinkling can or power sprayer as suspension in water.

After application, mylone should be thoroughly mixed with the soil to a depth of 13 to 16 cm with a rake or rotary cultivator after which the soil should be wetted with irrigation water. A period of two to three weeks should elapse between treatment and planting. A longer period may be necessary if the temperature is low. It is effective also against weeds and nematodes.

Formaldehyde : Formaldehyde should be used as 1 per cent solution at the rate of 4 litres per 10 sq m. After treatment the soil should preferably be watered and must be covered with cardboard, wet canvas or gunny bag for 24 hours. After removal of the cover, the soil should be raked for two to three days to allow vapour to escape. Normally a period of two to three weeks should elapse between the treatment and planting.

Allyl alcohol : Allyl alcohol has been recommended as soil fumigant but it has not been found to be effective against soil-borne fungi, though nematodes and weed seeds can be largely controlled.

Carbon disulphide (CS_2) : It was first introduced in 1872 for treatment against *Phylloxera* of grape vine in France. It is cheap, volatile and has good insecticidal properties. Carbon disulphide enters the soil by a simple process of diffusion, not by gravitational flow as considered by many. It is rapidly lost by evaporation which cannot be checked even by covering the surface with wet sacks. Penetration is good in light soils with not too great moisture content but is poor in wet and heavy soils. To prevent rapid loss of evaporation, different methods have been tried, including emulsion in oil but much success has not been achieved.

Other volatile soil-treating chemicals

    (a) Ethylene dibromide
    (b) 1, 3-dichloropropene-*Telone*
    (c) 1, 3- dichloropropene plus 1-2-dichloropropane 'DD' mixture.
    (d) 1, 3-dichloropropene plus ethylene dibromide
    (e) 1, 2-dibromo-3-chloropropane-*nemagon*, '*Dorlone*'

These above mentioned chemicals are effective as nematicides, but are not effective against soil-borne fungi and bacteria. They are recommended for use where soil infesting nematodes have created problems.

Apart from these compounds, coal tar antiseptics particularly cresylic acid has been used with success. It breaks down in the soil with the formation of phenol and orthocresol which have pesticidal properties. It is to be used as a drench in 0.05-0.01 per cent solution of active ingredient. It has been found to be effective against a number of soil-borne fungi.

Non-volatile soil-treating chemicals

Carbamates (zineb, ziram, maneb) have been suggested as promising chemicals for giving protection against soil-borne fungi. These chemicals may be used as dust or drench—as dust, 15 per cent dust to be applied at the rate of 1 kg per 500 sq m, or liquid, 0.05 per cent solution at the rate of 4 litres per 10 sq m. Action of these fungicides to a large extent is protection.

Nitrobenzenes : Pentachloronitrobenzene has been found tc oe effective against a number of sclerotia forming fungi namely *Sclerotium* spp., *Rhizoctonia* spp. It may be applied as a soil-mix, surface application, transplant solution. The soil-mix method is frequently used by broadcasting the chemical in the soil and then discing and cross discing the soil so that the chemical reaches the desired depth. It may be used in dust or spray formulations. It is available as dust with active ingredient varying from 10-40 per cent and for spray as 75 per cent wettable powder or emulsified concentrate. Dosage for dust varies from 40-50 kg per hectare depending upon the strength of the active ingredient and the crop and disease concerned. Spray concentrations vary from 0.1 to 0.5 per cent depending on the time of application, crop and disease.

Copper compounds : Copper fungicides at the rate of 25 per cent dilution of spray strength have been found to be promising in control of Phycomycetous fungi when applied as drench. Such applications may be made even with standing crop in the soil, as the fungicides are not phytotoxic.

Trenches filled with lime, surrounding the dead stumps have been found to be useful in the prevention of spread of the root infecting fungi in tropical plantation crops.

Chlorinated Hydrocarbons : Gamma-BHC, DDT, aldrin etc. have been applied successfully for the control of soil-borne insects including termites in the form of dusts at the time of soil preparation prior to planting. While gamma-BHC and DDT do not persist for very long time, cyclodiene insecticides like aldrin or dieldrin decompose very slowly and there may be accumulation in the soil. Gamma-BHC preparations should not be used where root crops are involved because of the taint they may impart.

Control of termites poses a difficult problem in the tropics and subtropics. Termites, which are often called white ants cause serious injury to a number of crops, particularly, wheat, sugar cane, etc. They can also cause damage to horticultural plants particularly in areas which have been recently cleared of jungles and are still littered with fragments of wood and stumps in the comparatively dry and arid conditions. They can cause damage to timbers, wooden furnitures, various woodwork in the houses in such areas.

Termites may be of two distinct categories—those forming mounds above the ground, e.g., *Odontotermes redemanni*, *O. obseus* and *Coptotermes horni ;* and those that live underground, e.g., *Odontotermes assumthi* and *Microtermes obesi*. Subterranean species are more injurious to crops. Control of termites is difficult as the insects remain underground in nests spreading over extensive areas. Cultural methods can afford control to some extent, but are not enough and they need to be supplemented by chemical methods. To control attack of plants particularly in the early stages till they are sturdy enough to withstand the attack, irrigation water may be charged with crude-oil emulsion. But this method is cumbersome and expensive to be recommended for large-scale adoption. Application of 2.5 per cent gamma-BHC or 5 per cent of chlordane or toxaphene or 1.2 per cent aldrin dust at the rate of 25 kg per hectare prior planting of crops gives good results. These chemicals should be thoroughly mixed with soil.

In the case of above ground termites building mounds over the soil, the termite colony inside the mound can be destroyed by the application of chemicals. For this purpose 0.004 per cent aldrin emulsion (3 gms of ·|(per cent emulsifiable aldrin concentrate in 10 litres), at the rate of 40-70 litres per mound can be used. Five per cent DDT or 0.8 per cent gamma-BHC emulsion is also equally effective.

As already stated some systemic fungicides like thiabendazole, benomyl have shown promise for soil treatment.

Cultural and biological methods may also be adopted for getting rid of pests in the soil. These will be discussed in their respective chapters.

In Central America, a peculiar treatment is applied to free the soil from infection of Panama wilt organism (*Fusarium oxysporum* f. *cubense*). Plantation soil is kept under water for a period of four to six months during which period pathogen dies out.

QUESTIONS FOR DISCUSSION

1. What is meant by 'soil sickness' ? What factors appear to be responsible for sickness in soil ?
2. What are the basic considerations of soil treatment ?
3. What is signified by "partial sterilisation" of soil ? Why "partial sterilisation" is recommended ?
4. Do you consider that sterilisation of soil by physical agencies like heat is a practical proposition ? Which source of heat is considered to be comparatively more efficient ?
5. Why are most chemicals which are meant for sterilisation of soil needed to be applied before planting crops ?
6. Why is the chemical sterilisation of soil a comparatively simple phenomenon ?
7. Why are the majority of soil treating chemicals volatile ?
8. Why is the control of termites considered difficult ? How do you propose to effectively tackle the problem of termites ?
9. Apart from the sterilisation of soil, what other methods are adopted for the control of soil-borne pests ?

CHAPTER 12

CONTROL OF VERTEBRATE PESTS

Many vertebrates including even elephants at times, can be pests, but among the mammals by far the most important are rodents and to some extent rabbits when they are in abundant population. Rats are direct pests consuming and spoiling vast amounts of food, They pose persistent problems in the fields as well as in godowns and warehouses. The subject of rat control is a very topical one among farmers, professional agriculturists and laymen. Everyone has a method to suggest for the purpose, but no single method is universally applicable nor efficient enough to exterminate rats. While the main objective of any method is to reduce the damage caused by rats by direct killing yet before any method can be recommended for large-scale adoption, its practical feasibility, cost, toxic hazards to men and animals and possible side effects etc., have also to be considered.

Of the different genera of sub-family *Murinae* of family *Muridae* of Order *Rodenta*, *Rattus*, *Mus* and *Bandicota* along with several species and subspecies are important from the economic point of view. *Rattus rattus*—the Indian house rat, *R. norvegicus*—the Norweigian brown rat, *Bandicota bengalensis*—the mole rat, *Mus musculus*—house mouse along with three subspecies, *M. m. bactrianus*, *M. m. homourus* and *M. m. tyteri* are widely distributed and cause extensive damage. Sub-family *Gerbillinae* of family *Muridae*, genera *Tatera*, *Millardia* and *Gerbillus* are important. Of several species of these genera *Tatera indica*, the Indian gerbil and *Millardia meltada* are common. *Rattus rattus* lives above ground in the neighbourhood of human habitation. It is a good climber and can swim but it seldom moves through or migrates to uninhabited area and open fields. It is found in rural as well as urban areas in warehouses and godowns near the sources of food. *R. norvegicus* is a burrower and lives in drains and sewers, but does not store any food in the burrows.

The mole rat, *Bandicota bengalensis,* possesses strong claws and is highly efficient in making burrows where it stores food. These rats are the most destructive of all field rats and cause serious damage to cereals including paddy and wheat and other crops in the field. They cut the earheads of cereals and store them in their burrows. Under upland conditions, they may also damage the tillers though such effects may not be evident except on close examination as the damage is dispersed throughout the field. Burrows produced by these mole rats usually have two to twelve openings and two to five lanes some of which are blind and used for storage of food and breeding. The opening of the burrows may be hidden by loose soil and may not be detected easily. Mole rats live in fields and farms in villages. They may be present in urban areas. They are rather aggressive in nature. They migrate from harvested areas to the cropped area. They are nocturnal in habit and omnivorus. They are also efficient swimmers and can live in bundhs and dykes of even deep water paddy.

The Indian field mouse, *Mus booduga,* is found in the same habitat as that of the mole rat. They secure their food from the field as well as grains stored in the food chamber of the mole rat. They make small burrows which do not have any side branches. *M. plathrix* and *M. cervicolor* also can cause damage to crops in the field.

Indian gerbils or white rats are inhabitants of sandy areas. They make their burrows near the field crops. Burrows made by them are deeper, but shorter than those made by mole rats. Openings of these burrows are hidden under bushes, or vegetation. These rats are bad swimmers. Apart from plant parts, grains, stems, rhizomes, etc., whatever they can secure, they also feed on insects.

For the effective control of rodents, the knowledge of bio-ecology of the predominant species involved in the area of operation, is necessary with particular reference to habitat, area of operation, feeding rhythm, breeding behaviour, etc. Diurnal species have their peak feeding hours, early in the morning while the nocturnal ones prefer to have food immediately after dusk. Natality or rate of birth is minimum during the summer months (May-June) and in December. Control operations during the lean breeding seasons may be easier and less costly in terms of requirements of baits, rodenticides etc. The feasibility of carrying operations in relation to weather conditions has also to be considered. It may not be possible to undertake field measure in the wet months.

Methods of rodent control may be classified as follows :

(1) Mechanical/physical methods,
(2) Chemical methods.

Mechanical/physical methods

Trapping

The population of rats, mole rats and mice can be reduced by catching them in traps. Traps may be divided into two categories : (a) live traps, and (b) kill traps. In live traps, rats are caught alive and then killed by immersion in water whereas in kill traps, such as breakback guillotine, rats are immediately killed as soon as they come in contact with them. The basic principle of the trap, as the name suggests is to allure the rats to move into the contrivances from which they will not be able to escape or in which they will be killed.

Wonder traps and cage traps are the two common types of live traps. A wonder trap is usually made of wires and measures 50×25×20 cm. It is divided into two chambers, a smaller one with a tubular opening through which the rat enters, a larger one connected with the smaller one through another tubular passage. This tubular passage allows the rat to pass only when the base of the passage is drawn downwards under the weight of the rat and rat is rolled down into the bigger chamber containing baits. As soon as the rat enters the bigger chamber and the weight is released, the plate again moves upwards to its original position and the rat cannot come out. In this trap, more than one rat may be caught in one setting.

A cage trap is made of wood with the roof of the front

portion made of wires. The door of the trap is provided with a long handle and two springs. It is kept open by engaging a curved wire on the roof of the box which is fixed at the end of the handle. On the other end of the curved wire suspended inside the box, the bait is kept hanging. As soon as the rat entering the box tampers the bait, the curved wire or hook gets detached from the handle of the door and the lid falls off. In one setting only one rat can be caught.

The back-breaking trap consists of an iron plate usually 15×10 cm having toothed edges pointing upwards at its distal end. An U-shaped striker of an iron wire is attached with a spring on which the bait is fixed. As soon as the rat disturbs the bait for feeding, trigger is released, crushing the rat between the edges of the basal plate and retreating U-shaped striker. Only one rat can be killed at one setting.

The trapping method is not considered to be a very effective one in large-scale control of rodents. This can be useful in the elimination of survivors of poison baiting.

The traps should always be clean and bait should be attractive.

Traps should be placed in the natural pathways of rats. The number of traps that should be set has a relation to the normal home range of rats. Success of trapping is dependent on the number of traps set. In case of house mice, it should be one to three metres apart and in the case of rats not exceeding six metres.

Electrocuting

In many advanced countries, iron fences carrying electric charges are placed around the fields infested with rats so that rats when they come out of their burrows in search of food and move across and come in contact with the fence carrying electric charge get killed. This method may prove useful when there are no rats inside the fenced area. This has the inherent danger of electrocution to men and animals. Such a device is not practicable in India.

Ultrasonic devices

Equipments are now available which can produce sounds inaudible to human ears but can cause sounds which will cause pain and irritation in the ears of rats. Rats exposed to such sounds may eventually die. Such equipments may be used both indoors and outdoors. It has been claimed by the manufacturer of one such proprietory machine that rats can be eliminated within 72 hours. The practicability of these machines, the cost involved, harmful effect, if any, on other animals have to be found out before such measures can be adopted. Ultrasonic methods can be used only in godowns and warehouses with facilities of electricity.

Chemical methods

These methods may be broadly divided into two categories : poison baiting and fumigation.

Poison baiting

Baits for this purpose require : (a) base or bait material, (b) chemical concentrate which is actually the toxicant, and (c) bait containers Normally the bait material may consist of normal food of rats and mice—low quality milled rice, flour, bread crusts, fishmeals etc. Sugar may be added to make the bait more attractive.

While selecting a suitable poison for rats, apart from its toxicity against rodents, two other factors need to be considered, namely, the action of the poison on other animals and men and the result of its action on the rodent itself. Almost all poisons which can be used for the control of rats are poisonous to men and domestic animals, but there are a number of chemicals which show a very specific action against rodents. Even with generally poisonous materials, toxicity hazards to men and animals can be minimised by careful handling.

For sanitary reasons, the rodenticide to be used should induce the rats to emerge and die in the open, so that rotting carcasses are not left in inaccessible places. Rodenticide should have the property of being quickly degraded in the poisoned animal so that other animals eating the dead rats are not affected.

Acute poisons : The powdered inner flesh of bulbs of red squill (*Urginia (Squilla) maritima*)—a bulbous plant of the subtropics is known to have rodenticide property since the Middle

Ages and is still used in many parts of the world. The exact mechanism of its action and the active compounds involved are not known. Specificity of action of this substance lies in the emetic action which causes vomiting in men and animals other than rodents. So the material is not particularly hazardous to men and other animals besides it gets decomposed fairly quickly in the body of rats. It is non-poisonous to poultry. However, such materials are available in limited quantity.

Thallium sulphate : This chemical which is odourless is used as a rodenticide. It is slow in action and forces the rat taking this poison to leave its haunt and search for water in the open. The poisoned rat dies two to three days after intake of poison due to respiratory failure. Normal dosage is 25 mg/kg of body weight of rats. Thallium sulphate is extremely toxic to men. It can be absorbed through the skin and has a cummulative toxicity effect. It can also be used for control of jackal and bird pests. On account of its extreme toxicity, it should be handled only by experienced operators and adequate precautions have to be taken.

Barium carbonate : It is also an odourless chemical like thallium sulphate and induces the rat to die in open in search of water. It is also poisonous and should be handled with care.

Sodium flouroacetate (FCH$_2$COONa) : It is used in several countries as a rodenticide. It is extremely toxic to men and animals. The lethal dose for a dog is 0.1 mg/kg. Its sale in the U.K. as an aphicide has been banned. This compound should be handled only by experienced operators and should be used in sewers, ship's holds and closed warehouses and not in the open.

Antu : α-napthylthiourea introduced in 1946 is effective against the Norwegian brown rat *Rattus norvegicus*, but *R. rattus* and mice are more tolerant to this compound. Adult *R. norvegicus* can be killed by a single dose of 6-8 mg/kg. It is relatively nonpoisonous to men and animals. It induces extreme bait-shyness in sublethal dose.

Other acute rodenticides that are or have been used are zinc phosphide and arsenic compounds. Norbromide 5-(α-hydroxy-α-2-pyridylbenzyl)-7-(α-2-pyridylbenzylidene)-norbon-5-ene, 2,3-dicarboximide is specific to the genus *Ratttus*. The lethal doses for the Norwegian rat and black rat are 12 mg and 60 mg/kg oral, but a dose of 1000 mk/kg produces no harmful effect on cats, dogs, monkeys. This compound however has poor acceptability to rats and induces baitshyness.

These acute poisons are also known as single shot poisons because they are capable of killing rats when taken once only, i.e., by a single dose. Therefore the dose should be lethal and the rat should take the poison without any suspicion. To overcome the initial suspicion, prebaiting with unpoisoned baits should be provided prior to feeding with the poisoned bait.

Chronic or multidose poisons : Chronic or multidose poisons are those which kill rats when taken in several doses over a period of time. Normally these compounds do not induce any baitshyness. They provide means for carrying on a continuous rat control with very little cost of labour. Active ingredient for killing of rats is provided in low concentrations. Anticoagulants which appeared in the 1950s for rodent control are examples of toxicants of chronic or multidose poison. These compounds interfere with the action of vitamin K and reduce the coagulating powers of the blood so that small injuries may cause fatal haemorrhages or the rats may die from internal haemorrnage. A daily dose of 2 mg/kg for rats for several days is sufficient whereas dogs have been found to survive 50 mg and the therapeutic dose for human beings for treatment of coronary thrombosis is 200-300 mg.

Most of the anticoagulant rodenticides are hydroxy-coumarin derivatives, though some are indandiones. The most widely used is warfarin 3(α-acetoxy-benzyl)-4-hydroxy coumarin. Warfarin is an excellent rodenticide without any baitshyness on the part of rats. 3-(acetonyl-p-chlorobenzyl)-4-hydroxycoumarin marketed under the trade name tomorin, known now as coumachlor, has the same properties as warfarin. Other competitors are coumafuryl, and coumatetralyl.

Diphacinone, 2-diphenylacetyl indane-1,3-dione is used as rodenticide. It has been found to be more effective against *Rattus rattus*, while warfarin is more toxic to *R. norvegicus*. Use of 0.025 per cent of active ingredient is recommended for anticoagulant bait. However, resistance against anticoagulants has been reported in scattered populations.

For the control of rats by baiting, poison baits have to be placed in a protective container for prevention of their being used by non-target animals and protection against spoilage from rain, etc. Indigenous materials may be used for making bait stations. Empty cans, pieces of bamboo with one intact node, spathes, etc., can be used. Irrespective of the shape and the material used, there should be ample space for the rats to enter and feed on the poison bait and the opening for entrance of rats should not be too large. Bait stations should be placed near the source of food along their routes of travel and close to their habitats. In buildings they should be placed near the walls in corners. It is necessary to place enough bait stations so that all animals have an easy access to poison bait. Baits should be replenished regularly as rats consume them. The cooperative effort of farmers in a locality is expected to yield desirable results.

Fumigation

The control of rodents, particularly mole rats and others which form burrows in the field may be achieved by the process of application of toxicant in gaseous form. Fumigation for rodent control can be carried out in the open in the dry season. It should never be done inside any building nor in the rainy weather or wet ground. Hydrocyanic acid is the active principle involved in many cases. It is usually formulated with magnesium carbonate and anhydrous magnesium sulphate to increase storage properties. Alternatively the cyanogas process is adopted in which calcium cyanide is used. In the case of cymag, the compound is discharged in the form of fine powder inside the burrows by foot pump. For effective operation all openings of a burrow should be searched out and simultaneous pumping through all the openings should be carried out. After pumping of the powder, openings should be sealed with mud. Cymag on absorption of moisture slowly liberates poisonous hydrocyanic acid gas which kills the rodents. A teamwork with a number of foot pumps is necessary so that the rats cannot come out, the gas generated cannot escape and desired results may be achieved.

Calcium cyanide may be used without a foot pump by direct application inside the burrows. The powder is placed inside the burrows with a spoon. At the first instance entrances into the burrows are closed. On the second day, when a new burrow is located, the powder should again be placed with the help of a spoon fixed at the top of a long stick and the entrance should be sealed. The operator should take precautions against inhaling the powder. Approximately 7 gm of powder per cubic metre is required.

An alternative method is use of phosphine gas which may be released from aluminium phosphide. This is particularly useful for the control of mole rats in paddy fields. Tablets containing aluminium phosphide and ammonium carbonate are placed deep inside the burrow and the openings sealed with mud. Phosphine gas produced from the tablets kills the rats.

Mole rats can also be controlled by baiting with earthworms into which strychnine has been injected.

Control of birds

Birds are often considered to be beneficial for the control of insects as they often live on insects and reduce the population. Before large-scale use of chemicals came into vogue, in many cases, special perches used to be made in the field to minimise the incidence of insects. Nevertheless birds may cause damage to early season and off season crops. They have been found to be very destructive to the cobs of maize or the heads of sunflower or ripening fruits. They may cause damage to tender seedlings of vegetables.

They may prove a menace to the successful growth of field crops as well as harvest of fruits. Various methods that are employed include covering by nets, using scaring devices, reducing their population by shooting, trapping and use of chemicals. Control of bird damage by shooting is not a practicable proposition. Covering can only be used in a very limited way in the experimental fields or orchards, or by some enthusiastic small sized orchard owners.

Scaring devices using mechanical, acoustic and visual means are normally taken recourse to. Beating of drums either mechanically or manually to produce sounds is still in vogue in many parts of the country particularly in the harvest season in the field when the damage is maximum.

Fire crackers placed at regular intervals along a cotton rope in which the fuses are inserted can be placed in the field. The rope burns from one end and ignites the crackers at regular intervals which produce sounds and scare away the birds. Interval between explosion of fire crackers can be adjusted by placement of fire crackers at proper intervals and diameter of rope. Fire crackers may also be useful for the control of flying foxes which damage fruits.

Bird scaring acetylene exploders have been used with success. Loud sounds due to the burning of acetylene gas produced at intervals are utilised to scare away birds and small animals. A number of models are available for the purpose.

Birds may be scared by display of scarecrows, dead birds, visually attractive flags, etc.

Electrical perches have been adopted in many countries. But these are inoperative in the wet weather. Birds are often shy to sit on these perches and avoid them.

Control of birds by use of poisonous chemicals as baits has not been favoured in many places. Sometimes narcotics such as chlorolose may be used, effects of which will be temporary. With the use of such a substance harmful birds can be separated from the useful ones which may be replaced.

Control of snails and slugs

Snails can cause damage to crops, particularly at the early stages and thus be agricultural pests. Besides they may be vectors of diseases such as schistosomiasis. The best known chemical control of snails is the use of poison bait with metaldehyde, a polymer of acetaldehyde. It is toxic to snails by contact as well as by ingestion. The chemical immobilises snails and copious slime is exuded out of the body of snails and they die of dehydration. It is applied as a bait mixed with bran and the dosage is very low. About 400 gm of metaldehyde mixed with 30 kg of bran is sufficient per hectare. This will control slugs. Metaldehyde has very low mammlian toxicity.

DNOC and dinitro-O-cyclohexyphenol have been reported to be effective against snails when used as herbicide. Copper sulphate and N-trityl morpnoline (frescon) have been found to be useful against snails when they are spread on meadows harbouring these animals.

Rodent species (1)	Habit & habitat (2)	Distinctive characters (3)
1. *Bandicota bengalensis* (Lesser Bandicot rat)	Good burrower/ Crop fields/living in or about human dwellings.	Grey, black or blackish grey, creatures of large size sufficient to distinguish from house rats ; Size : Head & body 15-22 cm, tail : 12-18 cm.
2. *Tatera indica* (Indian Gerbil)	Burrowing type and fast runner ; noctural, Crop fields and grasslands.	Reddish brown to fawn or greyish fawn ; tail is clothed with hair and ends in a tassel ; hind feet very long which helps in taking long leaps. Size : head & body : 15-18 cm. Tail : 10-12 cm.
3. *Rattus meltada* (Soft furred fieldrat)	Lives in simple burrow, Nocturnal. Crop fields and grasslands.	Soft furred rat, Tail the same length as or shorter than head & body. Ears rounded. Sole with 4 or 5 plantar pads, Dorsum light grey. Feet and belly off-white. Tail bicolour, Mammae : 8
4. *Mus spp.* (Fieldmouse)	Nocturnal. Crop-fields.	Tail shorter than head and body. Dorsum pale, sandy. Underparts and feet off-white. Tail bicolour.
5. *Nesokia indica* (Short tailed Mole rat)	Nocturnal, specialised fossorial rodent, Extensive burrow at each opening the excavated soil is heaped into a pyramid like mole hill.	Solidly built rat. Tail shorter than head and body (about 70%). Ears small. Six plantar pads, four digits on hand, well developed claws ; five toes on foot, Dorsum rufous brown or tawny ; belly murkey grey. Incisors broad. Mammae : 8.

Rodent species	Habit & habitat	Distinctive characters
(1)	(2)	(3)
6. *Bandicota indica* (Large Bandicot rat)	Nocturnal, fossorial Omnivorous, Rural environment.	Bigger & robust built than *B. bengalensis*. Ears rounded. Dorsal hair blackish brown, long hairs with tips; grizzled appearance. Ventral hair greyish brown. Mammae : 12.
7. *Meriones hurrianae* (Desert gerbil)	Burrowing type. Diurnal; gregarious, Crop fields and grass lands.	Sandy yellow on the back, dirty white on the belly; smaller than the Indian gerbil. Length : 5-8 cm.
8. *Rattus norvegious*	Can burrow & climb; Lives in sewers.	Brown, darkest on the back; whitish or light brown on the belly; tail shorter than the body; snout blunt; droppings in groups generally ellipsoidal or spindle shaped; Head & body : 18-20 cm. Tail : 12-18 cm.
9. *Hystrix indica* (Indian-crested porcupine)	Nocturnal & terrestrial. Orchards and tuberous crops near hills.	Body covered with quils with alternate deep brownish black and white bands. A long crest of bristles from forehead to back of head. Tail short and covered with short white quills.
10. *Rattus rattus* (House rat)	Not a burrowing type, good climber; rare in sewers. Rural and urban residential places.	Grey, black, brown or fawny on the dorsal side, may have white belly; tail is longer than the head and body; snout pointed, droppings scattered always, sausage or banana shaped. Head & body : 12-15 cm. Tail : 20-25 cm.
11. *Mus musculus*	Climber; burrower; unknown in sewers. Warehouses and godowns.	Brownish grey and lighter shades; tail is longer than the body; snout pointed; droppings small, scattered, thin and spindle shaped

Rodent species	Habit & habitat	Distinctive characters
(1)	(2)	(3)
		Head and body : 3-5 cm Tail : 10-12 cm.

RODENT PEST OF MAJOR AGRICULTURAL CROPS
The predominant rodent pests associated with the major agricultural crops in India are indicated below :

Important crops	Predominant rodent pests

A. KHARIF CROPS

Pearl Milet,	Meriones hurrianae, Tatera indica,
Sorghum, Maize	Bandicota bengalensis, Ratius meltada,
Rice	R. meltada, T. Indica, M. spp, * Mus muculus,
Ragi	Mus booduga, Rattus meltada, Meriones
Groundnut	hurrianae, Ratus meltada,
Cotton	R. meltada, M. hurrianae, T. indica
Oilseeds	Mus spp., R. meltada, T. indica.
Pulses	

B. RABI CROPS

Wheat, Barley,	Tatera indica, Rattus meltada,
Gram & Mustard	Bondicota bengalensis, Nesokia indica, Mus booduga.

C. MISCELANE- OUS CROPS

Tuber crops	Hystrix indica, M. musculus, Tatera indica.
Vegetables	R. meltada, Mus spp. Funambulus pennanti, F. palmarum, Hystric indica.
Orchards	Funambulus pennanti, F. palmarum, T. indica Nesokia indica, Hystrix indica.
Coconut	Rattus rattus, Tatera indica.

---

* Mus musculus, Mus booduga, Mus platythrix and Mus cervicolor.

QUESTIONS FOR DISCUSSION

1. Why of all vertebrate pests, are rodents considered to be the most important ?
2. What are the important species of *Rodenta* causing damage in the fields and warehouses ?
3. Why is the knowledge of bioecology important in the control of rodents ?
4. Of the mechanical/physical methods, which one is the most widely adopted ?
5. What are the different methods of trapping ? How can the efficiency of traps be improved ?
6. What are the requirements of a bait ? What factors need to be considered apart from toxicity in baiting ?
7. What are single dose and multidose poisons in rodent control ? Which one is preferable ?
8. What is the mechanism of anticoagulants in rat control ?
9. Where is fumigation to be done for the control of rats ? What steps are to be taken for the successful operation of fumigation ?
10 What are the different methods of control of bird menace ? Which do you consider to be the most effective ?
11. What is the mechanism of the action of metaldehyde on snails and slugs ?

CHAPTER 13

PROTECTION OF STORED PRODUCTS

Losses of food grains and other stored products of agriculture due to the attack of pests are not new phenomena. They have been observed by farmers since they became food gatherers from food hunters. Various indigenous methods of storage and protection against losses still in vogue in different countries of the world bear testimony to the fact that men have been aware of the depredations due to pests and the necessity of proper storage. When food was in comparative plentiness and population was much less, such losses did not make much impact except when there had been unusual spell of adverse conditions affecting yield. The effect of such losses was however begun to be felt with increasing population pressure and to meet the inadequacies in supply within the country, food grains had to be imported. Besides, to offset fluctuations in production which in many cases, under normal conditions is not enough or barely sufficient to meet the requirements, with the consequent threat of soaring prices operating under the laws of demand and supply and famine, many countries conceived of the maintenance of a buffer stock of food grains. Hence there has been a gradual development of storage facilities to keep a surplus stock, at different places, either from the produce of the country or imported from other countries, to maintain an equitable supply and fair price level in the lean years. In many developing countries, a comprehensive distributive system with a developed market structure has become a necessity for a number of reasons—social, economic and political.

Damages due to pests and the consequent losses both in quality and quantity of food grains always take place in storage due to a number of factors and their quantum is dependent on the length of storage. Hence reduction in the period of storage is expected to minimise losses. However, under prevailing conditions, food grains have to be stored for a fairly long period and measures will have to be taken to reduce losses to the minimum. Infestation may result in financial loss in terms of tonnage, besides there may be deterioration in quality rendering them not quite fit for consumption. Infestations of pests may fetch lower prices on account of consumer rejection. In the developing countries, where chronic shortage of food and poverty prevail, emphasis is naturally placed on the financial aspects, and other considerations are often lost sight of.

Organisms directly responsible for causing loss in stored products are insects, mites, rodents, fungi and bacteria. It has not yet been possible to ascertain how far the individual category of pests is involved, to what extent the effects are cumulative and infestation by one category aggravates infestation by another. Damages due to fungi and bacteria are normally attributed to unhygenic conditions of storage and are associated with the initial high moisture content of the stored products or absorption of moisture during storage due to defects in the system of storage. Deterioration due to fungi can be detected on account of visible mouldy growth and formation of lumps, but the same due to bacteria often remains unnoticed and passes under normal deterioration. Rodents are also responsible for losses in quantity, particularly when they are allowed to multiply in abundance and defects in storage structures and conditions permit such infestation. Insect and mites are responsible for reduction in tonnage along with deterioration in quality. In infestations with fungi like *Aspergillus flavus* and the related spp., there may be production of mycotoxins (aflatoxin) which may cause injury to health of men and animals, when such grains are consumed as food.

In dealing with the control of infestation or infection of stored products, the stages at which such infestation may take place have to be taken into consideration. It may occur at the point of initial storage immediately after harvest. Such condition of the commodity will depend on the degree and type of endemic infestation in the locality, which is again dependent on a number of factors including climate and harvesting conditions and methods. It is desirable that treatment should be effected at the initial stage. If the commodity is meant for consumption after a short period of storage, then of course, it has to be considered whether the cost of such treatment would commensurate with the extent of possible loss. Storage for a short period in humid tropical countries for consumption may cause heavy loss, but very often treatment for a short period is usually ignored, particularly when there is no possibility of these commodities being offered for sale. If the stored products are to be consumed elsewhere, either in or outside the country very often the treatment is left to the secondary or final point of storage after transportation from the initial point and the treatment at the initial point, however desirable it may be, is often considered unnecessary. The quantum and nature of infestation after transport to the secondary or final storage godowns again depend on whether the transport system is reasonably free from chances of infection or infestation or does not allow absorption of moisture from the atmosphere or get wetted so as to favour infection of micro-organisms. Besides, infestation or reinfestation may again take place in storage while trading through retail channels where a commodity may be stored for a short time under unhygenic conditions in infested godowns. Sometimes the total damage due to infestation or infection may be due to the cumulative effect of neglect at different stages. It may not be too much to state that problems of stored pests to a large extent can be controlled easily if systematic care is taken at different stages. It is desirable to effect such control for long-term storage.

In the protection of stored products two basic principles should be followed :—(a) the commodity must be free from infestation or infection when it is to be stored, and (b) it must be stored under such conditions which will not normally allow infestation or infection during storage. Evidently it may not be possible to meet these ideal conditions or prerequisites. There is, however, no single measure which may be made applicable or acceptable for the prevention of loss. Storage at low temperature under conditions of controlled low humidity no doubt offers the best solution. Under such conditions, pests and micro-organisms cannot thrive. But on account of costs involved, storage at low temperature is restricted to more valuable perishable commodities in which deterioration results in heavy financial loss.

Damage of stored materials particularly cereals and other grains, pulses, etc., caused by insects, mites and micro-organisms depends mainly on three factors : (a) moisture content of the stored materials, cereals or other grains, pulses etc., (b) availability of oxygen in storage, and (c) development of temperature inside the bulk stored materials. The design and construction of the storage structures and storage practices influence these factors.

Except probably *Trogoderma granarium*, very few insects thrive below nine per cent moisture content of grains. Low moisture content is not conducive to infection by micro-organisms and infestation by mites. Unless direct contact with the atmosphere is not prevented, stored grains adjust their moisture in accordance with the atmospheric relative humidity, even if they are dried thoroughly when put into storage. Deterioration or spoilage with increasing moisture level and higher relative humidity is dependent to some extent on the nature of the grains. When temperature is comparatively low and relative humidity high, moisture content of the grain may increase by 1 per cent in a day. Corresponding rise in temperature may bring about a drop in relative humidity. Increase in moisture content above 10 per cent will tend to increase infestation by insects and mites and infection by micro-organisms. So it is essential that before grains are put into storage, the moisture content should preferably be 8 per cent and absorption of moisture from the atmosphere in storage should be prevented. It is also necessary to see that moist or highly humid conditions are not created due to defects in construction in storage structures.

Under conditions of high moisture, grains produce heat due to increased respiration consequent on increase in moisture resulting in deterioration of the quality of grain. This condition is also favourable for infection by different saprophytic fungi which, as stated earlier, may bring further deterioration and render them unfit for consumption due to production of aflatoxin. If dry conditions prevail inside the storage structures and grains are properly dried, then heat and moisture produced are negligible. In cases of bulk storage particularly in flat godowns, there is likelihood of migration of moisture produced from warmer regions to cooler regions The condensation of moisture at certain points leads to localised growth of micro-organisms, infestation by insects with 'the formation of cakes or lumps. Hence circulation of air may be essential to maintain an equilibrium of temperature and moisture.

Under conditions of suitable relative humidity and appropriate moisture content of grains, temperature becomes a limiting factor in the development and multiplication of insects and mites infesting the grains. There may be development of heatspots due to growth of micro-organisms, or infestation of insects, or there may be fluctuations in temperature in metallic silos. Oxygen content inside the container under airtight conditions has also a deciding influence on the infestation of grains and multipli cation of insects.

Basing on the above considerations, methods which are generally adopted may be classified as : (a) physical, which aim at the control of physical factors influencing materials under storage ; (b) chemical, in which chemicals are applied so as to eliminate insect pests and mites and if possible micro-organisms ; (c) biological, in which steps may be taken to control or minimise infestation by use of other organisms, in contrast to the adoption of the physical and chemical methods.

Physical Methods of control

While adopting measures for control of pests in stored products, the guiding principle, should be that the materials or commodities treated in storage must not suffer in quality. In recent years with the availability of newer compounds, emphasis has been laid on chemical methods of control rather than on physical methods. Simultaneously refrigeration or storage at low temperature is gradually gaining importance and its scope is extending beyond convential commodities. In the Belgian Congo, safe measure of storage of tobacco is refrigeration. Not only to mitigate the effect of high temperature, but also of micro-organisms and pests acting at high temperature, potato is stored under low temperature conditions in the tropical countries. Physical methods should be given due importance along with chemical methods of tackling pests of stored commodities. Physical control measures may comprise one or more of the following : (a) heat treatment, (b) refrigeration, (c) drying, (d) blowing of air, (e) radiation treatment, (f) radiasthesia treatment, (g) trapping, (h) sound, (i) percussion devices, and (j) self destruction or autocidal treatment.

Heat treatment

Insect pests in storage structures can be killed where the temperature of the storage structures is raised to and maintained at a temperature lethal to insect pests. For this purpose, the inside temperature of the storage structure should be raised to 50-70°C and maintained at that level at least for 10-12 hours. In flour mills and food processing plants, this device may be taken recourse to by arrangement of a well laid out heating system which will generate and radiate heat in an uniform manner throughout the storage space. It should be ensured that drop in temperature by cooling is done slowly and the possibility of condensation of moisture due to cooling should be reduced to the minimum, otherwise reinfestation or infection with micro-organisms may take place at a later stage.

Refrigeration

It has already been pointed out that valuable perishable commodities are stored at low temperature. Most fungal spores fail to germinate and the growth of fungi does not take place at 0-5° C, and at 0-10° C growth of mites is prevented. A relative humidity below 65 per cent does not encourage fungal growth and the growth of mite is inhibited at a relative humidity below 60 per cent. Most insects fail to multiply below 10° C. Hence considering all aspects, a low temperature combined with low moisture content and relative humidity would be very suitable for the prevention of loss in storage from different organisms. Cost of these methods in relation to gains has always to be assessed. Practicability of uniform cooling conditions together with low moisture in large storage structures may have to be ensured. In many advanced temperate countries, this method may be advantageous.

Drying

Moisture content of the grain is the determining factor in the protection of stored grain from the attack of insect pests, fungus and spoilage by other micro-organisms. High moisture content in the grain encourages the same while maintenance of grain moisture below 11 per cent will ensure a fair amount of safety or protection against infestation with insect pests and destruction by micro-organisms including fungi.

In the tropics, sun-drying is still the cheapest and popular method in the rural areas, because virtually no investment is made though the process is slow and labour-intensive. Sun-drying can be taken recourse to in comparatively smaller quantities of harvested products meant for home consumption. In bulk storage, particularly when harvesting is done in the wet season with frequent spells of rains and high relative humidity, sun-drying is not a feasible proposition. On the other hand, slow drying may enhance the problem of aflatoxin due to encouragement of infection. It has been observed that some grains at the time of harvest may contain very high moisture, 20 per cent or more, which is not safe for storage. Hence grains will have to be dried by some other means, even if they are intended for storage for a comparatively short period. Mechanical driers may be used in such cases in which grains inside the storage structures can be dried by a continuous flow of air at low humidity forced through the bulk stock till the desired moisture level is attained. If necessary, heated air may be reinforced with atmospheric air to enhance the process of drying. Circulation of air of 50 per cent relative humidity is sufficient enough for bringing down the moisture to 12 per cent within a reasonable time. In many cases atmospheric air of low relative humidity reinforced with hot air may not suffice for the purpose. Besides this type of drying may not be possible in wet weather. Hence to achieve the desirable moisture level, hot air at 60° C or above may have to be circulated at a desired volume through the grains for heating of grains and other food articles at any time of the year regardless of weather.

Driers depending on the source of heat, low relative humidity, quantity of grains handled, and effectiveness of purpose may be of diverse types. The type is to be chosen in consideration of the job requirements.

Aeration or blowing

Grains in storage respire so also insects and micro-organisms infesting or infecting them. In the process of respiration heat is liberated and may raise the temperature of grains as well as inside the storage structure. This increase in temperature may result in liberation of water, at least in small pockets, in case the moisture content of the grains is comparatively high. Such increase in moisture is normally accompanied with increased infestation of insect pests, infection by micro-organisms, loss of viability, etc. This undesirable condition may be removed by blowing of air through the grains in the storage structures. This method can be successfully operated when the ambient air has a low moisture content. Aeration of grains with cool air is practised in many countries and mechanical devices have been adopted for the purpose.

Irradiation

(a) *Radiofrequency energy* : Dielectric heating of materials exposed to particular radio frequency (RF) energy which is converted to heat inside the materials may be useful in control of insect pests. In this process insects acquire or absorb more heat than the grains, hence this method is considered suitable because insects are killed whereas stored materials are not damaged. Besides, heat being uniformly distributed ensures uniform killing. It may, however, be pointed out that this method, in spite of its potentialities is still in the experimental stage. Capital investment and recurring costs are much higher as compared to chemical methods of control.

(b) *Infrared energy* : Exposure of the materials to infrared energy will bring about a rise of temperature of the insects to the level at which they will be killed. Infrared radiation however, has a low penetrability, hence grains at the surface in bulk storage only will be heated and acquire higher temperature whereas the grains in the inter and lower layers will not be affected. Hence grains are passed through over a conveyer belt in thin layer when this treatment is adopted on a commercial scale. Initial capital cost is comparatively low, but recurring cost is high.

(c) *Ultraviolet radiation energy* : Exposure to ultraviolet energy at high doses is lethal to insects, whereas in low doses results in aberrant behaviour of the insects including phototactic movement. This energy, like infrared acts on the surface material, though the power of penetration is higher in this case. It has potentiality, but has not yet become commercially feasible.

All the three above sources of physical energy have one advantageous feature—they do not leave any residue toxic to men and animals. X-rays cannot be used for treatment against infestation of grains, for cost and difficulty of bulk exposure.

(d) *Radiation from sources of atomic energy* : Sources of radiation from atomic energy are now being used for treatment in bulk storage, of grains particularly in port terminals. It is a cold disinfectant. Higher the dose, the more effective is the treatment and higher the cost involved. Dosages vary with individual species of insect pests. Doses of 250-500 krad (rad— unit of radiation with energy equivalent of 100 ergs per second. krad = 10^{-3} rad) can kill stored grain pests within 24 hours of treatment, 150 krad within a week and a dose of 50 krad within one to four weeks depending on insect species. Sterility in insects can, however, be caused at a much lower dose. Increasing attention is being paid to use of gamma rays at doses below 50 krad for controlling most of the insects. Cobalt irradiators are preferred for use in grain and food stuff. Irradiated food has been approved for consumption in different countries, though the permissible level of krad to be used varies. Irradiation is more effective against immature insects than on mature insects which are more susceptible to fumigation. Seeds cannot be treated with radiation from atomic energy, as radiations from gamma rays may have mutagenic effect.

It has been claimed that combination of gamma rays and infrared radiation is more effective than the individual application of either. Similarly, exposure to gamma radiation followed by fumigation particularly with malathion gives better results than either of these methods.

Radioasthesia

This treatment is still in the experimental stage. It is very economical if controllable and without ill-effects on commodities treated.

Trapping

It is normally considered to be effective against rats. It has been already dealt with separately under rodent control.

Sound

Ultrasonic waves have an effect on insects particularly on number of progeny from eggs, and longevity of adults. Ultrasonic waves though found to be effective under experimental conditions, yet have not been found to be of much practical utility in the control of infestation of insect pests in bulk storage due to the dampening of sound waves by the resistance of materials in storage, particularly grains, flour, etc.

Percussion

Centrifugal types of machines are used to produce percussion effect on particles of free flowing materials namely grains, flour, etc., and the insects contaminating them are also exposed. The impact of these machines which consist of revolving hard steel discs is sufficient to kill the insects.

Self-destruction treatment

When an infested material is kept in a storage bin or silo under airtight conditions, carbon dioxide produced by the respiring insects, also by the grains eventually reaches a concentration lethal to insects and sometimes micro-organisms. Attainment of this level of concentration may take sometime during which insects may already cause damage. Hence the reduction in time needed to reach a lethal concentration may be provided by the addition of carbon dioxide from an outside source in the storage structure before sealing the container.

Storage structures

Though attention has been paid to the storage structures and bins, yet there is ample scope for improvement. Considerably less attention has been paid to warehouse conditions. It is true that substantial progress has been made in the tropical and subtropical countries, but more research work in this context is desirable. The warehouse has to be viewed as a large package. An ideal storage structure should be damp-proof, airtight, resistant to heat, and as far as practicable be rat, and termite-proof as well. The silo (above ground steel or concrete bins) is now being recommended for bulk storage. Information is now available on a large number of storage structures of varying capacities which may be used for proper storage.

Chemical Methods of control

Insecticides of plant origin : Of the different insecticides of plant origin, only pyrethrum can be used. It can be applied as such in finely ground dust form, or its active constituent after extraction may be suitably formulated to be used as dust or spray, or aerosol as the case may be. Pyrethrum and its active constituents leave no toxic residue, hence are very safe for treatment of grains. The products are comparatively costlier. Cost can be minimised by addition of synergists. Under tropical conditions, dust applications are preferred than wet sprays. Synergists include piperonyl butoxide, bicarpolate, sulphoxide, sesoxane, etc. Recently synthetic analogues of active constituents of pyrethrum are available.

Organic Insecticides : DDT, gamma-BHC, dieldrin, malathion, chlordane, and lethane are used to a large extent in control of insect pests of stored grains. It has been reported that strains of insects have appeared which have shown resistance to these pesticides.

Inorganic Insecticides : No inorganic insecticide is used for control of pests of stored grains.

Fumigants : Commonly used fumigants are hydrogen cyanide, methyl bromide, carbon disulphide, ethylene dibromide, chloropicirin, mixture of ethylene dichloride and carbon tetrachloride.

Infested stored grains can be disinfected by dusting, spraying and fumigation. Fumigation destroys larvae and adults, but eggs of insects are not destroyed except with special fumigants like methyl bromide. Radiation from nuclear sources can destroy eggs also. Residual sprays and admixed dusts are used to supplement fumigation. Such measures yield good results even at low dosages and prevent reinfestation. Such treatments at the recommended dosages are not attended with toxic hazards. It is difficult to substitute fumigation with any other method of chemical treatment. Storage structures whether they are small bins or receptacles or big warehouses should be so constructed as to permit fumigation in an effective manner. Fumigation is a convenient method of tackling insect pest of stored grains.

Fumigation should be done under expert supervision only where proper conditions are present because many fumigants are highly toxic. Under average conditions, a mixture of ethylene dichloride and carbon tetrachloride at the rate of 70 kg per 100 cu m space with 72 hours of exposure should be done. The mixture is comparatively safe. If warehouse conditions do not permit fumigation, then operation may be done in the open under gastight polythene covers. Proper care should be taken to weigh down the lower edges to prevent leakage of gas. The fumigant is introduced through some holes specially provided for the purpose.

Methyl bromide is highly toxic to insects, and less costly. The rate of application is 3.5 kg per 100 cu m for a period of 10-12 hours. Ethylene dibromide has a slow penetrating power and needs exposure of six days, the dosage being 18.5 kg per 100 cu m. Aluminium phosphide is also in use. It can be applied in the form of tablets, usually one or two per metric ton depending on the airtightness of storage structures, exposure period being five days.

Many agricultural commodities other than grains are also stored in warehouses. These include a wide range of products including oil seeds, fibres, pulses, spices, fruits, tea, coffee, etc. They are stored in bags or in bulk, as may be convenient for their protection against pests. Periodic dusting with pyrethrum of the stock is beneficial. Disinfection with fumigants as is done in the case of cereals may be done.

Premises associated with stored products may be disinfected by chemicals which may be applied as dusts, smokes, gases, aerosols or surface sprays. Satisfactory results are usually obtained with aerosol spray of pyrethrum or synergised pyrethrin followed by a surface spraying of non-tainting organic insecticide with residual properties effective for at least two weeks.

Disinfection of used bags which harbour insects in their meshes and seams should be done by exposing them to a mixture of ethylene dichloride and carbon tetrachloride in an airtight container at the rate of 70 kg per 100 cu m or other fumigants at the prescribed dose. Impregnation of gunny bags with gamma-BHC at the rate of 163 mgm per sqm is a good preventive meausre.

QUESTIONS FOR DISCUSSION

1. Why is protection of stored products important ?
2. Which organisms are mainly involved in causing losses in stored products ? Can you ascertain damages caused by different categories ?
3. Which points need to be considered in the control of infestation or infection of stored products ?
4. What are the basic principles in protection of stored products ? Primarily on what factors does damage of stored products depend ?
5. Why is the moisture content of grains considered to be of prime importance in the spoilage of grains ?
6. What different categories of methods may be used for the control of stored grain pests ? What should be the guiding principle in such cases ?
7. What are the different physical methods employed for the treatment of stored grains ? Which do you consider to be practical and economic ?
8. What are the basic considerations in the fumigation of grains ?
9. Why should attention be paid to storage structures in the protection of stored products ? What should be the specifications of ideal storage structures ?
10. Why is the use of pyrethrum in grain storage advocated ? How can the cost of pyrethrum be reduced ?

Chapter 14

TOXIC HAZARDS OF PESTICIDES

Chemicals are now widely used as pesticides in agriculture for the protection of plants against the ravages of pathogens, insects and other pests and weeds, etc. Application of them has helped at least to guarantee the yield, if not increase the same. Many chemicals used in the control of pests are highly toxic and may be hazardous to the health of men and animals, if proper care is not taken. Problems of toxicity and the hazards they may cause to the health may often be serious and must be fully appreciated. To cite an example, treatment of seeds with organomercurial seed dressings has been in vogue since the early 1920s and when properly handled, these chemicals do not pose any danger. Nevertheless a serious outbreak of organomercury poisoning due to human consumption of treated seed grain occurred in Iraq in 1971 and 1972. The number of cases officially reported to be admitted in the hospital was 5500 and some 280 people died (as reported by Damulji and Tkriti, 1972). This particular episode was the largest, but there were previous instances of death or permanent disability due to ingestion of grains treated with organomercurials. Consumption of seeds treated with hexachlorobenzene was responsible for large casualty in Turkey between 1955 and 1959. In India, in 1958, a number of human deaths were reported in Kerala and Tamil Nadu following the consumption of imported wheat, which by accident was shipped together with a pesticide (methyl-parathion) and got contaminated during transit due to the leakage of the pesticides from the containers. Since then periodic reports have been made on death or disability of men and animals due to either the indiscriminate use or careless handling of pesticides or the ingestion of materials treated with pesticides without consideration of the hazards involved.

A distinction is, however, to be made between toxicity of a chemical and the hazards it may present to health. A chemical may be highly toxic, but may not present hazards under normal conditions of use, while a chemical of low toxicity may, under certain circumstances, present highly significant health hazards. The toxicity of a chemical is its innate ability to cause injury to living beings. It is the sum total of various untoward effects caused under varying conditions on the human or animal body. The toxic effect of a chemical may be different under a variety of circumstances. Toxicity of malathion has been estimated between 1 and 2 gm per kg of body weight of rats, but when injected into the peritoneal cavity, less than half of the amount will cause death. Rats fed with 250 mg of malathion in the diet for a period of two years do not show any untoward symptom. Hazards to living beings by any chemical depend not only on its inherent toxicity, but also the conditions under which it is used and whether or not such uses lead to an exposure significant enough to cause injury.

Hazard is best defined as the probability that a substance of a given toxicity is likely to cause damage or injury in a particular set of conditions under which it is normally used. While the toxicity of a pesticide is important no doubt, but other factors which affect the frequency period and manner of exposure are equally important. For example, organophosphorus insecticides and carbamates, which affect synaptic transmission in the nervous system are more toxic, but they do not persist in the environment as the organochlorines like DDT, BHC and cyclodienes which persist in the environment and result in pollution with untoward effects. Systemic organophosphorus compounds do not leave any toxic residue, whereas nonsystemic ones leave the same. While assessing toxic effects of pesticides, attention has been paid on the properties of these chemicals, but these pesticides are often used in a formulated state, mixed with other chemicals which act as solvents, emulsifiers, or synergists to enable the active ingredient to be used to the greatest advantage. Such chemicals may also have undesirable effects.

Hazard of a chemical is also determined by the manner in

which a man is likely to be exposed. During the manufacturing, formulating and application stages the chemical dust or vapour may be inhaled by workers. It might penetrate their skin, or get into their eyes and/or be accidentally ingested. Workers may also be contaminated with the chemical when carrying out other crop treatments or harvesting crop and skin contact may be the likely form of exposure. Consumers are exposed in a different way, to a small amount of toxic residues which may be present in the plants or in the meat, milk or egg of the treated animals. Workers involved in the process of manufacture, formulation or application are exposed to high concentrations for a comparatively short time, whereas a consumer is likely to be exposed to low concentrations for a long time. In the former, inhalation and contact with the skin or the eye may pose a threat, while in the latter ingestion creates the problem.

Some chemicals damage various surfaces of the body : skin, eyes, lungs, etc., by direct contact, but others, do not cause any injury on direct contact, but they may be absorbed and concentrated in certain internal organs before they can cause any damage. While many chemicals may be involved in inducing malformation or carcinogenic growth or causing damage to the genetic make up of the somatic or reproductive cells, there may be other types of injury namely dermatitis, asthma, urticaria, eczema, etc., which are generally grouped under irritation or allergic sensitisation.

Tests on acute toxicity need to be carried on several animal species by administration in single or multiple doses through different routes, mouth, skin, etc., to have an idea how much toxicant in which form is lethal to human beings. A figure called LD_{50} the dose at which 50 per cent of the test animals will be killed is calculated. The qualitative as well as quantitative aspects of the ill-effects need to be considered. This is particularly important in cases, where there is probability of ingestion of toxic residues in small quantities over a long period. Dangers to the applicator also need to be determined to formulate safety precautions for those engaged in the application of the chemical for agricultural use. The safeguards may range from restriction on use of the chemical to trained personnel who wear protective clothing ; labelling the container carefully with clear-cut specific

Instructions on how to use the chemicals ; precautions to be taken and what to do in the event of exposure.

Many countries have enacted legislative measures on the control of pesticides keeping in view many ways in which they can affect human health, for the purpose of prevention of hazards to human and animal life and the environment. Such legislation usually consists of a number of different items namely registration of pesticides, licensing of the manufacturers, quality control, supply and application, occupational health, aerial application or residues in food. It may also cover fumigation, dressing of seeds, restriction on the use of particular pesticides.

Control is exercised over both safety in use as well as efficacy. The protection of the operator is provided in the form of labels which are to be prepared according to instructions given. The labels are to be checked to ensure that adequate guidance is given on safe and efficacious use and symbols to indicate the degree of toxicity.

In India dangers from unregulated and indiscriminate use of pesticides became evident and were brought into focus in 1958, when a Commission of Enquiry was appointed to suggest remedial measures following, as stated earlier, a number of deaths in Kerala and Tamil Nadu by consumption of wheat grains which became poisoned during transhipment, due to leakage of toxic pesticides with which the grains were shipped. Following recommendations of an Expert Committee (1964-67) appointed by the Government of India which examined the whole question of pesticide use, the Insecticide Act was passed in 1968 to regulate the import, manufacture, sale, transport, distribution and use of insecticides to prevent risks to human beings or animals and for other matters connected therewith. Insecticide Rules were framed and brought into force with effect from 1st August, 1971.

In the Act and the Rules framed thereunder, registration of the product at the Central level is compulsory and licenses for manufacture, formulation and sale are dealt at the State level. Registration is not time-bound, but the license is time-bound, subject to renewal. The inter-department/ministerial/organisational coordination is achieved by the Central Insecticide Board constituted as per the provisions of the Act. The Registration Committee constituted under the Act looks after the registration of the pesticides. After the registration of a product, the license for manufacture, formulation and sale are to be procured from the appropriate authority of the State Government. The Act empowers the Government to prohibit the import, manufacture and sale of pesticides and confiscate stocks. Enforcement machinery is provided for in the Act, as Insecticide Inspectors, Insecticide Analysts, Central Insecticide Laboratory, etc. Offences are punishable and seizure and other penalties are prescribed. The Act also empowers the Central and State Governments to frame rules, prescribe forms, fees, etc., for the effective implementation of various provisions of the Act.

The Act does not require commercial operators to be licensed, but their activities are governed by provision of the Act and the Rules. Exemptions have been made to individuals for their own use for domestic purposes and to educational and research organisations for research work.

Details of salient provision of the Insecticides Act of 1968, as amended from time to time and the Insecticide Rules, 1971, are given in the Appendix.

Based on toxicity tests, Bailey and Swift (1968) have classified the chemicals into six broad categories as stated in Table 14-1

TABLE 14.1

Toxicity Rating	LD_{50} (mg/kg) oral-rats	by single dose dermal-rabbits	Possible lethal dose (man)
Extremely toxic	1 or less	20 or less	A taste to a grain
Highly toxic	1-50	21-200	A pinch to one tea-spoonful
Moderately toxic	51-500	201-1000	1 teaspoonful to 1 tablespoonful
Slightly toxic	501-5000	1001-2000	28-560 g
Practically non-toxic	5001-15000	2001-20000	560-1120 g
Relatively harmless	> 15,000	> 20,000	> 1120 g

In the Insecticides Act of Government of India, the chemicals have been put into four categories, detailed in Table 14.2.

TABLE 14 2

Classification of the insecticide	Medium lethal dose by the oral route (accurate toxicity) LD$_{50}$-mg/ kg of the body weight of test animals	Medium lethal dose by the dermal route (dermal toxicity) LD$_{50}$ mg/ kg of the body weight of the animals	Colour of the identification band on the label
Extremely toxic	1-50	1-200	Bright red
Highly toxic	51-500	201-2000	Bright yellow
Moderately toxic	501-5000	2001-20000	Bright blue
Slightly toxic	more than 5000	more than 20,000	Bright green

As per classification of Bailey and Swift (1968), categorisation of some common pesticides on the basis of toxicity hazards is given in Table 13.3

Skin irritation (after Bailey and Swift, 1962)

Mild skin irritation : Chlorfenson ; dalapon ; 2,4-D butyl ester ; dichlone ; diuron ; formaldehyde ; glyodin ; hexachlorophene ; linuron ; neburon ; paraquat ; sodium arsenite and sodium chlorate.

Severe skin irritation : Lime sulphur ; and phenyl mercury acetate.

Agriculture (Poisonous substances) Regulations under the Agriculture (Poisonous substances) Act of the Government of the U. K. classifies the toxic chemicals into the following categories from the safety of the operators :

Part I. Extremely toxic : demeton, dimefox, mazidox.

Part II. Severely toxic—amiton and its salts ; dinoseb and

TABLE 14.3 Classification of pesticides on the basis of toxicity (after Bindra and Hareharan Singh, 1977)

Extremely toxic 1	Highly toxic 2	Moderately toxic 3	Slightly toxic 4	Practically non-toxic 5	Relatively harmless 6
aldicarb, carbanolate, demeton, dimefox, disulfoton, endrin, fensulphothion, hydrocyanic acid, mevinphos, parathion, phorate, TEPP, thionazin	aldrin, carbofuran, carbophenothion, chlorophenamidine, DDVP, dichlorvos, dicrotophos, dieldrin, dinitrocyclohexyl phenol, dioxathion, endosulphan, EPN, ethion, mephospholan, mercuric chloride, methyl parathion, organomercury compounds, phosphamidon, schardan, sodium pentachlorophenate	azinophosethyl, azinophos methyl, BHC, binapacryl, calomel, chlordane, coumaphos, 2, 4-D sodium salt, 2, 4-D isopropyl ester, diazinon, dicamba, dichlorobenzil, dimethoate, dinitrobutylphenol, dinoseb, dinitrocresol, diquat, fenthion, formothion, leptophos,	arprocarb, Bordeaux Mixture, chloranil, derris, rotenone, 2, 4-D acid, dicapthon, dichlorprop, dicofol, dimetilan, fenchlorophos, ferbam, maneb, MCPA acid, MCPA amine, nitraline, zineb, ziram	allethrin, aramite, atratone, biphenyl, calcium arsenate, carbaryl, chlorfenvinphos, chloroneb, DDT, dicryl, dinocap, diphenamid, fenitrothion, lead arsenate, malathion, maleic hydrazide, menazon, methoxychlor, oxythio-quinox, Paris green, phosmet, piperalin, propanil	Amitrole, ammonium thiocyanate, captan, dichloral urea, dichloran, sabadilla

lindane,
MCPB sodium salt,
monocrotophos,
nicotine sulphate,
oxydemeton
methyl,
PCP,
phenthoate,
phosalone,
silvex,
2, 4-5-T acid,
thiometon,
thiram,
toxaphene.

ryania,
siduron,
TDE,
tetrachlorovinphos,
triallate,
trichlorophon,
trifuralin,
triphenyl-
tinhydroxide.

(After Bindra and Harcharan Singh, 1977)

its salts ; disulfoton ; DNOC and its salts ; endosulfan ; endothal and its salts ; endrin, flouroacetamide, medinoterb and its salts ; mevinphos ; phorate ; potassium arsenite ; schardan ; sodium arsenite ; sulfotep ; TEPP (including HETP) and thionazin.

Part III. Moderately toxic : Azinophos ethyl ; azinophos-methyl ; chlorovenfinphos ; demetonmethyl ; demeton-S-methyl ; dichlorovos ; ethion ; fentin acetate ; fentin hydroxide ; mecarbam ; nicotine and its salts ; oxydemetonmethyl ; phenkapton ; phosphomidon ; vamidothion.

Part IV. Mildly toxic : Organomercury compounds (only when used as aerosols).

As per provisions of the Insecticide Act of Government of India, persons handling insecticides during its manufacture, formulation, transport, distribution and application need to be adequately protected with appropriate clothing. The protective clothing need to be used wherever necessary in conjunction with appropriate respiratory devices. Protective clothing should be made of such materials which prevent or resist the penetration of any form of insecticide formulations. The materials should be washable to make it possible to remove toxic elements after each use.

The complete set of protective clothing should consist of : (a) protective outer garment, overalls, hood/hat ; (b) rubber gloves or such other protective gloves which should extend half way to the forearm, made of materials impermeable to liquids ; (c) dust proof goggles ; and (d) boots.

Besides, for prevention of inhalation of toxic dusts, vapours or gases, suitable respirators or gas masks are to be used. Respirators should be of chemical cartridge or supplied air or dem and flow type and gas masks should be full or half face with canister.

The Ministry of Agriculture, U. K., prescribes more rigid specifications of protective coverings and gas masks. (Table 13.4).

TABLE 14.4 Protective clothing requirements according to Agriculture (Poisonous Substance) Act of U.K. (Insecticide and Fungicide Handbook, Blackwell, 1969).

PART I EXTREMELY TOXIC

Jobs for which protective clothings must be worn	Clothings etc., to be worn
Opening a container or diluting or mixing or transferring from one container to another	Rubber gloves, rubber boots, respirator, either (a) an overall and rubber apron, or (b) a mackintosh
Washing or cleansing soil application apparatus	Rubber boots, face shield either (a) an overall and rubber apron, or (b) a mackintosh.
Soil application (other than green house) by (a) unaccompanied driver of tractor drawn apparatus,	Overall, rubber gloves and rubber boot
(b) Driver of tractor-mounted apparatus accompanied by operators on foot, (c) an operator on foot	Overall, rubber apron, rubber gloves and rubber boots

PART II SEVERELY TOXIC

Opening a container or diluting, mixing or transferring from one container to another except	Rubber gloves, rubber boots, face shield and either (a) an overall or (b) a mackintosh
1. Dinoseb or DNOC or its salt when used as insecticide	Rubber gloves, face-shield or eye shield
2. Granular formulation	Rubber gauntlet gloves and either an overall or mackintosh with sleeves worn over cuffs of the rubber gauntlet gloves
Washing or cleansing spraying apparatus, soil application apparatus or granule application apparatus	Rubber boots, face shield either (a) overall and rubber apron or (b) mackintosh

TABLE 14.4 (*Contd.*)

Jobs for which protective clothings must be worn	Clothings etc., to be worn
Spraying a ground crop	Overall, hood, rubber gloves, rubber boots and either a face shield or a dust musk
Spraying bushes, climbers or trees	Rubber coat, rubber gloves, rubber boots and face shield.
Placing granules by hand or hand-operated apparatus	Rubber gauntlet gloves and either (a) an overall or (b) mackintosh with the sleeves worn over the cuffs at the rubber gauntlet gloves
Placing granules by a mechanically operated machinery	Overall or mackintosh
Spraying in a green house (not an aerosol or smoke generator)	Rubber gloves, rubber boots, hood, face shield, either an overall or mackintosh
Applying aerosols in greenhouse	Overall, hoods, rubber gloves and respirator
Soil application in the field by mechanically drawn apparatus by (a) unaccompanied driver (b) or driver accompanied by on-foot operator	Overall rubber gloves and rubber boots
(c) Any operator on foot	Overall, rubber apron, rubber gloves and rubber boots
Soil application in green house	Overall, rubber apron, rubber gloves and rubber boots
Bulb dipping or stepping with thionzanin, handling wet bulbs disposing of the solution and washing dipping and steeping apparatus	Rubber gauntlet gloves, rubber boots, an overall and rubber apron

TABLE 14.4 (*Contd.*)

Jobs for which protective clothings must be worn	Clothings etc., to be worn

PART III MODERATELY TOXIC

Opening a container, or diluting, mixing or transferring from one container to another except	Rubber gloves and face shield
Granular formulation	rubber gloves
Applying aerosols in green house	Overall, hood, rubber gloves and respirator

PART IV SLIGHTLY TOXIC

Organomercury compounds when used as aerosols	Overall, hood, rubber gloves and respirator

There is no record in this country of toxicity values to Indian fish. However, data are presented of foreign countries which will give an idea of the relative toxicity of the chemicals. (Table 14.5).

Safety to the consumer has to be considered in respect of commodities receiving deposit of toxic materials. It should be seen that no unacceptable or undesirable residues remain on the crop, or parts of the crop, when the crop is harvested after treatment. Hence sometimes a minimum time interval has to be imposed between the last application and harvest of the crop. There should be a limitation on the amount of active ingredient that may be applied. There are some chemicals which though effectively control the pest impart a taint to the sprayed plants or plant parts which may persist even after harvest.

Apart from the safety to consumers, pollution due to pesticides has drawn attention of environmental scientists and biologists. Studies in many cases have shown that use of pesticides is economically justified nevertheless in some cases there may be over-use, and the users are often led by the advice

TABLE 14 5 **Toxicity of insect to fishes**

Pesticide	Acute 24-h LC$_{50}$ values/ug/I for Rainbow trout	Bluegills
aldrin	14	10
carbaryl	3500	3400
chlordane	22	54
DDT	8	7
demeton	—	195
diazinon	380	54
dichlorvos	500	1000
dieldrin	6	14
dimethoate	20,000	28,000
endrin	0.7	0.8
fenthion	840	1800
heptachlor	15	35
gamma-BHC	30	61
malathion	100	120
methoxychlor	20	31
methyl parathion	—	8500
parathion	2,000	56
phosphamidon	4,500	—
pyrethrins	56	78
rotenone	32	24
toxaphene	7.6	7.2
trichlorfon	28,000	5 600

(From Bailey and Swift, 1968)

of the dealers in agricultural chemicals. It has been suggested by environmental scientists that it is desirable to replace broad spectrum persistent pesticides by selective and less persistent ones. But the latter ones present toxic hazards to operators, as well as to mammals, hence they need more caution in storage and handling.

The search for alternative methods and integrated pest control have been suggested and some progress has been achieved in this direction which will be dealt with separately.

Pesticides may be applied in such a manner that minimum disturbance is caused to honey bees and other beneficial insects. Pesticides which are safe to handle and apply and have low mammalian toxicity may be harmful to bees, e.g., malathion.

Domestic bees in apiaries may be either removed or provided
with protective covers during the application of pesticides, but
wild bees cannot be protected. Pesticides may be put into three
categories in respect of their toxicity to honey bee (*Apis
mellifera*-workers).

I. (LD$_{50}$ 0.001-1.99 µg/bee hazardous to bees if they are
present at the time of application of pesticides or within a few
days.) Aldrin ; arsenicals ; azinophosmethyl ; calcium arsenate ;
carbaryl ; diazinon ; dichlorvos ; dieldrin ; dimethoate ; dinitro-
butyl phenol ; fenitrothion ; fenthion ; heptachlor, isodrin ; lead
arsenate ; gamma-BHC ; malathion ; methyl parathion ; mevinphos ; parathion ; phosphomidon ; TEPP ; tetrachlorvinphos and
thiometon.

II. (LD$_{50}$ 2.0-10.99/µg/bee, hazardous to bees if they are
foraging. Can be used in the vicinity of bees under appropriate
dose, timing and methods of application.) Carbophenothion ;
chlordane ; chlorbenzilate, coumaphos ; DDT, demeton ; disulfoton ; endosulfan ; endothion ; endrin ; fenson ; demeton-S-
methyl, and phorate.

III. (LD$_{50}$ above 11/µg/bee—relatively nonhazardous to
bees.) Aramite, binapacryl, Bordeaux Mixture, captan, chloro-
benside, copper compounds, 2,4-D, dalapon, diquat, ethion, ferbam,
maneb, MCPA, methoxychlor, nabam, nicotine sulphate, paraquat,
pyrethrum, rotenone, schardan, simazine sulphur, 2,4,5-T, thiram,
toxaphene, zineb and ziram.

Symptoms of Poisoning due to Toxic Action of Pesticides and First Aid Measures

Symptoms of pesticide poisoning should be known not only
to the operators, but also to the members of the family even
though they may not handle the pesticides. Symptoms in all
cases are not the same, as each chemical group has its own
characteristic symptomatology,

Early symptoms : Early or warning symptoms of mild poisoning are headache, fatigue, dizziness, weakness, restlessness,
nervousness, perspiration, nausea, diarrhoea, loss of apetite,
thirstiness, moodiness, soreness in joints, irritation of skin, eyes,
nose and throat.

Moderately acute poisoning : Nausea, diarrhoea, excess salivation, cramps in the stomach, excess perspiration, tremors, muscle twitching, extreme weakness, mental confusion, blurred vision, difficulty in breathing, cough, rapid pulse, weeping, flushed or yellow skin.

Severe or acute poisoning : Fever, intense thirst, increased rate of breathing, vomiting, uncontrollable muscular twitching, pin-point pupils, convulsions, inability to breathe, unconsciousness.

Whenever symptoms of poisoning would be detected, doctor should immediately be called. When the doctor arrives, he should be told of the pesticide involved. The container and the label should be shown to the doctor.

While waiting for the doctor, some assistance can be given to the victim. The first aid measures may prevent death in some cases. It is to be carefully seen that no further exposure to the pesticide occurs to the victim.

(i) *Poison on the skin :* Pesticides should be quickly washed off the skin. Clothing should be removed and skin should be cleaned with detergent which in such cases is better than soap. The patient should be well dried and wrapped up in a blanket.

(ii) *Chemical burns on the skin :* Contaminated clothing should be removed. Skin should be washed with running water, and covered loosely with clean soft cloth. Ointments, powders, drugs normally used for treatment of burns must be avoided.

(iii) *Poison in the eye :* Eye should be immediately washed in running water for this purpose, eyelids should be held open and eyes should be washed gently with spout of clean running water. Washing should be continued for nearly 10-15 minutes. No drug or chemical should be given to the eye or running water.

(iv) *Inhaled poisons :* Victim should be carried to the fresh air. He should not be made to walk. Doors and windows in the room should be kept open. Garments should be loosened. Artifical respiration should be tried if considered necessary. The patient should be kept as quiet as possible. In cases of convulsion, care should be exercised to see that the head does not strike against hard object. To facilitate better breathing, chin should be kept up. Patient should be covered to prevent exposure to cold.

(v) *Swallowed poisons :* It is always desirable to get rid of the swallowed poisons very quickly. Hence the victim should be made to vomit if possible. On this his life may depend.

Vomiting must not be induced when the patient is unconscious or convalescing. Vomiting must not be induced in cases of corrosive materials being swallowed. Vomiting must not be induced when the swallowed poison is a petroleum product, as concentrated petroleum products may cause severe burns when they come out. He can be forced to vomit if diluted forms of these products are taken. In cases of swllowing of emulsified concentrates or solutions, vomiting must not be induced.

To induce vomiting victim should lie with face down wards or kneeling forward. When inducing vomiting he must not lie on back. By tickling the back of the throat with clean fingers or blunt end of a clean spoon he may be made to vomit. One cup of soapy water or salt solution may be helpful to induce vomiting. In case the victim has swallowed acid or alkali based pesticides, milk may be given, 250 ml for victims under five and nearly 1 litre for above five. If the poision is an acid, milk of magnesia, baking soda or chalk in water may be administered. Activated charcoal may be useful and can be given in the form of a thick slurry in water.

(vi) *Shock* : Sometimes the victim may suffer from a shock, which, if ignored, may be lethal. In such cases, the skin of the victim is pale, moist and cold, eyes lustreless with dilated pupils, breathing shallow and irregular. The victim may be unconscious or conscious. Victim should be kept flat with face up. If he is not vomiting, then legs may be raised ¼ metre above his head level.

Antidotes for Pesticide Poisoning

Group I—Organophosphates—azodrin, trithion, dasanit, DDVP, demeton, dimethoate, dursban, ethion, fenthion, metasystox, methyl parathion, monitor, phorate, phosphamidon,

> Antidotes :
> 1. Atropine sulphate is used. Injections should be repeated as symptoms recur.
> 2. Protopam chloride (2-PM) should be injected intravenously.

Morphine, theophyllin, aminophyllin or barbiturates must not be administered.

Group II—Carbamates—Aldicarb, Carbofuran, Propoxur, Methomyl, Carbaryl.

Antidotes :
1. Atrophine sulphate.
2. Protopam chloride (2—PM) must not be used.

Group III—Chlorinated hydrocarbons—Endrin, dieldrin, aldrin, lindane, endosulfan, BHC, DDT, Toxaphane.

Antidotes :
1. Barbiturates for convulsions or restlessness.
2. Calcium gluconate given intravenously.
3. Epinephrine (adrenalin) should not be used.

Group IV—Inorganic arsenicals—Sodium arsenite, Paris green.

Antidotes :
1. BAL (dimercaprol) is specific for arsenic poisons. Intramuscular injection.

Group V—Cyanides—HCN or Cyanogas.

Antidotes :
1. Amyl nitrite through inhalation.
2. Sodium nitrite given intravenously.
3. Sodium thiosulphate given intravenously.

Group VI—Anticoagulants—Warfarin, Valone, Pival, Diphacin.

Antidotes :
1. Vitamin K by mouth, intravenously or intramuscularly.
2. Vitamin C is a useful adjunct.

Group VII—Fluoroacetates—For poisons such as sodium fluoroacetate.

Antidotes :
1. Monacetin (Glycerol monoacetale) intramuscularly.

Group VIII - Dinitrophenols—DNOC, Dinoseb.

Antidotes :
1. Atropine sulphate must not be used.
2. Life supports should be maintained.
3. Sodium methyl thiouracil may be used to reduce basal metabolic rate.

Group IX—Bromides—FDB, MB or mixtures of EDB+MB.

Antidotes :
1. BAL (dimercaprol) may be given before symptoms appear.
2. Barbiturates for convulsions.

Group X—Chlorophenoxy herbicides, Urea & miscellaneous—
2, 4-D, 2,4,5-T, Monuron, Diuron, Bromacil, Paraquat, Diquat.
   Antidotes :
   1.   None available.
   2.   Life supports should be maintained.
Safety measures to be taken in connection with application of pesticides.

Most of the pesticide poisoning results from careless handling. Accidents can be avoided, if safety measures are scrupulously followed.

A.   **Before application of pesticides :**
   1.   Pest concerned should be identified and the extent of damage, to be ascertained.
   2.   Application of pesticides should be need based.
   3.   Advice should be sought from competent persons.
   4.   Pesticides recommended for the purpose should only be used. If a number of pesticides are recommended, then the least toxic ones are to be chosen.
   5.   Labels and instructions enclosed with the containers should be carefully read.
   6.   Appropriate protective clothing should be made available.
   7.   Application equipments should be checked for leaks or any other defects and they should be kept in proper working condition.
   8.   Plenty of clean water and soap should be made available, as well as clean clothing for change.
   9.   Pesticides should be stored dry under locked conditions.
   10.   Neighbours particularly those owning apiaries should be notified if the operation is undertaken near the households.
   11.   Only sufficient pesticides for one day's application should be taken from the store.

B.   **While mixing pesticides and during application :**
   1.   Appropriate clothing should be used.
   2.   Only required quantities should be mixed by measuring. Pesticides should not be indiscriminately mixed. Spills should be avoided. In case of any spills, they

should be immediately cleaned whether on floor or clothing with soap and water.

3. Mixing should not be done by hands if not protected by rubber or polythene or plastic gloves. It should preferably be done by some instruments.

4. Children or domestic pets should not be allowed near the mixing.

5. Instructions on the lables should be rechecked.

6. Liquid formulations should be carefully poured to avoid splashing. Dust formulations should be prevented from 'puffing up' into the face. If contaminated with the concentrate, immediate washing should be done.

7. Eating, drinking and smoking must be avoided while handling pesticides and during spraying and dusting operations. Hands and face must be thoroughly washed with soap and water before drinking, eating and smoking.

8. Spraying or dusting must not be done when the wind is high. Spraying or dusting should not be done against the wind.

9. Clogged nozzles or hoses must not be blown out with mouth for cleaning.

10. Pesticides must not be left unattended in the field.

11. All the pesticides in the sprayer or duster must be used. They should not be stored for future use. If some is left, it should be used on other labelled crops.

12. Application of pesticides near the beehives, pastures, houses, schools or areas with sensitive crops to the pesticide concerned should be avoided. Near living areas, application should be done in the morning or afternoon when people and pets are least likely to be exposed.

13. Spraying should not be done across the streams and cannals.

14. Any unusual symptoms on operators should be watched. In case the operator is sick, attempt should not be made to finish the job.

ESTIMATED RELATIVE ACUTE TOXIC HAZARDS OF PESTICIDES TO SPRAYMEN*

The estimates of hazards in this table are based primarily on the observed acute dermal and to a less extent oral toxicity of these compounds to experimental animals. Where it is available, use experience has also been considered. It should be noted that the classification into toxicity groups is both approximate and relative. These toxicity categories are not related to specific categories spelled out for label requirements.

**Most Dangerous	Dangerous	Less Dangerous
aldicarb (C)	aldrin (CH)	azinphosmethyl, (OP)
carbanolate (M)	bidrin (OP)	BHC (CH)
demeton, (OP)	carbophenothion, (OP)	binapacryl, (N)
disulfoton, (OP)	DDVP, dichlorvos (OP)	chlordane (CH)
mevinphos, (OP)	dieldrin (CH)	coumaphos, (OP)
parathion (OP)	dioxathion, (OP)	diazinon (OP)
phorate (OP)	DNOC (N)	dichloroethylether (M)
schradan, OMPA(OP)	DNOSBP (N)	dimethoate, (OP)
TEPP (OP)	endrin (CH)	endosulfan, (CH)
thionazin, (OP)	EPN (OP)	fenthion, (OP)
	ethion, (OP)	heptachlor (CH)
	methyl parathion (OP)	lead arsenate (M)
	nicotine (M)	lindane (CH)
	pentachlorophenol (M)	naled, (OP)
	phosphamidon, (OP)	oxydemetonmethyl, (OP)
	sodium arsenite (M)	SMDC, (M)

Less Dangerous (continued):

aramite (M)	DDD, TDE (CH)
captan (M)	dicofol, (CH)
carbaryl, (C)	dilan, (CHN)
chlorobenzilate (CH)	dinocap, (N)
2, 4-D (CH)	diquat (M)
	IPC, Propham (M)
	malathion (OP)
	maneb (M)
	methoxychlor (CH)
	mirex (CH)
	morestan (M)

**Most Dangerous	Dangerous	Less Dangerous
	zectran (C)	trichlorfon, (OP)
		toxaphene (CH)
		VC-13 (OP
		NAA (M)
		Perthaner (CH)
		piperonyl butoxide (M)
		ronnel, (OP)
		rotenone (M)
		simazine (M)
		2, 4, 5-T (CH)
		tetradifon, (CH)
		thiram (M)
		zineb (M)
		ziram (M)

* Source of Data : Safety in the Use of Pesticides, Homer R. Wolfe, William F. Durham, Proc. 2nd. Estern Wash. Fertilizer and Pesticide Conf. Pullman; Washington State University, pp. 14-21, 1965.

** The chemical class to which the pesticide belongs is designated as follows : C, carbamate, CH, chlorinated hydrocarbon; M, miscellaneous ; N, nitro ; and OP, organic phosphorus.

C. After Use :

1. Unused pesticide formulation should be returned to the store.

2. Equipments should be thoroughly cleaned, after the use for the day. In case of sprayers, washing should be done in clean water. In case of dusters, dusts should be removed by gentle tapping.

3. All empty containers should be safely disposed of. As it may be difficult to bury empty containers after each days' application, they should be kept in the store until a convenient number are ready for disposal. IT IS ABSOLUTELY IMPOSSIBLE to clean out a container sufficiently well to make it safe for use of storage of food, water etc. If containers are burnt. smoke should be avoided.

4. After the days' work, thorough cleaning with soap and water and bathing is necessary. Scalp, groins, face, hand and feet should be thoroughly washed. Washing or bathing should not be done in cannals or ponds, as the pesticides may be toxic to aquatic animals.

5. If clothes are exposed to pesticides, they should be kept away from other uncontaminated clothes, and family laundry, and washed separately. Washing should not be done in cannals or ponds. Babies must not be handled when the operator is wearing clothes contaminated with pesticides.

6. Children, pets or farm workers must not enter the sprayed fields for the specific period mentioned by the manufacturer of the pesticide sprayed. Even with safe pesticides, at least spray deposits should be allowed to dry up before entering the field.

7. Operators using large quantities should have periodic medical check up and blood tests.

QUESTIONS FOR DISCUSSION

1 Why should there be proper appreciation of the problems of toxic hazards in the use of pesticides ?

2. Why is distinction necessary between innate toxicity of a chemical and the hazards they present to men and animals when used ?

3. What would you define as a hazard due to the use of a toxic chemical ?

4. How do tests on acute toxicity need to be carried ? What considerations need to be taken in this context ?

5. How is control to be exercised in respect of toxic hazards ? How is the same being done in different countries ?

6. How would you rate or categorise toxic effects ?

7. What are the main provisions in the Insecticide Act of Government of India as far as toxic hazards are concerned ?

8. How can the safety to operators be ensured ? What provisions need to be taken in this context ?

CHAPTER 15

PROBLEMS OF PESTICIDE RESIDUE

A large number of pesticides are now used in agriculture for protection against diverse pests. Pesticides being poisonous in nature not only to the target pests, but also to warm blooded animals including men, their residues left over the sprayed surface of the crop, or in the soil have become a matter of concern in respect of health hazards to men and animals and environmental pollution. Presence of toxicants in food and fodder is of considerable importance to every one. Residue level of any pesticide, apart from the physico-chemical properties depends on a number of factors, namely :- (1) crop and its variety with particular reference to morphological features of the leaf, stem, fruit etc. (2) climatic conditions which include temperature and rainfall, the degradation being faster, in many cases under hot and sunny condition and precipitation causing more repaid removal from the aerial parts of the plants to a much lower level; (3) doses of pesticides applied including frequency of application and time interval between two applications which have a direct bearing on the amount deposited and residue left; (4) method of application of pesticides-formulation, nature of applicances used have a direc tinfluence, e.g. granular forms are much less attended with the problems of toxicity residue; with low volume particularly ultra-low volume spraying, problem of toxicity residue is much leas ; (5) treatment of the crop produce, namely washing, drying or cooking etc. is known to decrease the residues considerably; and (6) soil type-PH and texture most pesticides are unstable under alkaline conditions, persistence is longer in acidic and clayey soils than in sandy and alkaline soils.

Thus problem of pesticide residue varies from country to country depending on agroecological conditions, as well as cropping conditions and intensity. An excellent review of work done in India on pesticide residues has been given by Agnihothrudu and Mithyantha of Rallis India Limited.

A brief account of some of the findings in India is given in this chapter.

A. Chlorinated hydrocarbons

Being highly persistent in the environment, in recent times there has been a popular clamour agianst use of these insecticides, but they are still in use in many countries of the world including India for their effectiveness against a wide range of insects, comparatively low cost and much less acute toxicity as compared to many pesticides and virtually no danger being involved in handling them. Deshmukh *et al* (1977) and Gupta and Gupta (1977) are of the opinion that inspite of its persistent and residual action, usefulness of DDT exceedes its harmful effects. Hence it is recommended to be safe for use in India, not only in agriculture, but also for public health and veterinary purposes.

Agnihotri *et al* (1974) after a study of vegetables and high fat food stuffs in Delhi market observed that most samples examined by them contained either BHC or DDT, about 10 per cent having both. About 25 per cent showed close to or above tolerance limit. Most of the milk samples contained DDT higher than tolerance limits. Concentration of DDT varied from 1.1 to 0.8 ppm in butter and 5.01 to 25.7 in oils. In eggs only thirty per cent was found to contain DDT, but the concentrations were less than tolerance limits.

Similar studies at Hyderabad by Laxminarayan and Menon (1975) showed wide variety of food stuffs to contain either BHC or DDT. Out of 1316 samples, 65 samples contained DDT and 235 BHC, many of which exceeded tolerance limits prescribed by WHO.

Jain and Gupta sprayed brinjal (*Solanum melongena*) fruits with 0.01 per cent lindane solution at 350 l/ha. Analysis of fruits by bioassay for residues showed that the initial deposit of 10 to 11 ppm decreased to level below tolerance limit in one day.

Studies on deposit of BHC residues on leaves of yellow sarsoon (*Brassica campestris*) for control of saw fly (*Athalia proxima*) showed that the residues fell below tolerance limit of 5 ppm by the third day. Washing the plants with water reduced the residues nearly 50 per cent.

In a series of experiments conducted over three years under two different agroclimatic conditions in Rajasthan on carrot, radish and beet grown in soil in which BHC was incorporated at varying doses, it was observed that the crops raised with normal doses of BHC were fit for human and cattle consumption. In all cases insecticide concentration inside the roots was below the tolerance limits. Study on spraying of spinach leaves (Spinacea oleracea) with 0.1 per cent BHC and DDT showed that the residue level of DDT and BHC decreased appreciably in 3-7 days. Washing reduced the residues by fifty per cent or more.

Soil application of 5 per cent BHC dust to cowpea and green gram showed that 97.5 percent was lost in 100 days and residues in the seeds at harvest was below the tolerance limit.

Soil drenching of sugarcane field with 0.1 per cent lindane at the rate of 1 Kg a.i./ha before planting did not result in any detectable residue in the sugarcane juice.

Studies on persistence of DDT and lindane in the soils of Delhi as well as on BHC in clay loam soil in Udaipur had shown rapid degradation of the toxicant.

Spraying of pea (Pisum sativum) with 0.1 per cent lindane twice in the formation and development of pods showed increase in residue of toxicant in the fruit but decrease in the plants as a whole.

Cyclodienes

Cyclodienes are known to be persistent and many members of this group have high mammalian toxicity.

Application of 5 per cent aldrin on crops like green gram, cowpea, maize, bajra showed rapid degradation in the soil as well as inside the plants into which it was translocated as dieldrin. More insecticide was removed from clay loam than in sandy loam. Repeated application however resulted in increase of residues.

Endosulphan

Spraying of endosulphan on okra fruits, cauliflower curds, showed that there was a steady decline in the level of residue with time. In majority of cases, the residue level fell below the tolerance limit, washing and cooking further reduced the quantity, but some workers reported that the residues were not reduced below tolerance limit in many cases even with washing and cooking. In some cases high volume spraying was found to be responsible for greater deposit and better results in respect of residue problem were noticed with low volume spraying. Consistent results were not obtained by different workers on the residue on the fruits below tolerance limit at the time of harvest.

Organo-phosphorus

Organophosphorus compounds are usually readily degraded on the sprayed surface, inside plants, soil and water and pose much less problems of toxicity residue and environmental pollution. But many of them have high mammalian toxicity, improper use of these chemicals may cause problems.

Chlorfenvinphos

Application of chlorfenvinphos in the soil to the potato crop at the rate of 4 kg a.i./ha did not leave any residue on the tubers at the time of harvest.

Diazinon

Application of diazinon either as granules in the soil or spray formulation in paddy crop at different days after transplantation showed presistence of the insecticide above tolerance limit even after 50 days of application.

Granular application in eggplant and cowpea however showed that the residues were less than tolerance limit 10 days later.

Disulphoton

Granular application of this chemical at the recommended dose in egg plant and cowpea crops was observed to result in accumulation in the plant residues higher than prescribed tolerance limit. Similar application to mustard crop showed that the plants had tendency to accumulate the chemical in the apical portions, the insecticide, however, could not be detected in the seed. It dissipated faster under irrigated condition both in the plant and soil. Similar observations were noticed on soybean crop treated with five per cent granules wherein chemical could not be detected in seeds at the time of harvest the residues were maximum and above tolerance level at 50-65 days after application.

DDVP

Applied as grain protectant, the chemical was found to be absorbed only on the surface of the grains. Residual toxicity on treated wheat and gram lasted for a short period ranging from 17 to 20 days.

Dimethoate

In cabbage and cauliflower receiving drenching spray of 0.03 and 0.04 per cent emulsions, residues were higher on leaves, particularly wrapper in cabbage than on heads and curds. Dissipation which was independent of the spray solution concentration, but varied with plant part and temperature was faster in heads and curds than on leaves. Residues got reduced to below WHO tolerance limit of 2 ppm in 5-9 days in heads and curds and 9-14 days in leaves. On mustard residue fell below tolerance limits within 5-10 days and it could not be detected on seeds. In guava and peach sprayed with 0.1 and 0.05 emulsions of dimethoate it took 3 days on peach and 1 day on guava for the residues to fall below tolerance limits. Washing the fruits under running tap removed the residues considerably.

With soil application of 5 per cent granules at the rate of 2 kg a.i./ha to cowpea, the residues were less than tolerance limits 10 days after application. It was noticed that the effect of dimethoate on insects lasted upto 10 days in cases, where spraying of the chemical was undertaken. Four sprays of 0.03 per cent dimethoate on grapes left a residue much below the tolerance limits, 15 days after the last spray.

Estimation of dimethoate in green pods of French bean sprayed regularly with 0.06 per cent dimethoate once in 10 days showed that the residues fell to a narrow range within 2 days and got decreased below tolerance limit in 10 days. Boiling the fruits reduced the residues considerably. Similarly green chill fruits and tomato sprayed with dimethoate once in 10 days showed fast degradation below tolerance limit within 10 days.

Fenitrothion

Where low volume spray was undertaken, initial deposit was low and a large portion of it was lost within 7 days of application. In case of high volume sprays its persistence above tolerance limits was noticed upto 2 weeks after application and a waiting period of 2 weeks was suggested.

In cauliflowers sprayed with 0.95 per cent solution, residues were more than tolerance limit of 0.2 ppm on curds for 9.5 days (by bioassay method) and 14.8 days (colorimetric method) and on leaves 13.8 days (by bioassay) and 16.0 days (as per colorimetric method). Washing and cooking reduced the residues upto 60 per cent. Similar treatment on cabbage revealed that the residues were not more than tolerance limit on hearts for 3 days by bioassay method and 6 days on leaves. Washing and cooking further reduced the amount.

Malathion

Studies on residues on cole crops showed that the type of formulation did not have much differential effect, degradation was faster under higher temperatures, low humidity and bright sunshine. On okra, residues were observed to fall belowt he tolerance limit within 2 days after treatment. Washing the fruits removed 74.49 to 89.15 of the residues, open cooking 58.3 to 86.82 per cent, steam cooking 57.82 to 75.97 per cent and dehydration 91.79 to 91.86 per cent. Studies on a number of crop plants showed that the initial deposit was below tolerance limit or fell below the tolerance limit in 2 to 3 days.

Monocrotophos

Residual toxicity of monocrotophos was found to last for 9-11 days. Considerable reduction in residues occurred due to washing and boiling.

Ethyl and Methyl Parathion

Results with ethyl parathion broadly indicated that the residues fell below tolerance limits after 7 days when applied on lower doses (0.05 per cent), but in high doses, it took nearly 11 days for the same. Residues were more on the leaves than on the pods or fruits or curds in cauliflower. Washing, rubbing and cooking considerably reduce the spray deposit to the extent of 84 to 87 per cent. In market samples in many cases, no parathion could be detected on cabbage. Methyl parathion has been found to be of comparatively less persistence in case of cauliflower, washing and cooking resulted in 40 to 65 per cent loss of residues in both ethyl and methyl parathion. In case of spray applications on bhindi, tomato and brinjal, degradation below detectable limits was noticed by 10 days. Both washing and cooking considerably reduced the residues.

Phorate

On brinjal and cowpea treated with phorate granules in the soil at the rate of 2 kg a.i./ha, it was observed that the residue level on both crops was above the tolerance limit for more than 3 weeks and it decreased to levels below tolerance limit only after one month of treatment. The insecticide, when applied to the soil with mustard crop, dissipated faster under irrigated conditions (45 days) as compared to unirrigated conditions (60 days). In soybean, phorate persisted in leaves and pods at 65 days after soil treatment and persistence might be upto 80 days when applied at the time of sowing. However at the time of harvest, no residue of phorate could be detected. It required 50 days on moong and 60 days on arhar for residues to fall below tolerance limits. Two applications of 1.25 kg a.i./ha to rice at 15 and 45 days after transplanting resulted in residues of 0.07 to 0.08 ppm in hulled grain, upto 0.15 ppm in straw and nil in bran.

Phosphamidon

Residue work on phosphamidon on cabbage, cowpea, mustard and bhindi showed presence of 0.5 per cent residues subsequent to 9 to 12 days after spraying 225 to 400 g. a.i./ha. Studies on bhindi by other workers showed that it took 10 days for the deposits to fall below the tolerance limit of 0.5 ppm. Similar results were obtained in cowpea. Foliar application on mustard showed accumulation on top leaves than on lower portions.

Quinalphos

Residual toxicity was found to last for 4 to 6 days. Waiting period of 9 days is usually suggested for cabbage, cauliflower and knolkhol. Dissipation of quinalphos is faster on cauliflower

Carbamate

Carbaryl

A number of studies conducted on the initial deposit of residues and the level of degradation clearly showed that in majority of cases, the initial deposit immediately after spraying was in many cases below tolerance limit. In all cases, there had been fast dissipation and even where the initial deposit was above the tolerance limit, the residue concentration fell below the tolerance limit within 10 days, actual time varied in individual cases. Granular application however in many cases resulted in greater concentration and effect persisted for longer period upto 20 days. Leaves showed greater deposit in comparison to fruits. Uptake as well as dissipation was not influenced by doses applied. Washing the fruits even after one hour of spraying was found to bring about considerable reduction in residues which may vary from 50-70 per cent, Cooking reduced from 64 to 100 per cent. Four spray applications of 0.1 per cent carbaryl to fodder maize left a residue of 52.4 ppm initially which was dissipated to levels below detectable limits after 7 days. Feeding the contaminated fodder to milch cows showed a residue of less than 0.05 per cent in milk, which is much below the WHO tolerance limit of 0.2 ppm for milk. In all cases of application of carbaryl to grain or vegetable crops residues were not detected in the grains or fruits.

Carbofuran

Carbofuran, a broad spectrum systemic and contact insecticide and a nematicide is used extensively for soil treatment and seed treatment to some extent. A number of crops ranging from cereals, millets to vegetables and fruits have been treated. The treatment usually gives a good protection upto a period of 4 to 6 weeks. Metabolites of carbofuran such as 3-hydroxy carbofuran, 3-keto carbofuran and carbofuran phenols are less toxic than parent compound. It is known from the literature that the harvest time toxic residues are negligible and well below tolerance limits, even though it has a fairly low LD 50 value.

Schedules of carbofuran applied to the soil as granules at the rates of 0.40 to 1.33 kg. a.i. per hectare, or as seed treatment at the rate of 5 per cent a.i. as showed no build up of residues in the grain or straw in the sorghum crop.

Analysis of leaves, stems and pods of soybean exposed to soil application of granular carbofuran (Furadan 3 G) did not show any detectable amount of insecticide or its metabolite after 65 days in all parts in the kharif season and 85 days in the spring season, indicating that no residue would be present in the shoot or grain at the time of harvest.

Studies on carbofuran residues in fruits of brinjal plants treated with Furadan 3 G at the rate of 1, 3 and 10 g per cent (Corresponding to 0.26, 0.78 and 2.61 kg a.i./acre) harvested at varying periods from 7 to 75 days after application showed that the residues ere 0.088 to 0.171 ppm in the fruits after 7 days and it decreased to traces to 0.027 ppm after 75 days. Residues were higher with higher doses of treatment and decreased with advent of time. Boiling the fruits in salt water removed large portion of the residues.

A large number of crop samples including paddy (grain and straw), sorghum (grain and straw), potato (tubers), tobacco (leaves), maize (grain and straw), brinjal (fruits), banana (fruit), cotton (seed and lint) receiving varying levels of carbofuran either as soil application 0.21 to 1.4 kg a.i./acre or as seed treatment at the rate of 5 per cent on seed weight basis have been analysed for residues. It was observed that even with application of high doses and under a wide range of agroclimatic and soil conditions and on a number of crops, residues at harvest were less than the tolerance limits.

Fumigation and grain storage pesticides

Based on residue levels of fenitrothion, malathion, iodofenphos and tetrachlorvinphos in wheat flour, it was concluded that the samples treated with fenitrothion at the rates of 4 and 6 ppm, iodofenphos at 15 ppm and malathion at 30 ppm could safely be consumed after 84, 112, 42 and 84 days of treatment respectively. Based on the laboratory tests to study persistence of pesticides against pulse beetle on gram seed, it was recorded that the efficacy of malathion. DDVP, methoxychlor, thanite. pyrethrin were u to 41, 17, 44, 31 and 27 days respectively.

CHAPTER 16

LIMITATIONS OF CHEMICAL METHODS OF PEST CONTROL

With progress in the knowledge of chemistry from the beginning of the twentieth century, there has been a gradual shift towards chemical methods of pest control, as the conventional methods, were not found to be adequate enough to meet the situations particularly in cases of serious outbreaks of pests. With the discovery of DDT, gamma-BHC and other chlorinated hydrocarbons from 1940 onwards and their phenomenal success in tackling some of the serious pests affecting agriculture and health of men and animals, chemical methods of control began to be used almost exclusively and still continue to be so. With the progress in synthetic organic chemistry, new products are gradually appearing in the market with more spectacular performances. In many countries, crop protection methods are still confined to the use of pesticides.

Chemical methods of control have been found to be extremely effective in controlling pests, particularly insects in the last three decades. A very rapid progress has been achieved in the development of more target-specific, stable, systemic and persistent toxicants which have been profitably utilised in the protection of crop plants against serious pests which has not only stabilised the yield, but also increased production. Unfortunately the use of chemical pesticides has created a number of problems which have indicated their limitations for pest control. One of the major problems pertinent to the use of chemicals is the development of strains or species resistant to various pesticides. The other major problem is the toxicity hazards resulting in environmental pollution and affecting the health of human beings and livestock.

According to Brown and Pal (1971), some 130 species of arthropods of agricultural and veterinary importance and 102 species of importance to human health have been found to be resistant to chemical insecticides. In fact there are very few major insects that have not shown resistance to one group or other of insecticide in some part of the world. Once a particular pest or pests begin to show resistance to any particular category or group of pesticides, the normal customary practice has been to switch over to another group of pesticides. Sometimes the change over has been so frequent that alternatives may be exhausted, as insects can show resistance to many insecticides simultaneously.

It is not quite correct to state that resistance of insects to insecticides was unknown before DDT came into existence in 1940 and this phenomenon has been observed following large-scale use of DDT and other chemical pesticides. Insect resistance to hydrogen cyanide and lead arsenate was known to exist as early as 1908. The problem has been focussed on account of the rapidity with which resistance of insects against these potent chemicals, has developed. Insects showing resistance to DDT were found to be markedly resistant to the cyclodiene group of insecticides, gamma-BHC and later to organophosphorus compounds and carbamates. Naturally the situation is serious. The resurgence of malaria due to failure to control mosquitoes by DDT and other pesticides in many countries has caused concern.

Insecticides do not confer any resistance to the insects, nor do the insects become adapted to the insecticides. Insecticides also do not act as mutagens in developing resistant strains, the exception being in the case of rodenticide warfarin attributed to be responsible in the production of resistant mutant strains of *Rattus norvegicus*. Insects do not exist in nature as a homogenous population, but a mixture of strains. Strains or species susceptible to the action of insecticides are rapidly el ninated from the mixed population with the result that the resistant strains which constituted a very insignificant per cent of the population begin to multiply very quickly due to the elimination to a large extent of intraspecific competition. Besides resistance to insecticides being genetically controlled, subsequent generations of resistant strains, are produced. Resistance to many insecticides, e.g., DDT, dieldrin, organophosphorus compounds, carbamates is genetically controlled. A single gene, usually dominant, or recessive in some, appears to be involved in these cases and modifying affect of minor genes on the principal genes may also be present.

Georghiou (1965) based on his studies on inheritance pattern of resistance of insects to DDT, dieldrin, organophosphorus compounds reported that 13 cases out of 17 on inheritance of resistance to DDT was monogenic, in respect of dieldrin 20 out of 23 and organophosphorous three out of seven. Brown and Pal (1971), basing on similar studies observed monogenic inheritance of resistance to DDT in 13 species (dominant two, recessive eight, intermediate three) ; to dieldrin in 16 species (one dominant, 15 intermediate) ; to organophosphorus in five species (four dominant, one intermediate) and to carbamate in two species, dominance being observed in both the cases. Strains resistant to DDT and dieldrin appear quickly in comparison to organophosphorus compounds.

Different mechanisms are involved in detoxification of different groups of insecticides. Cross resistance has also been noticed where a single mechanism is able to detoxify two or more related groups. Mechanisms of resistance to DDT is associated with that of methoxychlor, similarly resistance to gamma-BHC and cyclodiene group of insecticides is often linked together. There are cases of multiple resistance where insect species can show resistance to a number of groups of insecticides.

The limitation or handicap in the use of organic insecticides due to the development or appearance of strains of insects resistant to action of one or more groups of insecticides may be counteracted for the present in two ways, namely use of : (a) two different groups of insecticides alternately, or (b) formulations consisting of two different groups of insecticides. In case of the house fly, use of DDT and malathion alternately for a number of generations did not yield substantial results, on the other hand, strains resistant to both DDT and malathion appeared. Similar results were obtained when a mixture of DDT and malathion was used. Addition of synergistic compounds has been suggested to reduce the problem

17

of development of resistance to insecticides. Incorporation of synergists like sesamin, piperonyl butoxide has been found to be useful in respect of DDT and carbamates.

In recent times emphasis is being laid on nonchemical methods of control to alleviate the probem of resistance. Usefulness of these methods will be discussed in subsequent chapters. Till such time chemical pesticides can be completely replaced to a large extent, only a cogent solution appears to lie in the development of newer chemicals. These chemicals need not be more toxic, but will operate through different pathways so that build-up of resistance may be delayed.

In respect of a few systemic fungicides, the appearance of resistant strains of fungi has been reported. In many such cases, point mutation appears to be involved. Strains of *Candida* spp.—dermatophyte showing increased tolerance to amphotericin B and nystacin have appeared. They are characterised by decreased sterol content which contributes to their resistance. This condition has arisen out of slight changes in the gene mechanism responsible for sterol production. Strains of yeast *Saccharomyces cerivisae* have appeared which are tolerant to cycloheximide which acts by inhibiting protein synthesis on the ribosomes. In tolerant strains, in the process of gene mutation, the composition of ribosome has been slightly altered without interfering with protein synthesis. Similarly in *Coprinus lagopus* an agaric, in relation to ethionine an antimetabolite of methionine, strains have arisen with resistance to ethionine due to mutation in the structural gene which forms messenger RNA coding for this enzyme. Acquired resistance of a number of fungi to a number of systemic fungicides, namely benomyl, thiophanates, oxathiins, kasugamycin, blasticidin-S, chloroneb, kitazin, ethirimol, dimethyrimol have been reported.

Resistance of micro organisms to antibiotics have been known for a long time. Plant pathogenic bacteria are no exceptions. *Stellaria media*, a weed has been reported to develop resistance to 2, 4-D compounds.

Apart from the appearance of strains resistant to action of insecticides, application of chemical pesticides is attended with the danger of appearance of new pests which have been of minor importance and resurgence of pests in more severe form due to the effect of insecticides on parasites and predators which normally take a longer time to multiply in comparison to pests. Pesticides applied to control harmful pests are applied in an environment where besides destructive ones, beneficial ones are present. Such applications often upset agro-ecosystem and the effect is reflected in many ways. When chemical pesticides appeared since 1940 onwards, and made rapid progress within a very short time in the control of pests of importance to both agriculture and health by their quick action, cheap cost, convenient formulation, etc., undesirable effects produced or likely to be produced were largely overlooked in view of the benefits derived. Gradually, more potent pesticides—systemic in action have appeared in the market. Use of many of them is attended with hazards to health of men and animals including fish and honey bees. Environmental pollution and hazards to the health of men and animals, due to large-scale use of chemical pesticides are causing concern to biologists, environmental scientists and those concerned with the health and hygiene of the people.

Widespread death of fish has been reported in inland water reservoirs, including ponds, tanks, rivers, estuaries due to the direct contamination of the water with pesticides either in mixing of pesticides for spray preparations, washing of equipments and clothes contaminated with pesticides or runoff from agricultural lands where the chemicals have been applied. Chlorinated hydrocarbons and a number of other compounds are highly toxic to fish. Contamination by copper fungicides with arsenic compounds in minute quantities as impurities has caused mortality of fish.

Persistence of pesticides in the soil after their direct application as control measures ; or deposition due to dusting, spraying or drift of the pesticide is likely to cause problems. This aspect has not been properly looked into. The main reason for not taking the harmful accumulation of pesticides in the soil into consideration is the assumption that pesticides are degraded in the soil. Pesticides may be decomposed or degraded in the soil by microbial action, epioxidation, or may be inactivated due to the binding action of soil colloid. Fate of pesticide in the soil apart from its composition is also influenced by ; soil importance and resurgence of pests in more severe form due to the effect of insecticides on parasites and predators which normally take a longer time to multiply in comparison to pests. Pesticides applied to control harmful pests are applied in an environment where besides destructive ones, beneficial ones are present. Such applications often upset agro-ecosystem and the effect is reflected in many ways. When chemical pesticides appeared since 1940 onwards, and made rapid progress within a very short time in the control of pests of importance to both agriculture and health by their quick action, cheap cost, convenient formulation, etc., undesirable effects produced or likely to be produced were largely overlooked in view of the benefits derived. Gradually, more potent pesticides—systemic in action have appeared in the market. Use of many of them is attended with hazards to health of men and animals including fish and honey bees. Environmental pollution and hazards to the health of men and animals, due to large-scale use of chemical pesticides are causing concern to biologists, environmental scientists and those concerned with the health and hygiene of the people.

Widespread death of fish has been reported in inland water reservoirs, including ponds, tanks, rivers, estuaries due to the direct contamination of the water with pesticides either in mixing of pesticides for spray preparations, washing of equipments and clothes contaminated with pesticides or runoff from agricultural lands where the chemicals have been applied. Chlorinated hydrocarbons and a number of other compounds are highly toxic to fish. Contamination by copper fungicides with arsenic compounds in minute quantities as impurities has caused mortality of fish.

Persistence of pesticides in the soil after their direct application as control measures ; or deposition due to dusting, spraying or drift of the pesticide is likely to cause problems. This aspect has not been properly looked into. The main reason for not taking the harmful accumulation of pesticides in the soil into consideration is the assumption that pesticides are degraded in the soil. Pesticides may be decomposed or degraded in the soil by microbial action, epioxidation, or may be inactivated due to the binding action of soil colloid. Fate of pesticide in the soil apart from its composition is also influenced by : soil ecosystem, resulted in the death or elimination of many organisms and the balance has been upset. A simpler ecosystem may emerge but that will tend to be unstable.

In view of the drawbacks or limitations of the use of chemical pesticides, considerable interest has been given in the past decade on the use of such non-chemical methods which are not attended with toxic hazards. Some of the methods are well established and have been used profitably in the past, but have now been relegated to the background; some are in the experimental stage; while others are still in the conceptual stage.

The use of chemical pesticides has been decried for their adverse effect on the ecosystem, human and animal health and environmental pollution. There has been a demand on a total ban on the use of certain categories of pesticides, particularly chlorinated hydrocarbons. This has already been done in a few countries. Yet concept of wholesale banning of the use of chlorinated hydrocarbons has not been accepted. The application of chemical pesticides will continue to be a very effective method of tackling pests in spite of the various limitations.

QUESTIONS FOR DISCUSSION

1. Why has there been a shift towards chemical methods of control in recent years?
2. Which are the major drawbacks of the chemical methods of control? Which of them is the most important?
3. How does the application of chemical pesticides adversely affect an agroecosystem?
4. What conditions do you consider responsible for the appearance of resistant strains following the application of some insecticides?
5. How far is the resistance to pesticides in insects genetically controlled? How can the problem of the appearance of strains resistant to insecticides be overcome?
6. What other adverse effects are attended with large scale

application of pesticides apart from the appearance of resistant strains ?

7. Do you recommend the banning of the use of certain categories of insecticides ? What reasons would you adduce in favour of or against the banning of the use of insecticides ?

8. Why will chemical methods continue to be used in large scale in spite of limitations ?

Chapter 17

BIOLOGICAL CONTROL

For a number of reasons, particularly due to the awareness in recent years of the effects of chemical pesticides on agro-ecosystem and environment, the biological control of pests-insects and pathogens, has gained increased momentum and acceptance. Biological control is essentially based on the exploitation of mutual antagonism that exists in nature amongst different organisms influencing the dynamics of natural population. The concept of biological suppression of control rests on the findings that the density of many pest species is subject to reduction by ecological manipulation of suitable biological or environmental processes so that the pests do not find conditions suitable for their multiplication and thriving. Though the practical application of biological control is of comparatively recent origin, interactions between different organisms in an ecosystem must have been existent for the maintenance of a stable population in an environment. Insect predation was observed and recognised for a long time no doubt, but this knowledge has not been utilised for pest control till recently.

The first known use of the method which may be termed as biological control in the modern sense was the practice of the Chinese citrus growers of the introduction large yellow ants with long legs (probably *Oecophylla smaragdina*) into orange groves to protect oranges from "wormy fruits". A similar method of pest control was used by date-growers of Yemen who used to bring colonies of ants from the mountains and put them into date-palm trees.

The idea of biological control gradually germinated in the sixteenth to eighteenth centuries when insect predation began to be recorded in a scientific manner. During this period observations and descriptions of occurrence of parasitism and predation in insects were noted. Various suggestions on utilisation of natural enemies and the natural control of insect pests gradually emerged. Successes were claimed in a number of cases to illustrate : the use of lady beetles for the control of aphids and other phytophagous pests ; and the use of the predatory beetle *Calosoma sycophanta* in suppressing the gypsy moth. Transfer of beneficial insects—parasites or predators—was done from one country to another on an international level with the object of introducing insects for biological control.

Success of the control of insect pests by biological method was achieved in the case of the control of the cottony cushion scale (*Icerya purchasi*), which was threatening citrus plantations in California, by Vedalia beetle *Rodolia cardinalis* introduced from Australia by Koebele in the last decade of the nineteenth century. Quick success in California led to the adoption of this method in Florida, Hawaii in the U.S.A., South Africa and other countries.

In Hawaii in 1905, the control of leaf hopper of sugar cane *Perkinsiella saccharicida* was attempted with a series of beneficial insect introductions starting from egg parasites from Australia. Ultimate success was achieved in 1920 with the mirid bug *Cyrtorhinus mundulus.*

Attempts were made in the U.S.A. to control the gypsy moth *Porthetria dispar* and European corn borer *Ostrinia nubilalis* by biological methods with the introduction of parasites. During this period, chemical methods of control were not very efficient, hence considerable attention was paid to biological control.

Towards the end of the nineteenth century, when control of insects with other insect parasites was gaining importance, considerable interest was generated in pathogens of insects particularly entomophagous fungi. *Beauveria globulifera* was tried for this purpose against a number of insects and spore masses were discharged in the neighbourhood of insects for the purpose. These attempts met with ambiguous success and were eventually abandoned in scepticism.

The programme of control of insect pests by the use of biological methods received a setback from 1940 onwards with the discovery of DDT and other organic chemical insecticides and their massive use with spectacular success. However, with the gradual consciousness of the undesirable effects of use of chemical insecticides, there has again been a sudden development in the interest for biological control.

Biological control has been broadly defined as the encouragement of beneficial organisms already existing in a locality or of the introduction of suitable new species of exotic parasitic organisms, which are parasites on harmful insects in a locality where the pest is thriving with a view to control the pest. It has been observed that very often with the introduction of improved types of exotic plants, pests have also been introduced usually without the parasites normally associated with the pest in its original locality. Such pests, in the absence of parasites, multiply and reach such a density, without natural control of population and they pose serious problems. Hence the introduction of parasites from the original locality of the plant may be a practical proposition of pest control.

Whatever may be the case, whether any indigenous or exotic species is proposed to be utilised for a biological control programme, it should possess certain desirable attributes. The beneficial organism in question must have ecological requirements similar to those of the target insect. Closely linked with the ecological compatibility which ensures that the pest and parasite can thrive in the same habitat, there should also be temporal compatibility or synchronisation, which will bring the two contemporaneously. Other desirable attributes are : (1) density responsiveness—both numerical and functional ; (2) high reproductive potential through either short generation time or high fecundity or both ; (3) capacity to search for target insect, which is again determined by its power of locomotion, preception to its host, survival and aggressiveness and persistence, particularly in cases of low density residual pest population ; (4) ability to disperse in space so as to be in close proximity to the target insect ; (5) host specificity and compatibility ; (6) unexacting food requirements ; (7) culturability ; and (8) lack of susceptibility to hyperparasitism or attack of predators.

Once the parasites or beneficial organisms have been identified, the next step is their propagation or multiplication. Propagative increase of these insects will require complete knowledge of their life cycle, biology, behaviour, proper rearing conditions, suitable laboratory host or artificial diet, in case the target insect or other insect is not amenable to laboratory testing or not available. Extensive investment in rearing facilities involving both equipments and labour is necessary. After identification of the appropriate beneficial organism, and arrangements for rearing and multiplication, the next step is release of the same in nature so that colonisation in the environment may take place. Normally an early establishment with minimum colonisation will reduce the cost. Sites for introduction should be carefully selected which will provide for the requirements of the beneficial organism. Extremes of climates should be avoided in the new environment so that physical conditions may not limit the effectiveness of the beneficial organism. There should also be synchronisation of the introduction of the parasite with the appropriate phase of the life cycle of the host. It is difficult to determine the biotic potential needed for colonisation. However, the release of larger colonies at a few points rather than lesser number at larger points is likely to meet with success. Multiple releases may be necessary to find out the available conditions for selection. Ready availability of the artificially reared individuals may also be a limiting factor. Beirne very recently has concluded that greater the number of release, greater is the likelihood or possibility of establishment and if the number is small—below 5000, then the success likely to be achieved will not be significant. If the environment is stable and discrete, as in orchards, then greater success is likely to be achieved with smaller numbers, but the environment of field crops, where particularly monoculture is practised, is comparatively unstable, hence to achieve success under such conditions, a larger number has to be released. With smaller numbers, the rate of success will be smaller. Whenever any release is made, voucher specimens which will be representative samples of the beneficial organisms released should be kept for comparison in follow-up recoveries. To determine the success or otherwise of the introduction of beneficial insects, regarding their establishment, colonisation and eventual suppression of pests, studies on follow-up recoveries should be made. It should begin at the end of a year of introduction and be continued at least for three successive years.

Any programme of biological control of pests needs very careful evaluation regarding its effectiveness. This can be done satisfactorily in three ways : (1) qualitative analysis, (2) experimental exclusion procedure, and (3) quantitative evaluation. Qualitative analysis essentially relies on the observations on the progress of the beneficial organism introduced regarding its establishment, multiplication and spread as would be determined from extensive samples frequently drawn. In this approach, however, normal mortality and fluctuations in pest population as well as of parasite in relation to environment have to be discounted. Qualitative methods may lead to inconclusive or misleading results, if conclusions are not properly drawn. In the experimental exclusion approach, a comparative study is made of the population density of the pest in the absence and presence of the introduced parasites, in two adjacent plots—one receiving parasite and other without the same and in retrospect when parasite was not introduced. The best method is the quantitative mathematical analysis of the extensive life system information or data of pest population before and after the introduction of the parasite. With advances in computer technology, such methods can be adopted for critical evaluation of the programme.

The biological control programme has its limitations. It cannot be the only basis of protection of plants against injurious pests in all cases. It has to be suitably integrated with other measures, the aim being to develop an overall approach in tackling pests. It has been suggested that biological control is likely to be more useful in controlling indirect pests which can tolerate high pest density and have a high economic injury level. It may not give the desired effect in controlling direct pests which may cause injury at low density. Pests in the tropics and geographically restricted areas or islands are more liable to suppression by biological means as would be evident from successful examples of such control in these areas. Successes have been achieved also in the temperate regions and in continental areas. There are two schools of thought whether an "individual" species considered to be the best should be used or multiple species of potentially beneficial organisms should be introduced. In general the multiple species approach has been accepted, because it has been found to be practical and safe as supported by most population models. It is generally accepted that biological control is likely to be useful in respect of monophagous or narrowly oligophagous insects as is seen from records of successful instances and in view of the general belief that highly specific entomophage is biologically well adapted to host-parasite interaction.

The comparative value of parasite and predators in biological control programme has often been argued. It is felt that the role of predators is comparatively insignificant which are usually polyphagous and highly mobile, though both parasites and predators act in a similar fashion in relation to population density or in other words show similar density-dependent response. Considering their reproductive potential, searching capacity and dispersal ability, as well as adaptations, hardly any preference can be given to predators, on the other hand predators may not be easily available as compared to parasites. Parasites usually have been found to be more effective than predators because of their specifity in host choice and life cycles synchronous with the target insects.

Agents in Biological control

Agents for biological control programme can be any type of suitable living organism—These include vertebrates—mammals and birds; invertebrates—insects and mites and microbial agents:—bacteria, viruses, fungi and nematodes. Amongst the vertebrate agents, the earliest ones used are cats to control rodents. The degree of success using vertebrates is, however, fairly low, as they are not sufficiently specific and specialised in their feeding habits. Amongst the invertebrates, insects are the most commonly used biological control agents. They are usually very highly specialised feeders. Insects act on the host in two ways either as predators or parasitoids. Predators generally attack both larval and adult stages. Parasitoids commonly only kill one host during the life-cycle. They tend to be more specialised in their host range than predators. There are more parasitoids available for use than predators. Most insect pests have at least one parasitoid. Microbial agents are made up of bacteria, viruses and fungi. Most information available refers to pathogens which attack on insects. Some nematodes attack insect pests. They are usually parasitic for part of the life cycle within the body cavity of the host causing death on departure.

Most work on biological control has been on the control of insect pests of crops. Control agents are generally insect predators and parasitoids or pathogens. To be successful at controlling pests, predators must have high specificity, high searching efficiency, greater reproductive capacity than prelongevity and temperature/activity response similar to that of prey. Examples of control of insect pest by predator is suppression of the cottony cushion scale, *Icerya purchasi* by the coccinellid *Rodolia cardinalis*, coconut scale *Aspidiotus destructor* by the coccinellid beetle *Cryptoganths nodiceps*, glass house red spider mite *Tetranychus urticae* by the mite *Phytosciulus perssimilis*. Successful parasitoid characters apply to the egg-laying female. These qualities include: host specificity; high searching effeciency; development timed to coincide with the susceptible stages of the host life cycle; and oviposition discrimination, to prevent multiple parasitism of the host. Examples may be cited of control of woolly apple aphid *Eriosma lanigerum* by *Aphelimus mali*, glass house white fly *Trialeurodes vaporarium* by parasitic wasp *Encarisa farmosa*, Mexican bean beetle *Epilachna varivetis* by *Pedobius foveolatus*.

There are different ways in which a biological control agent can be used; namely introduction, inoculation, inundation, augmentation and conservation.

Introduction. When a pest organism is introduced from another country, it is usually not accompanied by its natural enemies. Classical biological control involves the visiting of the country of origin to obtain specimens of natural enemies, which, after screening is introduced into the country to 'control' the new pests and to maintain the reduction so that pest is no longer causing the damage. It normally aims at permanent solution of the problem.

Inoculation involves the seasonal release of the control agent. The method is useful when the introduction method fails due to control agent dying out after reduction in pest numbers. The aim is to establish control over a shorter period of time than in case of introduction, such as seasonal requirements e.g. against glass house pests such as redspider mite and whitefly.

Inundation. Involves the mass release of the control agent to control a particular pest outbreak at a particular time. It is once- and for-all treatments. There is no possibility of establishment of control agent, the function is like a pesticide. This method is most commonly used with viruses and bacteria e.g. *Bacillus thuringensis.*

Augmentation. Under certain conditions natural control agents, although present, achieve control too late in the growing season, because of slow population build up. Therefore, by adding other control agents early in the season control is achieved in time to prevent pest populations which are maintained at low level until the natural enemy populations are sufficient to take overcontrol.

Conservation—involves the main manipulation of the agro ecosystem to make it more beneficial to control agents, and enable them to perform better e.g. reduce the use of pesticides to the minimum.

Insects. Five orders in which parasites have been mainly recognised are the *Coleoptera, Diptera, Hymenoptera, Lepidoptera* and *Strepsiptera.* Of these, for insect pest suppression, insects belonging to *Diptera* and *Hymenoptera* have been widely used. Sixty-six per cent of the successes in biological control has been attributed to the introduction of Hymenopterous insects.

Among the successful cases besides the already reported : cottony cushion scale of citrus *Icerya purchasi* by *Rodolla cardinalis* California, U.S.A. there are more examples : control of hispid beetle *Promecotheca reichei* a serious pest of coconut palm in Fiji by an allied hispid beetle *Pediobius parvulus* ; black scale of citrus (*Saissetia oleae*) by *Metaphycus lounsburyi* in South Africa; citrus mealy bug *Pseudococcus* spp. by lady bird beetle *Cryptolaemus montrouzieri* in California, U.S.A.; control of leaf hopper of sugar cane *Perkinsiella saccharicida* by mirid bug *Cyrtorhinus mundulus* in Hawaii; glass house red spider mite *Tetranychus urticae* by predacious mite *Phytoseiulus persimilis*; banana skipper *Erionota thrax* by egg parasite *'Ooencyrtus erionotae*; and larval parasite *Apanteles erinotae* in Hawaii, suppression of greenhouse white fly *Trialeurodes vaporariorum* by *Encarsia formosa.*

Trichogramma spp. which are egg parasites are being produced on mass scale and are liberated in many countries of the world particularly in the U.S.S.R.; Mexico; Mainland China; Europe; California, and Texas in the U.S.A.

The control of myrmecine ant (*Pheidole punctulata*) and ant belonging to genus *Crematogaster* and arboreal ant *Oecophylla longinoda* has resulted in the suppression of mealy bug of coffee *Pseudococcus lilacinus* and *Pseudococcus njalensis*, the vector of swollen shoot of cacao respectively. Ants protect these insects for the honey dews they secrete and once this protection is removed, occurrence of mealy bugs is only sporadic.

Control of *Lantana camara* by insects in Hawaii, prickly pear (*Opuntia* spp.) by cochineal insect *Dactylopius opuntiae*, moth *Cactoblastis cactorum*, and red spider mite *Tetranychus opuntiae* are excellent examples of biological control of weeds where insects have been employed for their suppression.

Although no example of successful biological control of pests by animal organisms other than insects is available, there are possibilities that such organisms may help in suppression of insects under natural conditions, namely centipedes (*Lithobius fortificatus, Poabius bilabiatus*) upon the symphylid *Scutigella immaculata*, nematode *Monochus* sp. to sugar beet nematode *Heterodera schachtii, Tylenchinema oscinellae* against fruitfly *Oscinella frit* etc. The role of birds, frogs and other amphibians, in natural control of insects is also known.

Control of insects by pathogenic organism

Metchnikoff (1878) was probably the first to observe

bacterial disease of insects. He noted the presence of bacterium *Bacillus salutaris* in the diseased larvae of the beetle *Anisoplia austriaca*. d'Herelle may be called a pioneer in the microbial control of insects who used bacterium *Coccobacillus acridiorum* according to modern nomenclature *Cloaca cloaca* var. *acridiorum* isolated from desert locust *Schistocerca pallens* for the control of locusts. Findings of d'Herelle were however not confirmed by later workers.

Bacillus popilliae and *B. lentimorbus* have been found to incite two distinct types (A and B) of milky disease of Japanese beetle *Popillia japonica*, respectively. Of the two types of milky diseases of Japanese beetle, type A caused by *B. popilliae* has received attention because of its importance in the control of insects.

In recent years, bacteria which form toxic crystals at the time of spore formation have drawn considerable attention from the standpoint of insect control. Crystals, which have often been designated as parasporal bodies are formed at the time of formation of endospores. They are diamond shaped, but may be rhomboidal or cuboid. Chemically they are proteins containing at least seventeen per cent nitrogen and 17 amino acids, but no phosphorus. These crystals are highly toxic to certain insects, particularly some *Lepidoptera* but inocuous to others. The best known species is *Bacillus thuringiensis* which has been extensively used for the microbial control of insects. Efficiency of *B. thuringiensis* has been attributed to four toxic compounds found in the bacterial cell or in the cellular filtrate. Apart from endotoxin contained in the parasporal crystalline body other toxic compounds are lecithinase C, a thermostable toxin—a nucleotide linked complexly through ribose and glucose to all nucleic acids which appear to involve nucleotidases and DNA-dependent RNA polymerases, involved with ATP, and thus preventing RNA synthesis and the remaining one—an unidentified toxin, probably enzyme which clears egg yolk agar.

Mass production of *Bacillus popilliae* needs living organisms, for which only grubs of Japanese beetle *Popillia japonica* are used, although the bacterium is able to attack and thrive upon other scarabeid grubs. *B. thuringiensis* is readily culturable *in vitro*. For commercial mass production the bacterium is grown in bran-based semisolid medium or in submerged cultures using a powdered formulation which contains spores, crystals and exotoxins. The medium is inoculated with the bacterium in its logarithmic or active vegetative phase, and is provided with mechanical agitation to provide aeration. Temperature for optimal growth is 30°C at neutral pH. After 24-48 hours of inoculation collection of spores and toxins is made from the liquid medium by centrifugation. The concentrated residue or precipitate is mixed with stabilising agents and marketed as flowable concentrate. In case of semisolid preparations, brancake is airdried and powdered for use. The final product needs to be standardised, by suitable bioassay techniques which involves comparison with standard preparations or test organism such as larvae of housefly. Products of *B. thuringiensis* can be marketed as wettable powder, dusts, granular preparations and flowable concentrates. Adverse effects of use of product of *B. thuringiensis* on the environment are minimal. There is no toxic residue, and resistance of target insects against it has not yet been reported. Commercial preparations of *B. thuringiensis* are being marketed for nearly quarter of a century and have been found useful in the suppression of pests belonging to *Orthoptera, Coleoptera, Lepidoptera, Diptera* and *Hymenoptera,* including many pests of vegetable crops. It has been found to be effective against many forest insects including gypsy moth.

The viruses have been found, in recent years, to be a promising group of pathogenic entities for use in the biological control of insects. The important role of viral diseases of insects as a natural control factor in population dynamics has been recognised in the past and attempts have been made to utilise this phenomenon for control of insect pests. Only in the recent years, viruses have been successfully used for the purpose. More than 450 viruses have been described from 500 *Arthropod* species—majority being *Lepidoptera* (83 per cent), *Hymenoptera* (10 per cent) and *Diptera* (4 per cent). Six different types of viruses have been recognised so far : (a) nuclear polyhedral viruses, (NPV) ; they develop in host nuclei and they are occluded in a polyhedral proteinaceous body containing one to many rod shaped virions, which are composed of double strands of DNA and capable of parasitising any cell nuclei and cause death ; (b) granulosis viruses, (GV)—which develop in either nucleus or cytoplasm of host fat, tracheal or epidermal cells and are occluded in oval shaped bodies, each containing a rod shaped virion similar to NPV virion, (c) Cytoplasmic polyhedrosis viruses (CPV) ; they develop only in the cytoplasm of host midgut epithelial cells and spherical virions of double stranded RNA are occluded in oval bodies and do not always cause death, but retard larval growth and reduce adult longivity and fecundity ; (d) entomopox viruses (EPV) recently discovered, and have a similarity to vertebrate pox viruses ; and (e) non-occluded irridescent viruses (IV) in which virions are not associated with the inclusion bodies.

Occluded viruses are often very stable outside the host and may have a 'shelf life' for many years. They can persist in the field for a long period. Non-occluded viruses are generally unstable outside the host. Nuclear polyhedral viruses have received the most attention for control of insect pests. They are host specific with few exceptions. Most of the efforts have been with NPV and GV. Much of the successful work on the introduction of viruses for control of insect pests has been carried out in Canada and the U.S.A. against sawflies and a number of forest and agricultural insects. Mass production of viruses, has not reached commercial level except in a few cases. Very recently a registered product under trade name 'Elcar' has been approved in U.S.A. for use against the corn ear worm *Heliothis zea* and tobacco bud worm *H. virescens*. Application of virus at the rate of 2.9×10^{11} Polyhedral inclusion bodies (PIB) per hectare has given good control of *H. zea* on sorghum ; on cotton eight to nine applications per season at 1.5×10^{12} PIB/ha has yielded good results comparable to the use of insecticides both from standpoints of efficacy and cost. One of the must successful attempts to control insects by polyhedrosis virus preparations is that of alfalfa caterpillar in California by spraying polyhedra at the rate of five million polyhedra per ml and 50 litre per hectare. Thus entomophagous viruses may prove to be useful in microbial control of insect pests, the limiting factors of mass production being solved by pilot plant production of larvae needed for multiplication of viruses and adoption of tissue culture technique.

Fungal diseases of insects have been recorded in the nineteenth century. Insect mycoses have been caused by fungi belonging to *Phycomycetes, Ascomycetes, Basidiomycetes* and *Fungi Imperfecti* of which genera in *Phycomycetes* and *Fungi Imperfecti* have been mostly considered for insect suppression. Genera *Beauveria, Metarrhizium* of *Fungi Imperfecti*; *Entomophthora* and *Coelomyces* of *Phycomycetes*; *Cordyceps* of *Ascomycetes* have received most attention. For nearly a century, attempts have been made to reduce insect populations through the introduction of fungi parasitic on insects, but continued success has not been achieved though fungi concerned are capable of producing resistant spores for carry over through the unfavourable periods and of rapid spread in an insect population. Uncertainties of environmental conditions play a decisive role in the spread, multiplication and parasitisation of the target species. Besides density of host population, congenial microclimate for the host and parasite, resistance of the host and level of pathogenicity of the fungus, also play important part.

Though theoretically entomogenous protozoa offer considerable promise in control of insects, as a large number of protozoa are parasitic on insects, nevertheless it has not been possible to utilise them in biological suppression of insects because of the production difficulties and of the fact that they produce chronic, but not acute symptoms.

Biological control of plant diseases

Biological control of plant diseases has been defined by Garrett as any condition under which, or practice whereby survival or activity of pathogen is reduced through the agency of any other living organism (except man himself) with the result that there is a reduction in the incidence of the disease caused by the pathogen. It has been postulated that biological control can be achieved either by the introduction or augmentation in number of one or more species of controlling organisms, or by a change in the environmental conditions designed to favour multiplication and activity of such organisms or by a combination of both methods.

Biological control has been mostly attempted in case of pathogens operating in the soil where a microbiological equilibrium is maintained in relation to a soil agro-ecosystem. That the activity of a pathogen could be inhibited by an accumulation of its own metabolic products was shown by Potter in 1908, but significance of Potter's observations was not then fully appreciated. Sanford in 1926, by his work on control of scab of potato (*Streptomyces scabies*) by green manuring showed the possibility of biological control of soil-borne diseases. Works of Sanford led to a concise hypothesis of the biological control of soil-borne plant diseases in which it has been postulated that saprophytic microorganisms can control the activity of pathogens and microbiological balance in the soil can be changed by altering soil conditions which may include the incorporation of organic matter. From 1930 onwards, work of Weindling and his associates on the parasitism by *Trichoderma viride* of other soil fungi particularly through the production of antibiotics further advanced the concept as well as practical feasibility of biological control of soil-borne diseases.

Biological control of plant diseases through a living organism mainly operates by its action on the pathogen, which is commonly designated as antagonism. Antagonistic action may be broadly divided into three categories : antibiosis, competition and exploitation. Antibiosis in the broad sense denotes suppression of pathogenic organisms due to the production of metabolic agents by other micro-organisms having a harmful effect on pathogens. Competition is due to active demand of nutrient over supply, a situation which results primarily from quicker and greater utilisation of available nutrients by saprophytic micro-organisms with the result that pathogens which are normally poor competitors face lysis or supression of activities due to starvation. Exploitation is the direct parasitism or predation of other organisms over pathogenic ones.

The mechanism of action of *Trichoderma viride* which has been reported to be capable of suppressing pathogens has been postulated as due to the production of antibiotics, gliotoxin and viridin. Nevertheless there have always been controversies as to whether antibiotic production in the soil has any significant effect on microbiological balance in the soil. Parmer while summing up the observations of many workers in this field observed a lack of convincing evidence to support the thesis that antibiotics produced under natural conditions are capable of exerting a significant and generalised effect. The inability to detect any significant quality has frequently been cited as the strongest argument against activity of antibiotics in the soil in suppression of pathogenic organisms. Brian (1949), doubted any significant role of toxin in control of soil-borne pathogens mainly due to two reasons, namely these toxins have been found to be produced by the organisms in question, *in vitro* in rich medium—a condition not likely to be present in soil and many of the toxins are fairly unstable. From the findings of a large number of workers it may, however, be concluded that antibiotic or toxin production can take place in the soil n the vicinity of organic substrates, which are rich carbon ources, like straw or seed coats ; antibiotics or toxins may be extracted from the soil ; ability to secrete antibiotic toxin *in vitro* does not signify antibiosis in soil ; metabolic products connected with antibiosis may be inactivated in a number of ways in which adsorption plays an important part ; antibiotics may be slowly released from the adsorbed condition. Evidences however are present that organisms capable of producing antibiotics *in vitro* may form a substantial portion of soil microflora varying quantitatively or qualitatively with crop sequence. Positive evidence of the role of antibiotics can be concluded on correlation between suppression in disease associated with increase in antibiotic producing organism. The inability to show antibiotics production *in vitro* does not indicate that the organisms are not capable of adversely affecting soil-borne pathogenic organisms. *Trichoderma* and some other fungi have not been found to show distinct antibiosis against *Verticillium in vitro,* but in gnotobiotic cultures with *Trichoderma* there was considerable reduction in wilt incited by *Verticillium.* Antibiosis may be due to the release of large quantities of carbon dioxide during decomposition of crop residue, which is likely to affect some pathogens like *Rhizoctonia, Fusarium oxysporu* f. *cubense.* Influence of volatiles of plant origin in the soil (soil-borne pathogens has been indicated in some recent works.

Biological control of soil-borne pathogens has also be attempted by exploiting competitive saprophytic activity of mic organisms in the soil. Factors determining good saproph' ability include high growth rate and rapid germinability of re productive structures ; good enzyme producing ability, production of antibiotic toxins ; and tolerance of antibiotics produced by other micro-organisms. Competitive saprophytic ability is often manifested in quicker growth of micro-organisms in comparison to pathogenes thereby depleting levels of available carbon, nitrogen and vitamins which are necessary for the germination of propagules of pathogenic organisms and their initial sapro- phytic growth prior to parasitism. *Fusarium solani* f. *phaseoli* is very sensitive to supply or availability of nutrients. Root disease incited by this pathogen can be reduced to a great extent by the incorporation of carbohydrate-rich material in the soil. In California this is achieved by turning in residues from a barley crop. Soil micro-organisms utilise much of the available nitrogen in decomposing the barley straw. Control by this method can be nullified by the addition of nutrient or by in- corporation of soybean or alfalfa residues which have a low carbon/nitrogen ratio. Many soil organisms, such as strains of *Streptomyces, Nocardia, Pesudomonas, Bacillus,* isolated from soil are capable of causing lysis of hyphae of *Fusarium* spp., *Penicillium chrysogenum.*

There have been evidences in the laboratory of direct parasitism of one or more fungi or other organism on other fungi or other organism. Parasitism by destructive mycoparasites by their direct penetration into or coiling round the host hyphae has been observed in *Rhizoctonia solani* on various other fungi and *Penicillium vermiculatum* parasitising *Rhizoctonia solani.* *Didymella exitialis* is able to penetrate and kill hyphal cells of *Ophiobolus graminis*—incitant of take all disease of wheat. *Gliocladium roseum.* parasitises and destroys conidia of many species of fungi.

Most of the evidence of natural occurrence of mycoparasi- tism in soil is circumstantial. Attempts to control root diseases through mycoparasitism have been so far not very encouraging.

Attempts at biological control of noxious nematodes with parasites or predators have been limited almost exclusively with predacious fungi. Results obtained so far by the use of predacious fungi for biological control have neither been impressive or consistent or in effectiveness are not of the same level as attained by fumigation. Though *Dactylaria* spp. have been noted to be good nematode-trapping fungi, but protection against nematodes by these fungi has been far from adequate. Activity of these fungi in soil has not been fully studied or understood. Amendment of soil with organic manures has been found to be beneficial in suppressing the reproduction of plant parasitic nematodes and has shown the possibility of protecting host crops against a possible attack from them. While there is ample indication that organic manure may stimulate biological control agents, as yet no definitely dependable method is available for large-scale adoption.

Methods that have been proposed for the biological control of plant diseases where positive results have been obtained include inoculation of soil or plant tissues in the soil with antagonistic micro-organisms. But its possible applications are limited. Change in microflora of soil can be effected for a temporary period by such introduction and the former equilibrium is restored quickly. Unlike the biological control of insects, where exotic species can be easily introduced, it is, however, very difficult to find a suitable exotic micro-organism in the soil for biological control. This can however be achieved by modification of soil environment. 'Wiendling and Fawcett (1934). showed that citrus seedlings could be protected against attack of *Rhizoctonia solani* through soil inoculation of *Trichoderma viride* by acidifying the soil, thereby making conditions suitable for establishment of *T. viride* at high population level. This modification can be brought about by the incorporation of organic matter in the soil. A number of diseases including root rot by *Fusarium solani*, potato scab, etc., have been controlled by such incorporation. This method is suitable against specialised parasites with restricted host range, but not against primitive parasites with a wide host range and remarkably well adaptations for saprophytic activity.

Cultivation of decoy crop illustrates the example of utilisation of higher plant species for biological control. A decoy crop for a particular pathogen may be defined as one the roots of which will stimulate germination of dormant propagules of that pathogen without themselves being sufficiently susceptible to allow fresh formation of infective propagules. *Datura stramonium* has been used as a successful decoy crop against *Spongospora subterranea* powdery scab organism of potato. Planting potatoes from plants of *Solanum tuberosum* sub sp. *andigena*, resistant to golden nematode *Heterodera rostochiensis* has resulted in eighty per cent reduction in a year and egg population to the level of ten per cent of the original value in four years.

Biological control measures may be operative in several ways. A direct effect on the host may result from inoculation with a non-pathogenic or mildly virulent organism. When wounded roots of tomato seedlings were exposed to *Cephalosporium* spp. vessels were obstructed by mycelial growth of very weakly parasitic *Cephalosporium* sp. and later infection by *Fusarium oxysporum* f. *lycopersici*—incitant of wilt was very limited, as the latter fungus could spread inside the tissues to a limited extent. Severity of wilt disease of tobacco incited by *Pseudomonas solanacearum* was considerably reduced when a large proportion of cells of an avirulent strain was introduced along with the virulent ones. These results, however spectacular they might be, have not been put into practical use at yet.

It has already been stated that the addition of organic manures can control root rots incited by many fungi, as under such conditions, growth of saprophytic organisms is encouraged with the depletion of nutrients resulting in starvation of the pathogenic organism.

Many fungi have chitinous cell walls and laminarin or closely related polysaccharide is also present. When laminarin or chitin is added to the soil, there is a marked increase in the population of *Actinomycetes* which are thought to be responsible for lysis of fungal cell walls in such amended soils. By chitin amendment of soil, the severity of several diseases including *Fusarium* wilt of peas, *Rhizoctonia* stem rot of beans has been reported to be reduced. But damping off caused by *Pythium debaryanum* cell walls of which do not contain any chitin cannot be controlled by this method.

Control of potato ssab (*Streptomyces scabies*) by green manuring has often been successful. In many cases, control for the future years has been achieved by growing a leguminous crop after potato and turning them in whilst still green.

Legumes may result in control in a different way. When legumes grow under good conditions of illumination, they take up more nitrogen from the soil than they add. As nitrogen is depleted from the soil, the pathogen dies out rapidly in the infected stubbles or plant parts in the soil. Hence sometimes a legume mixture may be helpful in reducing diseases caused by soil-borne pathogens.

Effect of saprophytic micro-organism on pathogenic ones may sometimes be enhanced following selective chemical or heat treatment of soil. Partial sterilisation of soil by some agent which will favour selective survival or recolonisation of the treated soil by a micro-organism having the desired antagonistic properties has been found to be attended with biological control. Following the fumigation of soil with carbon disulphide to eliminate *Armillaria mellea*, it was observed that the killing of *A. mellea* was not due to direct fungicidal action of the fumigant, but to the lethal effect of a much augmented soil population of *Trichoderma viride* due to selective action fumigation on soil microflora.

It has been observed that long established cultural methods continue to give excellent results, which are largely due to the encouragement of microbial growth. Ir spite of accumulation of considerable knowledge on diverse aspects of soil-borne diseases in plants, the biological suppression of pathogenic organisms has been achieved only partially and has not been found to be acceptable as the only method. It may have to be integrated along with other methods. Where a biological buffering does not pose much problems, greater success is likely to be achieved with microbial antagonism. Probably there is ample scope of an integrated system of control combining both chemical and biological methods.

QUESTION FOR DISCUSSION

1. Why has biological control of pests gained momentum in recent years ?
2. What is meant by biological control of pests ?
3. Why is the introduction of a parasite from the original locality of the pest recommended ?
4. What should be the desirable attributes of a species to be utilised for biological control ?
5. What are the different steps to be taken in the introduction of insects for biological control ?
6. Why is the introduction of lesser number of beneficial insects more likely to give desired results in orchards rather than in field crops ?
7. Why is the evaluation of a programme of biological control considered essential ? How is it attempted ?
8. Why are pests in the Tropics and geographically isolated areas likely to be more easily suppressed ?
9. Against which types of insects or in which cases would biological control be more successful ?
10. What do you consider to be the limitations of biological control ? Why is it suggested that biological control should be integrated with other methods of control ?
11. Which do you consider to be more efficient in biological control—parasite or predator ?
12. Apart from suppression of other insects, have insects been employed for the control of other pests ?
13. Can insects be controlled by pathogenic organisms ?
14. How can and how has *Bacillus thuringiensis* been utilised for the control of insects ?
15. Can entomophagous viruses be utilised for the control of insects ?
16. Do you think that the control of insects by entomogenous fungi is a possibility ?
17. What do you understand by the biological control of plant diseases ?
18. In which ecological group of fungi has biological control been attempted ?

19. How does biological control through living organisms operate ? What are the mechanisms of such antagonistic actions ?

20. How far is production of antibiotics responsible for the control of pathogenic fungi ?

21. What do you understand by competitive saprophytic activity ? How can competitive saprophytic activity be utilised for biological control of plant pathogens ?

22. Do you think that control of nematodes by fungi is a practical proposition ?

23. How can the modification of soil environment be helpful in the biological control of plant diseases ?

24. What is a decoy crop ? How can decoy crops be utilised in the biological control of plant diseases ?

25. How can inoculation with a non-pathogenic or mildly virulent organism be helpful in the control of plant diseases ?

CHAPTER 18

AUTOCIDAL AND OTHER METHODS OF CONTROL

ιne biological methods of control or suppression of insects deal with the reduction or elimination of the population by management or manipulation of external biotic components of the environment operating under normal principles of population dynamics. The internal constitution of the biological environment of insects can be changed to the detriment of the insects or in other words genetic make up of the insects may be so manipulated as to reduce its fertility or reproductive ability with a demonstrable effect on population.

These methods may be concerned with the release of sterile, sterilised or incompatible insects into a wild population with the object of producing a progressively increasing proportion of sterile matings with gradual reduction in population of wild insects with the possibility of its ultimate extinction. Male sterile populations can be secured artificially by exposure of insects to ionising radiations or to certain chemicals.

Apart from the induction of sterility in males, hybridisation among closely related species may result in the production of sterile insects. Cytoplasmic incompatibility that may be present may also be taken recourse to. The result will be that one sex of certain species can be released capable of compatible mating with the production of fertile offspring with its own opposite sex, but incompatible with the opposite sex of another population of the same species and there may be replacement of population if not extinction.

Whatever may be the approach or methods adopted, the principal aim is to produce a sufficient number of healthy, competitive and sexually aggressive insects which may be of genetically different types. They may be released at appropriate time and place to enable them to mate successfully with wild insects. Methods of mass rearing, mass · sterilisation or other genetically manipulative techniques and the knowledge of ecology and population dynamics of the insects concerned are the prerequisites of the success of this method.

Male sterile method

This method of biological control was first suggested and developed by Knipling (1960, 1967) in respect of the screw worm *Cochliomyia hominovorax*. The hypothesis on which this method has been developed is that the reproductive capacity of many insects is barely enough for the perpetuation of the species under normal condition. Significant reduction in their successful reproductive ability will result in decrease of the population. If male sterile individuals could be produced and released in sufficient number among the wild populations of the species over a period of several generations, then there will be a progressive reduction in reproduction with the possibility of the ultimate disappearance of the natural population. In theory, more rapid control of insect population can be achieved by male sterile techniques than by application of insecticides.

Male sterility has been achieved by exposure of the pupae of a fly to a predetermined dose of gamma radiation from a cobalt-60 source. Ionising radiations have now been replaced by powerful chemosterilants which are principally either strong alkylating agents or derivatives of aziridine. The alkylating agents commonly used are TEPA [tris (1-aziridinyl) phosphine oxide], METEPA [tris (2-methyl-1-aziridinyl) phosphine oxide] and apholate [2, 2, 4, 4, 6, 6-hexahydro-2, 2, 4, 4, 6, 6-hexakis (1-aziridinyl)-1, 3, 5, 2, 4, 6-triazatriphosphorine]. Other aziridine derivates used experimentally are mostly sulphonyl derivatives namely 1, 1'-sulphonyldiaziridine ; 1,1'-sulphonyl bis (2-methyl aziridine) ; 1, 1'-dithiodiaziridine ; 1-1'-sulphonyl bis (2-methyl aziridine) 1,1'-sulphonyl diaziridine, bis (2-methly-1-aziridine) sulphide. These compounds have shown acute mammalian toxicity to mice when aqueous solutions were introduced intravenously. Sulphonyl derivatives of aziridine show toxicity values of LD_{50} less than 6 mg per kg in some cases to 20 mg per kg. They are more toxic than TEPA which has LD_{50} value greater than 300 mg per kg. Irradiated insects usually exhibit reduced sexual competitiveness, but chemosterilants are likely to have less adverse effect on longevity and mating aggressiveness. On the other hand use of chemosterilants results in more sexual activity. The chemicals are cheap and can conveniently be applied. Costly and laborious mass-rearing technique can be avoided as chemosterilants can be directly used in nature. Uniformity of results as obtained by ionising radiations may not be achieved in these cases because toxicity problems have to be encountered. Both the methods of ionising radiations and use of chemosterilants produce the same result, namely, dominant lethal mutation with sexual sterility in the treated individuals. However the most important factor for the achievement of success in the sterile insect technique is the ability of the natural population to respond to deliberate reductions in its fertility by a compensatory increased likelihood of individual survival of the remaining fertile insects.

Introduction of males sterilised by irradiation has been successfully employed in fruit flies ; oriental fruit fly (*Dacus dorsalis*) and Mediterranean fruit fly (*Ceratitis capitata*), to some extent in melon fruit fly (*Dacus cucurbitae*), onion fly (*Hylemya antigua*), cotton boll weevil (*Anthonomus grandis*) and cockchafer (*Melolontha vulgaris*), cabbage looper (*Trichoplusia ni*), corn ear worm (*Heliothis zea*), gypsy moth, codling moth (*Laspeyresia pomonella*).

Chemosterilants do not appear to have been used as frequently, as in domestic insects, for the control of insects of agricultural importance. Mexican fruit fly (*Anastrepha ludens*) has been successfully controlled by TEPA sterilised male insects. Chemosterilants have been used in sterile male production for the control of cotton boll weevil (*Anthonomus grandis*), red boll worm (*Diparopsis castanea*), and pink boll worm (*Pectinophora gossypiella*).

Limited trials with chemosterilant-cum-bait combinations have been conducted with agricultural pests namely olive fruit fly (*Dacus oleae*), onion maggot (*Hylemya antigua*) and cabbage root fly (*Erioischia brassicae*) in which apholate 0.4 per cent) or TEPA (0.1 per cent) have been used with baits.

Apart from the induction of sterility in male population either by irradiation or by the use of chemosterilants, which is however not inherited, the introduction of partially sterile insects with capability of transmitting sterility to the offspring has been conceived of and put into trials in the laboratory and in very small scale in the field mainly with insects of public health importance. The methods involve reduction in reproductive ability by the induction of hybrid sterility, irregularities in chromosomal translocation and cytoplasmic sterility by suitable genetic manipulations. Theoretically these methods appear to have tremendous potentialities, but they are yet to be tried and assessed in the field scale particularly with pests of agricultural importance.

Pheromones

The term "pheromone" was used by Karlson and Butenandt to designate chemical substances produced by an organism exogenously to influence the behaviour or physiology of other members of the same species. They are exocrine products with a number of specific functions amongst the insects.

Pheromones which have been classified in a number of ways by different authors have been found to influence diverse activities in both social and nonsocial group of insects. They may be of potential use in the control of insects. Amongst them sex pheromones have been the subject matter of much attention in recent times. Sex pheromone is concerned with direct facilitation of mating. It is normally produced by one sex and affects the other sex ; sometimes both sexes may be affected. In some cases it may be produced by both sexes. In general, substances produced by females serve as attractants and those produced by male as aphrodiasiacs. Female sex pheromones appear to be more useful because they are effective over long distances.

Identification of the chemical nature of sex pheromones have been made in a few cases. Gypsy moth sex pheromones have been identified and synthesised as cis-7, 8-epoxy-2-methyloctadecane known as disparlure. Silk worm attractant has been identified as trans-10-cis-12-hexadecadienol (bombykol), pink boll worm as 10-propyl-trans-5, 9-tridecadienyl acetate, cabbage looper as cis-7-dodecenyl acetate and honey bee 9-keto-trans-2-decenoic acid. The compounds must be volatile, and specific. These are released at specific times and insects have been found to respond to an exposure of pheromone only for a short period.

Sex pheromones may be profitably used either for pest survey or pest control. Use of pheromones for the detection and survey of pest population has been adopted in various countries of the world for determination of the abundance of the pest as an aid to plan other control measures. Control of pests, through the use of pheromones is a preventive measure usually operative at low densities. Basing on the phenomenon that sexual behaviour may either be stimulated or inhibited, suppression of insects may be directed in two ways : (a) luring the insects in large number in some traps so as to remove a large percentage of the breeding population, and (b) disruption of the inter-sexual communication system to prevent normal mating behaviour. However, basing on intensive investigation for nearly twenty years on exploitation of pheromones for the control of insects, any practical system of pest suppression basing on pheromones which may be employed with success has yet to be achieved. Mass trapping of red banded leaf roller *Argyrotaenia velutinana* in apple orchards and cotton boll weevil, pink boll worm using appropriate pheromones has been claimed on an experimental pilot scale.

Antifeedants

An antifeeding compound is a chemical substance which prevents pests from feeding on a material treated with this substance without necessarily killing or repelling them. Anti-feeding compounds namely Eulan New ; sodium salt of bis-[α, α-(3, 4-dichloro-2-hydroxyphenyl)]-0-toulene-sulphonic acid, mitin FF, 5-chloro-2 [4-chloro--2-(3, 4-dichlorophenylureido)] benzene sulphonic acid have been known of for sometime and have been used primarily for the prevention of moth damage in fabrics, etc. Zinc salt of dimethyldithiocarbamate acid mixed with cyclohexylamine has been used on the bark of the trees to keep deer, rabbits and rodents from feeding on the bark. Use of antifeeding compounds for the protection of crops is a recent innovation. Natural pyrethrins which have been widely used in the past are known to exhibit antifeedant characteristics. The only compound which has been used extensively as antifeeding compound is 4'-(dimethytriazeno) acetianilide (DTA). Fentin acetate (Triphenyl tin acetate) a fungicide has been found to inhibit feeding of insects.

An antifeedant functions differently from other pesticides inasmuch as it prevents the insects from causing damage instead of killing the insects. It is operative against surface feeding insects, but not against piercing, sucking and penetrating insects. It affords protection only to the sprayed surface, while new growth, as it expands, is not protected. Beneficial insects, parasites and predators are not affected. Mammalian toxicity is usually very low. Antifeedants can be suitably integrated with other biological methods.

Juvenile hormones

The juvenile hormone which prevents the formation of an adult insect within the pupae, was accidentally found in the abdomen of male cecropia moths by Williams (1967). This hormone was found to be effective even if applied to the cuticle of the pupa. Hence it has been considered to be a potentially valuable agent in the control or suppression of insects. Juvenile hormone has been chemically identified as neotenin. A number of compounds, namely, juvabiones, farnesol are found to be naturally occurring. Synthetic chemicals which are structurally similar to them also exert the same influence. The basic principle involved in the use of juvenile hormone lies on its application at the appropriate time on the appropriate target species which will cause eventual disturbances in metabolisms leading to aberrations in reproduction and development and eventual mortality. They incite delayed action, not immediate mortality, hence cannot be employed for the eradication in the case of an outbreak. Besides they are somewhat specific and not universal in action. They need to be applied at an appropriate developmental stage. Hence their potentialities are yet to be brought into commercial realities.

Attractants and repellants

Evidence has been present for sometime that certain stimuli closely related to olfactory ones often attract or direct the insects towards food.

In their struggle for existence, very often in the uncongenial environment, insects have developed various abilities which include a highly specialised sense of smell. This trait is utilised by the insects in finding out sources of food, host plants, members of the opposite sex and places for laying eggs. There are no powerful general insect attractants. The potent chemical attractants are often highly specific being confined to one or a small number of species. These attractants can be used, as in the case of pheromones for detection and survey and pest suppression, either alone or in combination with insecticides. Decomposition products of organic matter or decomposing fruits often serve as oviposition or food lures. But such substances, which are often chemically not properly defined are non-specific and inconsistent in their action. The search for such compounds has resulted in the discovery of methyl eugenol in citronella oil. A number of synthetic compounds have been developed which act as lures namely cuelure, siglure (α-acetoxy derivatives of benzyl acetone), Sec-butyl-6-methyl-3-cyclohexene-1-carboxylate, anisyl-acetone, medlure and trimedlure (secondary and tertiary butyl esters of 4 (or 5)-chloro 6-methyl cyclohexane carboxylic acid, methyl linolenate, butyl sorbate. Most of the compounds are specific as is evident from Table 18.1.

Volatility of the compound is an essential attribute because it imparts effectiveness over a distance. There should also be an optimum concentration for each compound which is related to the volatility. Attractant properties of essential oils have been used to trap the Japanese beetle (*Popillia japonica*), geraniol and eugenol being the most effective compounds. Siglure has been used extensively for the control of the Mediterranean fruit fly by trapping. To be more useful for detection and control operations, the efficiency of insect attractants should be increased so that they can compete with other methods.

Repelling of the organisms by the actions of certain chemical substances has been practised for sometime. Smoke is generated to repel various domestic insects. Naphthalene, p-dichloro-

TABLE 18.1

Name	Species attracted	Other species attracted
1. Methyleugenol	Oriental fruit fly (*Dacus dorsalis*)	(*Dacus umbrosus*)
2. Anisylacteone	Melonfly (*Dacus cucurbitae*)	Queensland fruit fly (*D. tryoni, D. ochrosiae*)
3. Siglure	Mediterranean fruit fiy (*Ceratitis capitata*)	Walnut husk fly (*Rhagoletis completa*)
4. Medlure	,,	—
5. Trimedlure	,,	—
6. Butyl sorbate	European chafer (*Amphimallon majalis*)	—
7. Methyl linolenate	Bark beetles (*Ips typographus*)	—

benzene are examples of moth repellants. During World War II, a standard mixture of repellant consisting of six parts of dimethyl phthalate, two parts of butopyronoxyl, better known as "Indalone" (2,2-dimethyl-6-carbobutoxy-2,3-dihydro-4-pyrone) and two parts of ethyhexanediol (3-hydroxymethyl-n-heptan-4-ol) has been used. For the protection of cattle butoxy-polyprop-elene of glycols C_4H_9 [O. $CH(CH_3)$. CH_2]$_n$ are used. Possibly the most useful repellant for human use at the present is N, N-diethyl-m-toluamide 2-hydroxy-n-octyl sulphide. It is used as repellant for cockroaches. As would be seen that work on synthetic repellants has been carried on exclusively on insects concerned with diseases of men and animals.

There are, however, indications that plants themselves may act as repellants, as certain crop mixtures do not favour the incidence of pests. *Pieris brassicae* never oviposits on cabbage surrounded by tomatoes, carrot fly *Psila rosae* does not attack carrot in a carrot-onion crop mixture, a few plants of *Datura* can repel white flies from tomatoes. Attractant and repellant

19

properties of plants have not been fully investigated. After a thorough study of the same, it may be possible to breed plants for greater attraction towards or repellancy against insects as a measure of control against pests.

Juvenile hormones, repellants, sex attractants, and other biologically active natural or synthetic substances which are able to quite specifically destroy particular pests have been termed as "third generation pesticides". The "first generation pesticides" include chlorinated hydrocarbons with broad spectrum activity and less acute toxicity and organophosphoric esters with a very acute toxicity and broad spectrum. The compounds which have appeared later with less toxicity and more selective effect on target organisms are named as "second generation pesticides". Third generation pesticides appear to work in a manner different from the others, are target specific and virtually non-toxic.

So far physical attractants are concerned, light traps have been used particularly against nocturnal insect pests which are positively drawn towards the source of light. Attractiveness of light towards insect is dependent on wave length, and intensity besides the environmental conditions and position of light. Light traps have been found to be useful for survey and detection of pests. But they have not been successful for the direct control of insect pests. Use of sounds for the control of insects is also not a feasible proposition, because of cost and operational difficulties.

QUESTIONS FOR DISCUSSION

1. How do autocidal methods of control differ from biological control ?
2. What are the advantages of the male sterile technique over other methods of control ?
3. How can male sterility be achieved ? Can male sterility be transmitted to the offspring ?
4. What do you understand by chemosterilants ? Which is more effective in inducing male sterility—ionising radiations or chemosterilants ?

5. What are pheromones ? Which pheromones appear to be more effective in the control of insects ?
6. How can pheromones be used for the suppression of insects ?
7. What are antifeedants ? How do they differ from other pesticides in action ?
8. How can juvenile hormones be effective for the control of insects ? What are the possible limitations of the use of juvenile hormones ?
9. How can attractants be used for the control of insects ? Can they be as effective as chemical pesticides ?
10. Can plants themselves act as repellants to some insects ?
11. What are known as third generation pesticides ? Why are they called so ?

CHAPTER 19

USE OF RESISTANT VARIETIES

Methods for controlling pests so far described are related to their eradication or suppression by the adoption of measures that are directed towards them. Another line of approach which has been found to be very useful in many cases is the genetic improvement of the host plants so that they become resistant to the attack of the pests. Use of varieties which are less amenable to the attack of the pests is not entirely new. Recognition and selection of such varieties showing some degree of resistance in the field must have been coexistent with the improvement of domesticated plants in various ways, though references to such work date only from the eighteenth century onwards when definite findings were recorded. In the pre-Mendelian era, the importance of resistance was recognised. Steps were taken to control late blight of potato, downy mildews and *Phylloxera* of grape vines by the introduction of resistant root stock from the U.S.A. The science of plant breeding which includes resistance against pests really made a beginning after the rediscovery of Mendel's law of inheritance and findings of Biffen (1906) with yellow rust of wheat that resistance to disease is a heritable character.

In recent years, breeding for resistance in plants has assumed greater importance for a number of reasons which may be as follows : (a) Greater homogenity or less diversity in cultivars of major crop plants. Cultivation over wide areas of cultivars of high yield potential with somewhat similar genetic background may create hazardous situations in which build-up of pests can easily be made with the possibility of serious epiphytotics or epizootics. Very frequently these varieties with high yield potential possess such characters namely succulence, richness in available food, abundant foliage that the pests find more congenial conditions for their establishment and quick multiplication. (b) Intensive and multiple agricultural cropping pattern with the availability of more facilities for cultivation in the Tropics and sub-Tropics, where weather is not a limiting factor. (c) Inadequacy of direct control measures in terms of efficacy, availability, cost, toxic hazards and development of resistance towards pesticides.

Use of varieties resistant to diseases and pests is considered to be a permanent solution to a large extent particularly in low value crops grown over an extensive area where direct control measures may not be economical or feasible, or both. In developing countries where the average holdings are comparatively small and very often fragmented, such measures appear to be ideal. Breeding for resistance is considered to be of primary importance apart from control of disease and pests, in reducing hazards of environmental pollution and man made ecosystems. In the present integrated approach on control of pests, varieties even with moderate degrees of resistance or field resistance are expected to play an important role.

The resistance of a plant to pathogenic organisms and insect pests may be defined as the heritable characteristics by which most plants are capable of nullifying or reducing probability of infection or infestation by the pests, their races or biotypes. In a large number of cases the host plant is capable of sustaining an insect population or shows infection at a somewhat moderate level, but the effect of such infestations or infections is not manifested in the reduction of growth, vigour and yield. Such cases are designated under "tolerance". The distinction is often made between tolerant and resistant varieties though similar results are likely to be derived in both the cases. There are also some cases in which the host plants do not possess any intrinsic resistance against pathogenic organism or insect pest, but apparently evade infestation or infection, because of susceptible age of very short duration, or build-up of inoculum or pest population at the appropriate stage is low. Early sowing of varieties sometimes results in escape of damages due to diseases and pests. These come under "pseudo-resistance" or "disease escape". Similarly plants may be susceptible at the young or seedling stage, but they are unaffected when they are fully grown and mature. Such cases are often termed as "mature-plant resistance". It may be pointed out that the use of resistant varieties is essentially a preventive measure. Very often particularly in plant diseases there has been a severe outbreak even with the use of resistant varieties due to one reason or other including the breakdown of resistance.

Resistance against diseases

Diseases in the field are normally held in check by : (a) presence of diversity of crop cultivars both at intraspecific and interspecific levels, (b) host resistance in plant populations, (c) stabilising tendencies in pathogen population, and (d) prophylactic measures which include both chemical and cultural. Interspecific diversity has been considerably minimised, if not eliminated altogether from early phases of agriculture when wild plants were domesticated for economic exploitation. With the hybridisation and selection of pure lines as cultivars, extent of intraspecific diversity has also been reduced to a large extent at least in some cases. Homogenisation of cultivars is being attempted for a number of reasons, instead of diversity. Stabilising tendencies in pathogen population have disappeared. On the other hand, there is always a continuous shift or change in the pathogen or its races or biotypes which is beyond control in most fungi. Hence increase in host resistance in plant populations is considered to be one of the practical methods of controlling diseases.

In developing a programme of breeding for disease resistance there should be a clear concept of the different stages in pathogenesis and the types of resistance that may operate (Table 19.1).

In breeding for disease resistance, the aim should be to develop such varieties that may have preformed characters which may be mechanical, or non-specific chemical substances as inhibitors or can induce enzymatic reactions resulting in inhibition of pathogens or may induce specific host-parasite reaction. Plant breeders and plant pathologists in the past concentrated primarily to evolve varieties with high degree of resistance against specific pathogens or few races and biotypes of pathogens, characterised by specific host-parasite interaction and this type is designated as "vertical resistance". In contrast to vertical resistance. another type is met

TABLE 19.1

Stage Prepenetration	Modes of performance	Resistance induced
(a) Contact	Pseudo resistance, exclusion, escape, hairyness	
(b) Germination of spores	Surface and composition of cell walls, unwettability	Induced excretions or substances detrimental to germination and growth of pathogens
(c) Penetration	Thickness and composition of cell walls	Induced thickness of cell walls
	Stomatal frequency and opening	Occlusion of stomata
(d) Establishment of infection	Physiologic state, acidity of all sap, presence of inhibitors	Specific interactions, hypersensitive reactions, primary reaction activating induced production of inhibitors or barriers, mechanical-tyloses, gums, corks
(e) Host-pathogen stabilisation	Inhibitors, deficiency of nutrients (non-specific)	Specific interactions, continued production of inhibitors (specific or non-specific or induced enzymatic responses polyphenol oxidases, peroxidases etc. or phytoalexins etc.)

which is fairly non-specific with moderate level of resistance against a large number of pathogens or races of pathogens which is commonly known as "field or horizontal resistance". In the former, differential reaction between varieties of host and races of pathogen is observed, whereas in the latter no such specific differential reaction is involved.

Vertical resistance conveys a very specific relationship. It is usually governed by a specific small number of major genes. In this context a gene for gene relationship which has been propounded by Flor working on rust of linseed *Melampsora lini* has gained general acceptance. Flor propounded the hypothesis that the disease reaction expressed in the infected host is dependent on whether or not host individual possesses individual specific genes for resistance and if it does, whether or not the pathogen possesses specific genes for virulence or avirluence and the specific interaction between these two sets of complementary genes one of the host and another of the pathogen will determine the reaction on the host. The hypothesis implies that for each gene conditioning resistance in the host, there is a reciprocal gene conditioning pathogenicity in the parasite. To be more precise, a gene for gene relationship exists when the presence of a gene in one population is contingent on the continued presence of a gene in another population and where the interaction between the two genes leads to phenotypic expression by which presence or absence of relevant genes in either organisation may be recognised.

Gene for gene relationship has been suggested or demonstrated in the following host-parasite systems (Table 19.2).

TABLE 19.2

Rust	
1. Linseed—*Linum usitatissimum*	*Melampsora lini*
2. Corn—*Zea mays*	*Puccinia sorghi*
3. Wheat—*Triticum* spp.	*Puccinia graminis* f. *tritici*
	P. *recondita*
	P. *striiformis*
4. Oat—*Avena sativa*	*Puccinia coronata*
5. Sunflower—*Helianthus annus*	*Puccinia helianthi*
6. Coffee—*Coffea* spp.	*Hemileia vastatrix*
Smut	
1. Wheat—*Triticum spp.*	*Ustilago nuda*
2. Oat—*Avena sativa*	*Ustilago avenae, U. kolleri*

TABLE 19.2 (*Contd.*)

Bunt

1. Wheat—*Triticum* spp.   *Tilletia caries,*
                                             Tilletia controversa

Powdery mildews

1. Wheat—*Triticum* spp.   *Erysiphe graminis* f. *tritici*
2. Barley—*Hordeum* spp.   *E. graminis* f. *hordei*

Other diseases

1. Potato—*Solanum tuberosum*   *Phytophthora infestans*
                                             Synchytrium endobioticum
2. Apple—*Pyrus malus*   *Venturia inaequalis*
3. Tomato—*Lycopersicon esculentum*   *Cladosporium fulvum*

Involvement of genes in different diseases are presented as in Table 19.3.

TABLE 19.3

Disease	Paper reviewed	Major vs Minor gene		Dominance vs Recessive		Interaction	Allelism
Rusts	301	291	19	292	36	43	27
Smuts	108	89	12	100	6	5	1
Mildews	85	85	2	89	10	6	7
Blight	45	43	3	43	4	2	3
Scab	25	24	2	25	0	0	6
Bunts	49	47	4	48	2	4	0
Viruses	75	76	1	63	15	5	6
Other fungal diseases	148	140	13	133	17	12	3
Other non-fungal diseases	82	80	4	78	10	5	2
	912	875	60	862	99	42	59

(Persons and Sidhu).

Regardless of the species that may be involved in the host-parasite interaction, factors for resistance to diseases generally segregate in simple Mendelian ratios and the alleles for resistance are usually found to be dominant over susceptibility. In a smaller number of cases, interactions of genes suggest modification of two factor hypothesis—presence of epistatic factors, etc.

Horizontal resistance is conditioned or governed by a large number of genes, which may be either major or minor genes. It may be termed as polygenic resistance in contrast to vertical resistance, commonly known as oligogenic resistance. Inheritance of polygenic type is rather vague and genes involved are too many to be identified. No single gene has any demonstrably large effect to be recognised. Some of the genes involved with resistance to diseases in polygenic types may also be concerned with the growth and development of the host and resistance may result indirectly or directly from these effects. To cite an example, some of the genes concerned in polygenic resistance to potato blight are also concerned with photoperiodism and duration of the growing period of the variety and resistance is indirectly related to them.

Extent, nature and mechanism of variability of the pathogens also need to be studied, as development of resistant varieties is closely linked with the pathogens some of which are variable in nature.

Though resistance is governed by genetic characters, nevertheless the expression of the same is liable to be conditioned by environmental factors. It has been observed that under adverse environmental conditions, resistant plants show susceptibility. This phenomenon is more explicitly evident in the case of horizontal or field resistance, where a large number of characters are responsible for the reaction of the plants. Expression of these characters take place under suitable environmental conditions. Besides, pathogens show parasitism under specific environmental conditions which are conducive to their germination, growth and establishment of infection. Hence study of environmental factors is necessary.

Development of resistant varieties has its own limitations. Recent works of Horsfall and his co-workers (Wood, 1967) have shown that some diseases are favoured by high sugar content of the host plant, namely, obligate parasites while many other

364 PRINCIPLES AND PROCEDURES OF PLANT PROTECTION

diseases are favoured by low sugar content, particularly leaf blights. It is not possible to combine both the characters, in one variety or cultivar hence many rust resistant varieties are often susceptible to leaf blights incited by *Helminthosporium* spp., *Alternaria* spp., and other related organisms. Southern corn blight caused by *Helminthosporium maydis* which appeared in epidemic form in recent years in the U.S.A. is an example of such a situation. High degree of vertical resistance of crops which has been aimed at to reduce the damages caused by the diseases has resulted in the rapid evolution of new races or biotypes over vast areas and resistant plants have shown susceptibility. Example may be cited of the development of varieties of potato with few dominant *R* genes which has resulted in the formation of newer virulent races, to which the varieties so far resistant succumbed easily. Appearance of races 56 in 1935 which attacked *Marquis 15A*, and *15B* in 1953 and 1954 which were not prevalent before and to which the varieties were not resistant created epiphytotics of stem rust in Northern America. Similar situation has also been observed in many other countries. One of the approaches has been to rely more on horizontal or field resistance. Horizontal resistance has a very useful role to play no doubt, but is governed by a large number of physiological characters which find their proper expression under congenial conditions. Under conditions of high density of inoculum and unfavourable to the host such varieties may not be in a position to show resistance. In case of rust diseases of wheat, it has been suggested by Borlaug that a multilineal approach should be made. A mixture of isogenic lines having similar agronomic characters may be released so that no single race would be able to cause damage. It has been pointed out that this may also lead to natural selection of races of pathogens which may be virulent to all the genes present in the multilines. A judicious combination of horizontal and vertical resistance is advocated so that epidemics over large areas can be avoided.

Introduction of varieties resistant to one pathogen may result in the appearance of other pathogens in a virulent form to cause concern to the growers. When dwarf *indica* varieties of rice were introduced, the most important pathogen was blast incited by *Pyricularia oryzae*. The varieties released from IRRI showed a high degree of resistance against blast disease, but new diseases, e.g., bacterial blight incided by *Xanthomonas oryzae*, bacterial streak incited by *X. oryzicola*, RTV (rice tungro virus) and other virus diseases have made their appearance to complicate the situation. This situation is likely to happen when the varieties have a narrow genetic base.

The development of resistant varieties involves the location of suitable genes which may not be available in the cultivars and has to be searched either in wild ancestors or related species. In the case of potato, resistant genes for late blight were available in *Solanum demissium, S. andigena, S. multidissectum, S. veruei.* For potato root celworm, *Heterodera rostochiensis*, a single dominant gene has been found in certain South American clones of *Solanum andigena.* For yellow rust *Puccinia striiformis* and leaf rust *P. recondita*, resistance is being attempted to be introduced into *Triticum* spp. from *Aegilops umbellulata* and *A. comosa.* Resistance in sugar cane (*Saccharum officinarum* and *S. barberi*) against a number of diseases has been secured by crossing with a perennial relative *Saccharum spontaneum.* Similarly in grassy stunt of rice, steps have been taken to introduce resistant genes into rice from *Oryza glaberissima, O. nivara.* These processes involve considerable time and energy and may not meet with the desired success.

It is easier to evolve varieties resistant to the pathogens which are host specific, but in the case of those pathogens which have a very wide host range the problem is difficult. Examples may be cited of some soil-borne pathogens, e.g., *Macrophomina phaseolina* (= *Rhizoctonia bataticola*), *Sclerotium rolfsii, Pseudomonas solanacearum* which have a very wide host range and cause extensive damage. In case of many host specific pathogens caused by facultative parasites, e.g., *Helminthosporium sativum* on wheat, *H. oryzae* on rice, resistance is conditioned by minor genes, hence complete resistance may not be achieved in these cases.

Widespread use of non-race-specific resistance has led to the more careful study of epidemiology of the diseases against which resistance is sought to find out the most effective way of selecting for resistance. Sometimes, a weak point in the life cycle of the pathogen may be helpful. In breeding for resistance to loose smut (*Ustilago nuda*) in barley, it was found that some varieties are pollinated apogamously, while in others lodicules expand, causing flowers to open at the time of anthesis. As infection is caused by the spores falling on the stigma, apogamous flowers escape infection.

Brian (1972) has stated that due to recent developments in the production of highly specific fungicides, systemic in action in many cases, emphasis need not be laid so much on the development of resistant varieties. Inadequacy of the use of such chemicals for the control of specific plant diseases is borne out by the findings that there has been breakdown of control by these chemicals, because of the evolution of new races of the pathogen concerned. It has been suggested that resistance to infection due to systemic fungicides can be broken down by the appearance of new races of pathogens which may either be adapted or possess specific mechanism to detoxify these fungicides.

In many cases, as in *Citrus* gummosis, compatibility between stock and scion results in control ; similar influence is also exerted in other diseases of citrus, namely, dieback, or tristeza, though the diseases are caused by pathogenic organisms including viruses.

In spite of different limitations, remarkable successes have been achieved in many cases namely wart disease of potato, cereal rusts and smuts, onion smudge, red rot and virus diseases of sugar cane, virus diseases of potato, anthracnose of beans, blast disease of rice, downy mildews and powdery mildews in millets and cereals, etc. While successes have been achieved more easily in the field crops and vegetables, in fruit crops due to obvious difficulties and limitations, spectacular achievements have not been made. New knowledge and recent advancements made in allied fields now show the possibility of having broad based resistance.

Resistance to insect pests

Though empirical knowledge on existence of varieties resistant to pest infestation was available from the later part of the eighteenth century, almost at the same time with information in respect of plant diseases, nevertheless factors affecting or controlling insect pests are much less understood or worked out.

There is a fundamental difference in the mode of parasitism of the disease-producing organisms and insect pests. Pathogens inciting diseases including viruses are typically endophytic except in some powdery mildews. They penetrate into the host tissue, establish themselves inside and incite specific host-parasitic interactions which ultimately result in expression of symptoms and damages caused to the host. In case of insects, the chewing types feed on the plant parts from the surface without entering inside the host tissue. In sucking types, a superficial resemblance with the pathogenesis is noted inasmuch as insects pierce through the host tissues and derive the nutrients from the host tissue, but a stable parasitic relationship, as noted in the host-pathogen combination in plant diseases is not observed.

Resistance to attack of insect pests may vary from complete immunity to complete susceptibility where damage caused is much greater than in an average case. Three basic mechanisms in resistance to insect pest attack have been recognised : (a) non-preference, (b) antibiosis, and (c) tolerance. In cases of non-preference, host plants possess such character or characters which do not encourage or adversely affect insect's response to the host for food, shelter or oviposition. Plants showing resistance due to antibiosis normally prevent damage by exerting an adverse influence on growth and reproduction of insects and disruption of their normal life cycles. As explained earlier, in cases of tolerance, inspite of heavy infestation, plants do not show decline in growth, vigour and yield.

Non-preference usually denotes the index of readiness with which an insect pest intends to utilise a particular host. It may be either relative or absolute. In the former, in the absence of a preferred host the insect may attack the non-preference hosts and in the latter, under no circumstances the host plants will be acceptable to the insects. In the previous chapters, it has been stated that the hosts may have some properties by virtue of which they can incite stimuli in insects and insects may choose these hosts for food, shelter and oviposition as the case may be or they may act as repellants. Failure to incite any stimulus which will attract the insects or possession of repellant properties may cause the basis of non-preference mechanism of host resistance. Chemical or bio-chemical basis of such properties of the host and response of the insects have not yet been fully explored.

The mechanism of action in antibiosis shows a wide range of

effect. There may be death in the early stage, egg, nymph, larvae or instar, or in the adult stage. The reproductive ability of female insects may be adversely affected. There may be shortening of the life period or lengthening of larval period, which may disrupt the normal life cycle of the insects.

Mechanism of antibiosis has been attempted to be explained on the basis of the presence of chemical substances inhibitory or toxic to the physiological activity of the insects. The presence of potent insecticides in many plants, e.g., pyrethrins, rotenones and nicotine lends support to this view. A number of plants contain precursors of potent insecticides, and juvenile hormones. Nutritional imbalance has been suggested in a few cases, e.g., pea. Varieties resistant to pea aphid *Acyrthosiphon pisum* have less free acids than the susceptible ones. Morphological characters may also be responsible, as in case of *I. R. 5* variety of rice which is resistant to stem borer *Scirpophaga incertulas* as the narrow lumen of the stem does not encourage the development of larvae, solid stem varieties of wheat resist attack of *Cephus cinctus*, because eggs are damaged when introduced inside and larvae have restricted growth due to narrow lumen.

Though resistance is governed by heritable characters, nevertheless climatic and edaphic factors play an important part in its expression and modification. This is particularly evident in cases of resistance to insect pests. In relation to uncongenial or nonpreference hosts high humidity facilitates detection of odour to which insects may be attracted or repelled. Similarly conflicting evidences are available to indicate that soil types or soil moisture and fertility status in the soil may have a modifying role particularly in relation to sucking insects. It may be pointed out that expression of such examples of resistance are evidently linked with the physiological processes of the host plants and presumably fall under the category of horizontal resistance. Hence environmental factors and age of the plant are likely to play an important part in such cases. Environmental conditions which encourage faster rate of growth of host plants or are not favourable for multiplication of insects may be indirectly responsible for more resistant reaction. Under no circumstances, however a resistant host plant becomes susceptible or vice versa.

As in the case of pathogens, biotypes exist among the insect pests, though occurrence of such biotypes in insects appears to be less frequent or common. The existence of biotypes in European corn borer *Ostrinia nubilalis* and fruit fly *Oscinella frit* has been indicated. In respect of spotted alfalfa aphid *Therioaphis maculata* where biotypes have been recognised, a gene for gene relationship has been postulated. As in plant diseases, the appearance of new biotypes of insects may complicate the matter. Varieties resistant to brown plant hopper *Nilaparvata lugens* in IRRI, the Phillippines have been found to be susceptible in India, because of the existence of different biotypes of the pest in India. Then the matter of adaptibility in insects has not been fully investigated. If resistance of a non-preference host is based on repellant odour, then it is quite likely that the insect may become adapted or habituated to the odour particularly under conditions of stress or non-availability of food. Varieties resistant to insect vectors of virus diseases may prove to be of great promise. A number of single dominant genes have been detected for some virus diseases of potato. Polygenic resistance in potato leaf roll and two viruses of beet namely Beet Yellow Virus and Beet Mosaic Yellow Virus has been noted. Similar approach has been made in the case of *Myzus persicae*, the most important aphid which is the vector of many virus diseases. Resistance to *M. persicae* has been reported in all crops where a thorough search has been made although levels of resistance are lower than recorded in oligophagous aphids. Steps are being taken to incorporate resistance to this aphid as well as viruses of which they are vectors.

Though the control of insects has been aimed to be achieved before the advent of chlorinated hydrocarbons by use of chemicals and the role of resistant varieties has not been projected as in plant diseases, nevertheless remarkable success has been achieved in many cases. To cite a few : development of grape vines resistant to *Phylloxera, P. vitifoliae* ; varieties of cotton resistant to leaf hopper *Empoasca fascialis* in Africa, *E. devastans* in India and Pakistan and *E. terrareginae* in Queensland, Australia (resistance is related to hairyness of plants) ; wheat to hessian fly *Mayetiola destructor* in North America and Europe. Some degrees of resistance to rice stem borers exists in most long-established rice varieties of tall indica group, which are now being utilised for breeding purposes.

There are however, general limitations in the use of resistant varieties for control of diseases and insect pests. It takes a fairly long time to evolve varieties resistant to diseases and pests. Besides, in order to be acceptable to farmers, such varieties should have a wide spectrum of resistance against a number of major diseases and pests. Such varieties should be decidedly superior in agronomic character including yield potential, otherwise their introduction and popularisation may cause problems. One of the limiting factors in the use of new resistant varieties is the availability of seeds. Multiplication of seeds may not be up to the level of requirement in many cases and there would be a gap. Cost : benefit ratio in resistant varieties is very high and has been estimated as high as 1 : 100 or even higher in some cases besides ensuring a minimum yield guarantee.

QUESTIONS FOR DISCUSSION

1. In which way is the use of resistant varieties considered to be different in approach from control or suppression of pests ?
2. Why has breeding for disease resistance in recent years assumed greater importance ?
3. Why is emphasis given to the use of resistant varieties ? What do you understand by resistance of plants to pathogenic organisms ?
4. Define the terms 'resistance', 'tolerance', 'pseudo resistance', and 'mature plant resistance'.
5. What are the different stages in pathogenesis where resistance to infection of disease producing organism may operate ?
6. What do you understand by 'horizontal resistance' and 'vertical resistance' ? What do they convey ? How do they differ ?
7. What do you understand by 'gene for gene' relationship ? Mention the cases where such relationship has been definitely established.
8. Do environmental factors play any part in the expression of the resistant character in plant diseases ?

9. What are the limitations in the development of resistant varieties ?

10. Can you cite some examples where remarkable success has been achieved in the development of resistant varieties ?

11. What are the three basic mechanisms of resistance in respect of insect attack ?

12. How do you explain the mechanism of antibiosis in insect resistant plants ?

13. How do climatic and edaphic factors play a part in the expression and modification in host plants in relation to insect resistance ?

14. What do you understand by 'multilineal approach' in breeding for disease resistance ? What and where has it been advocated ?

15. What are the general limitations in the use of resistant varieties for the control of diseases and insect pests ?

CHAPTER 20

MECHANICAL AND PHYSICAL METHODS
OF CONTROL

Apart from chemical methods, attempts are made to control pests by mechanical and physical methods. These methods aim in reducing the population of pests particularly insects by manual devices or manipulation of physical environment or employment of physical sources. Destruction of less mobile, large sized, conspicuous insects in the adult stage and egg masses by hand is practised even now where labour is still cheap, holdings are small and pests are relatively restricted in distribution. This method is still employed for the destruction of egg masses of top shoot borers of sugar cane, or borer of egg plant, egg clusters of cabbage butterfly or *Epilachna* beetles. A specially designed hopper doser is employed to collect grasshoppers and *Pyrilla* in some countries.

It is possible to prevent invasion of the nonflying insects in a limited area by preventing their migration through the use of physical barriers. Construction of deep trenches 30 to 60 cm deep and 60 cm wide is made in the villages to save crops from the crawling bands of hoppers of locust in gregarious phase or swarming caterpillar of rice.

Slippery or sticky bands may be used for preventing the movement of the crawling insects to the host. The wrapping of the trunks of mango trees with oil cloth or alkathene sheets (15-20 cm wide) prevents climbing of nymphs of mango mealy bug (*Drosicha mangiferae* var. *octocaudata*) which hatch out and begin crawling up the tree.

Corrugated card board bands with strips of waxed paper around the trunks have been used as traps for codling moth

(*Laspeyresia pomonella*) in apple. Encasement of the bands with wire mesh, impregnation with a chemical toxicant, but non-repellant improves the efficiency.

Placing of a greasy band containing 'Ostico', a proprietory product of Imperial Chemical Industries, around the tree trunks is practised in temperate countries to prevent geometrid moths from crawling up the trees depositing eggs on the buds and twigs.

Dragging of a rope previously wetted with kerosene can be effectively employed in dislodging rice hispa (*Hispa armigera*) from rice plants, which fall into water and cannot come up. Kerosene acting as repellant adds efficiency to this mechanical method.

Flame throwers are employed against locust hoppers and adults. They may also be used for burning weeds, scrub vegetations and localised sterilisation of the soil. A flame thrower is essentially a pneumatic sprayer in which the lance is modified into a burner. The tank is filled with kerosene, which on burning produces intense heat. The burner is so enclosed that the flame is usually thrown upwards. By providing a hood, it can also be thrown downwards.

Suction traps have been used for the control of cotton boll weevil. It works like a vacuum cleaner.

Many nocturnal insects can be attracted towards the source of light. Basing on this phenomenon, traps with light as a source of attraction, more popularly known as light traps, have been used primarily for the detection of insect pests and the determination of their population densities. However, often they have been employed as a control measure. The simplest light trap consists of source of light—an oil lamp placed on a brick or stone in a pan of kerosene water, or suspended over the pan. Insects attracted by light fall into the pan and get killed. In more sophisticated types, an electrical bulb of 100 watts or more is fixed at the top of a funnel shaped or trapezoid cone terminating in a bottle containing calcium cyanide. Insects attracted towards this powerful source of light which is placed in the field fall into the bottle and get killed.

Manipulation of the physical environment and use of physical means is possible in closed spaces, as in storage godowns.

These are employed for the control of stored grain pests and have been dealt with in Chapter 12.

QUESTIONS FOR DISCUSSION

1. What are the objectives of mechanical and physical methods of pest control ? How are they helpful in minimising pest incidence ?
2. Where are sticky bands used ?
3. How can the ability of insects being attracted towards a source of light be utilised in pest control ?

CHAPTER 21

LEGISLATIVE MEASURES : QUARANTINE

One of the methods of crop protection is to exclude the pests from entering into the area or sphere in which the host plants are growing so that the chance of attack is nil due to lack of exposure to the pests. This method of exclusion of the pests is enforced through certain legal measures commonly known as quarantine. Knowledge and methodology of exclusion are utilised and practised by a legally constituted authority to prevent the entry and thus spread of injurious crop pests in the public interest. Quarantine Regulations or Acts are of comparatively recent origin. The adoption of such steps by different countries of the world has arisen out of the fact that extensive damages, often sudden in nature, have been caused not by indigenous pests, but by exotic ones which have been introduced along with plants, plant parts or seeds in the normal channel of trade or individual interest. Instances may be cited of the introduction of grape *Phylloxera* (*Phylloxera vitifoliae*) from the U.S.A. to France which caused destruction of French vineyards ; Mexican boll weevil (*Anthonomus grandis*) whose original home was in Mexico or Central America, round about 1892 entered the U.S.A. and later to various countries in the world, causing extensive damage to cotton ; European corn borer (*Ostrinia nubilalis*) which reached North America probably through broom corn from Italy or Hungary and has since become a major pest there.

Pink boll worm (*Pectinophora gossypiella*) considered to be one of the six most destructive insects of the world probably a native of India is now established as a highly destructive pest in nearly all cotton growing areas of the world. Downy mildew of grape (*Plasmopara viticola*) introduced into France from the U.S.A. was responsible for the destruction of grape vines till the discovery of Bordeaux Mixture. Blight disease of chestnut (*Endothia parasitica*) introduced into the U.S.A. from Europe completely wiped out chestnut plants.

In India, the San Jose Scale (*Aspidiotus perniciosus*) is a pest of apple introduced about 60 years ago, now causing concern to apple growers in Himachal Pradesh, Jammu and Kashmir ; potato tuber moth (*Gnorimoschema opercullela*) which entered from Italy in 1900 is an established field and storehouse pest of potato all over the country ; wooly aphis (*Eriosoma lanigerum*) an introduced serious pest of apple ; fluted scale (*Icerya purchasi*), a native of Australia introduced through Ceylon in 1928 now a serious pest of *Citrus* spp ; leaf rust of coffee (*Hemileia vastatrix*) introduced from Ceylon in 1876 ; fire blight of apple and pear (*Erwinia amylovora*) introduced from England in 1940, now a serious disease in Uttar Pradesh ; flag smut of wheat (*Urocystis agropyri*) introduced from Australia now established in the Punjab, Rajasthan and Uttar Pradesh ; bunchy top of banana introduced from Ceylon in 1940 causing serious damage to dwarf Cavendish varieties in different parts of India ; wart of potato (*Synchytrium endobioticum*) introduced from Holland in 1952 ; golden nematode of potato (*Heterodera rostochienssis*) introduced in the last 20 years and onion smut (*Urocystis cepulae*) introduced recently are examples showing how many destructive diseases and pests have entered into this country and have established themselves causing extensive damage.

Plant quarantine regulations in order to be effective have to be based on sound scientific principles. The biology and ecology of the organism against which quarantine measure is proposed to be enforced should be known. Besides it has to be determined whether : (a) in the absence of any quarantine measure, the organism is likely to be introduced into the country ; (b) in the event of its introduction whether the organism is likely to be established and cause damage of any consequence ; (c) quarantine regulations can be framed on scientific lines and enforced satisfactorily ; and (d) it is economical to introduce the legislative measure in terms of benefit likely to be derived. Biological, legal and economical aspects of the problem have to be clearly understood to place the measures on a sound footing.

International quarantine regulations which aim to prevent entry of new pathogenic organisms and insect pests may; (a) completely prohibit entry of certain plants or plant materials; (b) allow import of certain plants or plant materials if they are certified to be free from certain specific insect pests and pathogenic organisms, by a competent authority of the country of origin; and (c) allow entry of plants and plant materials provided they are accompanied by certificates of freedom from pests and diseases by the competent authority of the country of origin. The importing countries may also impose restrictions on the mode of transport (air, ship or postal mail), and wrapping materials (soils etc.). They also have the right to examine the materials before they can be allowed to be introduced, even if accompanied by the certificate from the country of origin. Fumigation or any other treatment may be enforced. For this purpose the plants and plant materials to be brought into a country need to be channelised through certain specific ports of entry.

The enforcement of legislative measures to check the entry of destructive diseases and insect pests from other countries can be successfully done through the cooperation of Governments of different countries. Almost every country of the world has passed Quarantine Acts with specific provisions. Mutual respect of the provision of the Act is necessary for the successful promulgation. This calls for an understanding and agreement at the international level. Appreciating the need for such an effort at the international level, the Food and Agriculture Organisation (FAO) of the United Nations set up an International Plant Protection Convention in 1951 (details given in the appendix). Various member countries of the world were signatories to this convention. They are expected to respect the provisions of Quarantine Laws or Acts of different countries.

Under the International Plant Protection Convention member countries cooperate with one another in the prevention of introduction and spread of plant and plant products carrying destructive diseases and insect pests across international borders. Under this Convention Regional Bodies have been set up—one of which is meant for the countries of South East Asia and the Pacific Region, which includes India, Australia, Ceylon, the U.K.,

Laos, the Netherlands, Indonesia, Portugal and Vietnam. Some other regional Bodies are also existent namely the European and Mediterranean Plant Protection Organisation (E.P.P.O.), Courte Internationale Permanent Anticardien (C.I.P.A.), etc.

From time to time these Conventions and Regional Organisations meet and discuss the common problems pertaining to each organisation and the operation of plant protection and the involvement of international cooperation.

The International Plant Protection Convention defines the policies underlying import restrictions for plant protection and prescribes an organisational set up and functions at the national level by different governments.

Operation of plant quarantine in India

Various rules and regulations have been issued under the Destructive Insects and Pests Act, 1914 as amended from time to time, prohibiting or restricting the import of plants and plant materials, insects, fungi to India from foreign countries. This comes under foreign quarantine. Rules and regulations have been made prohibiting or restricting the movement of certain diseased and pest infested materials from one State to another in India. This comes under domestic quarantine.

For the purpose of quarantine, 'plant' means a living plant or parts thereof, including tubers, bulbs, rhizomes, corms, cuttings, buds, grafts, layers, suckers, fruits, vegetables, cut flowers and seeds.

The enforcement of plant quarantine regulations is carried out by the technical officers of the Directorate of Plant Protection and Quarantine, Ministry of Agriculture, Government of India, under the overall supervision of Plant Protection Adviser. There are a number of quarantine stations in seaports, such as Calcutta, Bombay Madras, Cochin, Bhavanagar, Tuticorin. Visakhapatnam and Rameswaram, besides in airports at Calcutta, Bombay, Madras, New Delhi, Amritsar, Tiruchipalli and Trivandrum ; and land frontiers of Attari-Wagha border (Amritsar), Sookeapookri, Kalimpong, Bongaon, Gede Road (West Bengal) and railway stations at Attari and Amritsar. In general plants and plant materials have to be imported through the prescribed ports, or routes as the case may be and must be accompanied by a proper phytosanitary certificate, as per International Convention, issued by the competent authority of the exporting countries. They must undergo inspection and fumigation if necessary. Import of plants by air is allowed only by permission of the Plant Protection Adviser. Government of India.

Complete prohibition is placed on import of unginned cotton, Mexican jumping beans, etc. Cotton imported from the U.S.A. and West Indies has to be fumigated in a specific manner at the specified airports to prevent entry of the Mexican boll weevil (*Anthonomus grandis*). Import of potato tubers from areas known to be infested with wart diseases and/or golden nematode is totally prohibited.

On account of certain destructive diseases and insect pests being prevalent in some countries and the possibility of their introduction and establishment in India, some host plants and pertaining plant materials from some specific countries are not allowed to be imported into India. These relate to : (a) sugar cane cuttings from Australia, New Guinea, Fiji, the Philippines, the West Indies ; (b) cocoa and other *Theobroma* spp. (plant and seeds) from Africa, Ceylon, and the West Indies ; (c) rubber (plant and seeds) from South America, the West Indies ; (d) sunflower from Argentina and Peru. Besides plants and plant parts *Citrus, Allium,* coffee, cotton, sugar cane, unmanufactured tobacco, seeds of flax, berseem, sunflower and cocoa, etc., are subject to various restrictions. Phytosanitary certificates must state specifically freedom from pathogens and pests as required under provisions of the Quarantine Act. Importation of living insects such as exotic parasites, predators, honeybees, silk worms and of living fungi is allowed with permission of the competent authority. Importation of plants, seeds, etc., by letter post is completely prohibited. Consignments must not contain any soil.

Besides prevention of the introduction of new insect pests and diseases into India, it is also expected that insect pests and diseases are not introduced from India to other countries. Hence plants and plant materials needing to be exported have to be examined and phytosanitary certificates be issued in the international form for the purpose. Apart from general health, the certificate is required to conform to the current phytosanitary regulations of the country to which the plants or plant materials would be exported. In addition to Plant Protection Adviser and certain Officers of the Directorate of Plant Protection, Government of India, State Entomologists and Plant Pathologists are empowered to inspect and issue phytosanitary certificates in respect of exportable plants and plant materials.

If the importing country requires the fumigation of a commodity to be imported, facilities available at port quarantine stations may be availed of.

Details of provisions of Destructive Insects and Pests Act, Government of India corrected up-to-date are given in the appendix.

Domestic quarantine

Apart from international quarantine which is aimed to check the entry of diseases and pests across international borders, quarantine regulations may be enforced on the movement of plants and plant materials from one part of the country to other parts of the same country with a view to restrict the diseases or pests in specified areas where they have been found to be endemic and to prevent their spread to other parts. Such regulations operating within a country fall under domestic quarantine. In India, domestic quarantine preventing inter-State movement of some plant or plant materials is in force against two insect pests and four plant diseases. These measures aim at preventing the introduction of these destructive pests from the infected areas to non-infected areas.

(1) *Fluted Scale (Icerya purchasi)* : Transport and movement of the host plants of *Icerya purchasi* from the States of Tamil Nadu, Karnataka and Kerala are prohibited by letter, sample or parcel post or by air, sea or road except through the prescribed routes, or by rail or inland steam vessels unless the consignment is accompanied with a certificate from the competent authority in respect of freedom from fluted scale.

(2) *Bunchy top of banana* (virus disease) : Rules prohibit the transport and movement of any part of the plant or the entire plant of genus *Musa*, including suckers, stems and also the materials of banana plant used for packing and wrapping

excepting fruits from the States of Assam, Kerala, Orissa, and West Bengal to other States.

(3) *Banana mosaic* (virus diseases) : Transport and movement of any part of the plant or plant of the Genus *Musa*, as stated in the case of bunchy top of banana are prohibited from the States of Gujrat and Maharashtra to other parts of India.

(4) *San Jose Scale* (*Aspidiotus perniciosus*) : Rules prohibit the transport and movement of host plants or parts of host plants and packing and wrapping materials of these plants, which are likely to carry San Jose Scale from the Punjab, Himachal Pradesh, Jammu and Kashmir, Uttar Pradesh, Tamil Nadu, West Bengal, Assam Orissa to other parts of India except through specified routes and with official certificates as required.

(5) *Potato wart* (*Synchytrium endobioticum*) : Movement and transport of potatoes grown in the Darjeeling district of West Bengal are prohibited to other parts of India due to the wart disease of potato being present in Darjeeling district. Potatoes grown elsewhere will be permitted to be transported through West Bengal, if accompanied by a transit permit from the Plant Protection Adviser, Government of India.

(6) *Golden nematode of potato* (*Heterodera rostochiensis*) : The movement and transport of potatoes grown in the Nilgiri hills, Tamil Nadu are prohibited to other parts of India due to the golden nematode disease of potato being present in that area.

Similar restrictions under domestic quarantine may be operative in other countries of the World.

The effectiveness of quarantine measures to check the entry of new pests has been questioned by many, particularly in relation to the cost of enforcement. They are of the opinion that these measures create inconvenience to the public, and have often been taken advantage of in putting up trade barriers. They consider that in spite of these steps, new exotic pests are gaining access to different countries of the world.

As a method of crop protection, legislative restrictions of the importation of plants and plant products are attended with economic and political involvements, but it is based on sound scientific principles. It is true that entry of new pests cannot be eliminated altogether, but it has slowed down the process and made people particularly tourists and travellers conscious of the danger of indiscriminate importation of plants and plant products. Besides these measures have given an opportunity to the scientists to study the consequences of newly introduced diseases and pests and restrict them. Quarantine measures, in spite of the criticisms, have been found to be largely successful, because they are handled by scientific personnel and constitute one of the important plant protection measures.

Certification of planting materials

Measures aimed to reduce the initial inoculum include use of clean planting materials including seeds. In many countries legislation aims to restrict the sale of infected planting materials in general by notifying that only certified planting stock of some crops are to be sold. These materials are inspected in the growing season and are expected to meet the standards prescribed for the purpose. In reality to make this measure successful, cooperation between the producers, users and government officials is necessary. The government is to fix minimum standards and lay down the procedure of certification keeping in view technical feasibility and labour involved and render assistance in inspection and certification, producers are to purchase only certified planting stock. As no compulsion is envisaged, the voluntary involvement of the producers and users can ensure the desired results.

For seed, certification is normally issued on the basis of purity, freedom from weeds and germination percentage for which standards are fixed for each crop. Seeds are treated with pesticides, though such treatment may not be obligatory. Freedom from diseases and pests is not specifically stated or certified. Difficulty has been experienced in the past in determining seed health for its freedom from infection because of the lack of reliable simple laboratory techniques which can be employed almost at the same time when germination and other criteria are determined. Largely due to the pioneering work done in seed pathology institutes in Holland and more recently in Denmark simple, but precise methodology is available to determine the health of seeds which can be gainfully adopted.

In respect of vegetative propagative stocks, tubers and other planting materials, use of certified material is essential. In potato, this has been in vogue in many countries of the world for sometime. The method also involves the production of nucleus virus-free materials, which is usually done by heat treatment, meristem culture or more conveniently by tuber indexing. Once nucleus virus free planting material is obtained, it is multiplied in areas under conditions unfavourable for virus infection. With potatoes, tubers from virus-free plants may be multiplied over three to five years depending on the freedom of the area from vectors. During this period, plants are examined in the growing season for the presence of virus infection and certified accordingly.

In virus diseases of fruit plants including *Citrus,* virus-free plants are located by suitable tests on local lesion hosts and buds or twigs of these trees are used as scions. Stocks are raised in virus-free areas and grafting is done also in virus-free areas. Planting materials after tests for freedom from viruses are certified to be used.

In sugar cane, special seed plot techniques are followed. Setts collected for healthy plants determined after regular tests are indexed and multiplied in special plots with intensive care. They are examined also in the growing season for the presence of disease. Heat treatment is also taken recourse to get rid of infection of viruses and nematodes in many cases.

Similar techniques are followed in banana for obtaining suckers of banana free from infection of Panama disease and bunchy top.

Seedling propagation is possible in some plants which are normally vegetatively propagated, and the same is advocated, because very few viruses are transmitted through true seeds. Seedlings however, need to be tested for the presence of viruses before they are allowed to be used.

The scheme for voluntary certificate, if properly worked out and implemented can serve a useful purpose in the control of plant diseases by exclusion of the pathogen.

QUESTIONS FOR DISCUSSION

1. In which way, are quarantine measures helpful in the control of pests and diseases ? Do you justify quarantine measures ?
2. What are the basic principles and scientific considerations in respect of quarantine measures ?
3. Why are quarantine measures called legislative measures ?
4. What is International Plant Protection Convention ? What significance has it with quarantine ?
5. What are the two different categories of quarantine ? State the cases of domestic quarantine.
6. What plants and plant materials are prohibited from entry into India ?
7. Is a certificate necessary for exporting plants and plant materials from India to other countries ?
8. What are the specified points of entry of imported plants and plant materials ? Why are plants and plant materials imported at certain points only ?
9. Why is the use of certified plant materials recommended ? How do you justify certification of plants, plant materials and seeds ?
10. Why is certification of planting material very useful for the control of virus diseases ?
11. Why is certification of seeds from the health stand point necessary ? What method is adopted for the purpose ?
12. What is seed plot technique ? How is it useful for raising healthy plants ?

Chapter 22

INTERNATIONAL PLANT PROTECTION

Protection of crops against the attack of pests in a country is the concern of the individual government which is expected to take measures in this context. With progress in agriculture, resulting in increasing production, there has been an augmentation of international trade in agricultural commodities. Consequently there has been greater awareness of importance of protecting crops against possible losses due to pests. Hence this situation has motivated different countries of the world to develop organisations at the international level for tackling common problems which include plant protection. Increased importance of plant protection has necessitated cooperation at the international level particularly in respect of pests which are capable of causing destruction in a number of countries. One of the classical examples in this context is desert locusts, which can breed in one country and develop swarms which will fly to other countries and cause extensive damage. The successful tackling of locust problem in a country needs cooperation from the contiguous ones, where locusts can breed and the swarms originate. This circumstance has prompted India, Pakistan, Afghanistan and Iran to form one group known as the West Asian Group with the Food and Agriculture Organisation of the United Nations acting as liason. Mention in this context, may be made of rust diseases of wheat in which propagules (uredospores) can be carried over long distances in air, often from one country to other. Hence the study of the rust disease situation in wheat needs the cooperation of Nepal where off-season cultivation may allow the pathogen to remain active and provide initial inoculum for the plains of North India.

Failure or success in the control of pests in one country is likely to have an effect on the neighbouring countries. Besides the study of diseases and pests which affect a particular crop in a number of countries is likely to give more detailed and precise information on the ecological conditions necessary for the appearance of the disease or pest in a serious form. There are many crops like rice, wheat, sugar cane, potato, maize, onion, etc., which are grown in many countries and may be considered international. The knowledge available in a number of countries with different agro-ecosystems is likely to be helpful in developing control of pests and protection of crop plants.

The aim of international plant protection is to limit the spread of pests and pathogens to different countries and prevent the development of epidemics. For this purpose, it is necessary to take advantage of all available information. To achieve this objective, it is necessary to : (1) organise survey and surveillance in a number of locations and exchange of information in respect of occurrence and distribution of pests, pathogens etc. ; (2) regular exchange of information regarding control measures including the distribution of resistant varieties or resistant lines for breeding ; and (3) cooperation and joint efforts in enforcing regulations involving quarantine measures.

Basing on these considerations, in recent years, a number of international organisations have come into existence. A brief account of a few important International Organisations is given as follows :

FOOD AND AGRICULTURE ORGANISATION OF THE UNITED NATIONS

The formation of the United Nations Organisation in 1945 opened avenues of cooperative effort at the international level on diverse problems of mutual interest including agriculture. Precisely with this objective, the Food and Agriculture Organisation was established in October, 1945. Amongst different problems which are being tackled at the international level under the auspices the F.A.O. plant protection and crop losses due to pests and diseases find an important place. The F.A.O. has taken leadership in diverse aspects of plant protection, which include estimation of crop losses ; organisation of locust control amongst different participating countries ; organisation of research in an international level in different countries on projects of global or regional importance ; training programme and seminars ; toxicity hazards and residue problems ; environmental pollution ; initiation of new concepts in pest control, e.g., integrated pest management ; provision of expertise to different countries ; dissemination of information through publication of a bulletin "Plant Protection Bulletin" ; plant quarantine ; etc.

INTERNATIONAL CONVENTION OF PLANT PROTECTION

The international Convention of Plant Protection which was organised under the auspices of the International Institute of Agriculture in 1919 was adopted by more than fifty member countries. Under this convention, several countries agreed to issue and accept phytosanitary certificates at an international level with the explicit purpose that future shipments of plants and plant materials from one country to another would be free from injurious pests. Agreements of the Convention came into operation in 1932. In 1949, the subject of a new International Convention for Plant Protection came before the Fifth Session of the Conference of Food and Agricultural Organisation. Accordingly the Second International Plant Protection Convention was drawn up and approved by the F.A.O. Conference in 1951. According to the terms of the Convention, an internationally agreed form and methodology of issuing phytosanitary certificates were devised which are being followed by member countries. Countries are also to report officially on the occurrence of new insect pests and diseases appearing in their countries for the first time.

PLANT PROTECTION AGREEMENT FOR SOUTH EAST ASIA AND THE PACIFIC REGION

A proposal for an Agreement was presented to the Council of the F.A.O. in 1955 to provide for measures to regulate importation of plants into this region as well as their movement within this region to prevent the introduction of dangerous pests and pathogens. The Region includes the following countries—

Bangladesh, Brunei, Sri Lanka, Fiji, the Gilbert and Ellies Islands Colony, India, Vietnam, Cambodia, Laos, Indonesia, Malayasia, New Guinea, the New Hebrides, North Borneo, Pakistan, Philippines, Samoa, Sarawak, Singapore, Solomon Islands, Timor, Tonga.

EUROPEAN PLANT PROTECTION ORGANISATION

The European Plant Protection Organisation was founded in 1951 with its headquarters in Paris. It came into existence as a result of an inter-governmental agreement amongst fifteen countries of Central and Western Europe. This organisation is supported by financial assistance from member countries. Its purposes are to advise member governments on technical, administrative and legislative matters related to the prevention, introduction and spread of pests and diseases of plants and plant products. The Organisation also assists the member countries in the control of pests and diseases of plants, the establishment of pest control campaigns, coordination of research in the field of plant protection, and exchange of information in matters relating to plant protection. The organisation regularly publishes progress reports and bulletins on international pest and disease problems.

INTERNATIONAL COMMISSION ON PLANT DISEASE LOSSES

This Commission was set up in 1952 by the International Union of Biological Sciences in cooperation with UNESCO. It first met in Paris in 1954.

Activities of this Commission include the assessment of diseases, compilation of information on losses caused by diseases in plants, methods of estimating losses caused by diseases, and compilation of statistical data on relevant matters. Reports are published in the Plant Protection Bulletin of the F.A.O. A number of important diseases have been dealt with.

INTERNATIONAL COMMISSION ON CHEMICAL CROP PROTECTION

Following deliberations of the International Congress of Chemical Crop Protection first held in Louvain in 1946, eventually

in 1952 an International Commission was set up for reviewing
recent progress and exchange of views on the chemical control
of pests.

COMMONWEALTH AGRICULTURE BUREAUX OF THE
UNITED KINGDOM

This Organisation came into existence in 1928 with its
headquarters in England. It is supported by financial assistance
from the Commonwealth countries. It is really the foremost
clearing house of worldwide interchange of information of value
to research workers in different branches of agriculture, animal
husbandry, fisheries and forestry. A large number of scientific
review journals are published by this Bureaux. Of them two
are of great relevance to plant protection, namely the Review of
Applied Mycology (now named as Review of Plant Diseases)
and Review of Applied Entomology. Besides these two review
journals, the Bureaux publishes bulletins from time to time which
are of importance to plant protection. The Commonwealth
Institute of Biological Control with headquarters in Ottawa,
Canada, and stations in various Commonwealth countries is under
the Commonwealth Agricultural Bureaux.

There is an organisation at the international level under the
auspices of the F.A.O. for control of locusts. It is involved in
the organisation of locust surveillance and control in countries,
normally affected by locusts.

Besides there are a number of organisations at the international level in different regions : (a) the Regional Plant
Protection Organisation operating in Mexico, Central American
countries, Panama aiming at the control of locusts ; pine pests
specially bark beetles, coffee diseases and pests, Mediterranean
fruitfly, and (b) Inter-American Conference on Plant Health
operating in Latin American countries in rendering assistance
in plant quarantine, advice on pests affecting agriculture, and
exchange of personnel. There may be other such organisations.

QUESTIONS FOR DISCUSSION

1. Why do you consider that collaboration at the international
   level is necessary in the field of plant protection ?

2. How has such collaboration been attempted ?
3. What is the aim of the International Plant Protection Convention ?
4. Why is an international approach necessary in the case of locust control ? How has this been achieved with reference to India ?
5. What is the role of the F.A.O. in international plant protection ?
6. How is the Commonwealth Agriculture Bureaux useful in international plant protection ?

INTEGRATED PEST MANAGEMENT

Prior to the development of a scientific basis of plant protection many cultural and physical methods were adopted by the growers for the control of pests. These methods were found to be useful by trial and error and experience. These have originated and been developed on empirical lines no doubt, but they have later been found to be scientifically valid. These practices which involve the destruction of the initial inoculum by sanitation measures, tillage practices to destroy insect pests and inoculum of disease, rotation of crops, circumvention of attack of diseases and pests, use of healthy disease-free seeds and planting materials, agronomic practices relating to depth of sowing and spacing; proper management of water and fertilisers, etc., are still being advocated. Whatever may be the cummulative advantages of these methods, they have not been found to be adequate for the control of sudden outbreaks of pests and diseases particularly in severe form. Being generalised in nature, these methods have not been found to be useful in tackling specific pests and diseases. Hence in the earlier part of the twentieth century varieties resistant to the attack of pathogens and insect pests and suitable biological methods for suppression of pests were evolved to meet such situations. But these methods have their limitations. It often takes a long period before varieties would be released or biological methods can be adopted on a large scale. Hence there has been a gradual shift towards chemical methods of control by the application of pesticides which received great stimulus since 1940 with the discovery of DDT and chlorinated hydrocarbons. Since then there has been a remarkable development in the chemical pesticides which have been put into wide use throughout the world. These pesticides have been found to be very potent and target specific in many cases. This situation is responsible for great technological advancement in plant protection which has made a strong impact. In the post World War II era, another technological advancement in crop breeding has been achieved and there has been a breakthrough in yield in many crop plants by the breeding of new varieties which have provided the base of the "green revolution". These two technological advances have been marked with other effects. Undesirable consequences of indiscriminate pesticide application have resulted in the development of strains of insect pests largely and plant pathogens to a limited extent resistant to pesticides; the rapid resurgence of pests following treatment due primarily to the destruction or reduction in population of parasites and predators ; and the appearance of other insect pests and diseases which have been of minor importance in more virulent form, completely upsetting the programme of control and taking the agriculturists and scientists by surprise who did not have any previous knowledge of control in many cases. Mention may be made of rice tungro virus (RTV), bacterial blight (*Xanthomonas oryzae*) and streak (*X. oryzicola*), brown plant hopper (*Nilaparvata lugens*), and gall midges (*Oresfia oryzae*) in severe form with appreciable reduction in yield following the introduction of high yielding varieties of rice. To meet these exigencies, greater use of more potent insecticides has been suggested.

This situation has evidently created environmental hazards apart from toxic effect on men and animals. Agricultural ecosystem has been upset with the use of more homogeneous narrow genetically based varieties and pesticides. It would be seen that *ad hoc* approaches to the problem by adoption of any method however effective has not met with the desired success. Hence a combination or integration of practices namely the use of resistant varieties, judicious use of mixture of pesticides and biological pest suppression has been considered necessary for effective pest management. More recently this integrated approach has been broadened to incorporate cultural practices and other such methods which are aimed to check the diseases and infestation of insect pests. According to the F.A.O. panel of experts (1968), integrated pest management has been defined as "pest management system that in the context of the associated environment and population dynamics of the pest species, utilises all suitable techniques and methods in as compatible manner as possible and maintains the pest populations below those causing economic injury." Integrated pest management is based on ecological principles operating in an unstable agro-ecosystem. Hence instead of being called pest control, the term 'pest management' has been used to convey a methodology of keeping pests below the injurious level. It is really a system of approach towards the suppression of pests by an integration of several methods, the purpose of which is not only to achieve the immediate object of preventing losses caused by pests, but also to keep a broader perspective. This perspective besides being confined to the population dynamics also takes into account the possible impact on society, environment and the financial aspect. It may be termed also as a management of available resources in dealing with pests to the best advantage in all possible aspects.

The importance of integrated pest management arises from the fact that modern agriculture which is based or tending to be based on extensive cropping programme with genetically uniform high yielding varieties which have been evolved in many cases without regard for its susceptibility to pests, combined with use of high doses of fertilisers, irrigation and pesticides together with rapid straightforward introduction in many countries avoiding requisite preliminaries, has created a serious situation by disrupting the natural balance. Pest attacks have consequently become more frequent and serious.

Empirical approach has been made in the past without much attention being paid to the broader perspective.

The concept of integrated pest management implies that the agro-ecosystem should be so adjusted or manipulated that pests are maintained at a sub-economic level of population so as not to cause any extensive injury. Hence various components or methods of control which are adopted or recommended to be adopted should be integrated and incorporated into a crop production system, but care should be taken so that a high level of yield is also maintained for the sustenance of the growing human population at the desirable level of nutrition. Some of the methods of control are based on natural phenomena namely host resistance and biological suppression of pests, while use of pesticides and cultural methods are inducted into the natural system. All these four methods together with autocidal methods of pest control form the basis of integrated pest management.

In case of diseases caused by pathogens, a somewhat integrated approach has been in vogue, because in many cases, a system approach has been advocated for achieving overall success. Since majority of the measures are preventive in nature in plant diseases and population dynamics of pathogen is very difficult to be assessed, many practices, namely, use of varieties at least moderately resistant or tolerant to infection, clean planting material, eradication measures including seed treatment, crop rotation and incorporation of organic matter in the soil to encourage microbial anatogonism have already found their place in overall planning of disease management. In case of pests, too much reliance has been placed in recent years on chemical methods because of quick success.

An integrated pest management, as ideally visualised, has its limitations. A number of diseases and pests may have to be encountered in a growing season and their ecological requirements may vary. Besides, as already pointed out varieties may not possess resistance to all, and resistance for all may not also possibly be incorporated into one; cultural methods particularly soil and water management suppressing one pest may encourage the other; and recommendations may not fit in well with agronomic methods aimed to achieve high yield.

Sufficient data on diverse pests relating to population dynamics and ecological requirements and conditions under which they may be kept at subeconomic level are not yet available. In tropical and subtropical countries where multiple cropping is practised often with same crop and is being advocated to increase production per acre per annum such knowledge is required. There cannot be an universal panacea for all pests, but system has to be worked out basing on the data in case of each pest-complex in different agro-ecological regions in relation to season, where cultivation is practised.

Emphasis has been given on natural methods, but in many cases to meet the exigencies of a situation or in view of the nature of pest or damage, reliance has to be placed on chemical methods of control, e.g., apple scab, late blight of potato, brown plant hopper of rice, sigatoka disease of banana, coffee rust, blister blight of tea, etc.

In many cases the ultimate effect of chemicals is not known. To mitigate losses due to loose smut benomyl compounds are being used, but enough is not known of the total effect of benomyl on various organisms, etc. But the use of such chemicals cannot be totally given up at this stage.

So far attention has been confined to field crops. For the protection of plantation crops, particularly tea, coffee, rubber, cocoa, chemical methods will have to be taken recourse to almost exclusively because of the investments made and risks involved.

Integrated pest management needs a multidisciplinary approach. The control measures that have been developed and adopted are essentially pragmatic in nature and have been developed in close connection with some disciplines, but unidirectional in nature. Hence a collaborative outlook has to be developed. It should not be viewed as an alternative to chemical methods of control, but may aim at the reduced usage or putting a brake on the excess use o chemicals, and incorporate the same with others.

Integrated pest management has been introduced with success in a number of crops, namely cotton in the Canete Valley, Peru ; spotted alfalfa in California ; apple and pear insects in Nova Scotia, glasshouse crops in the U.K ; tobacco pest management in North Carolina, U. S. A. In recent years, an integrated approach on pest management in rice and cotton is being attempted in India.

In the development of integrated pest management several phases have been conceived of. They may be stated as follows :

(1) *Single tactic phase* : In the initial stage, the best method for tackling or managing a pest is found out.

(2) *Multiple tactic phase* : In this phase, all the available methods cultural, biological, chemical and use of resistant varieties are worked out in relation to ecology, population dynamics of the pests and damages caused by them. This phase is the most critical one and needs to be worked out carefully.

(3) *Biological monitoring phase* : After the multiple

tactics have been worked out, they are to be carefully monitored particularly biological ones in relation to beneficial organisms and host plants so that control measures may be suitably timed and phased out. The main aim is to keep the pest at the subeconomic level by suitable adjustment of diverse control measures.

(4) *Modelling phase* : Data obtained from phases 1, 2 and 3 need to be very carefully worked out so that a model can be built pointing out the major steps that need to be taken regarding the application of control measures and the critical stages in the host plant with reference to diverse measures. The entire approach has to be systematised through critical analysis of data.

(5) *Management phase* : After the different methods of control have been worked out in an integrated manner, the entire process has to be incorporated into the overall production management techniques of the crop in question.

There may be different aspects which will necessitate careful integration. A grower is interested in maximisation of production often with a few tactics. Besides certain approaches or use of inputs which are essential for higher yield may aggravate the pest situation.

(6) *Acceptance phase* : The most critical stage is the development of systems approach which will be acceptable to growers. The approach has to be technologically sound, economically viable and not too cumbersome for adoption by the farmers.

Integrated pest management is a new system approach which has been necessitated primarily out of the growing concern about the undesirable side effects of large-scale use of organic insecticides and often failure of the same to provide for suppression of pests at economic level. The Food and Agriculture Organisation of the United Nations has been interested in 'Integrated Pest Management' since 1963. It has developed and implemented 13 field projects in different parts of the world. Global programme has been proposed to be undertaken. Criteria of priority have been developed which point out the necessity of integrated pest management. These are : (1) crop must be of vital national and regional importance, (2) serious losses are caused by pests, (3) inadequate control by use of organic pesticides, (4) use of pesticides is generating more problems, but the same cannot be given up otherwise food production will not be stepped up, and (6) an integrated approach can be developed which will yield desirable results and be acceptable to growers.

Emphasis has•been laid on major crops namely, rice, maize, sorghum, cotton, etc. Potato, sugar cane, grain legumes, tapioca, coconut, etc., have also been considered as second order of priority.

In pest management, normally the methods employed for control are judged mainly from the point of view of (a) mode of operation and action, (b) safety and biological efficiency. The objective is to achieve optimum control not always maximum control, but the density of pest population is reduced to such a level that the damage caused is not of economic significance.

The strategies for managing a pest problem may be put into three categories:

(i) *evasion* of the effects of a pest, such as growing crops for processing, when surface damage does not affect the yield or commercial value.

(ii) *elimination* of susceptibility of crops to attack of pests namely use of resistant varieties.

(iii) *reduction* of pest numbers to levels at which the damage caused is negligible.

In practice most IPM programmes concentrate on the reduction of pest numbers to below economically injurious levels, besides use of resistant varieties wherever such varieties with desirable agronomic characters are available.

There are several ways of limiting pest numbers, either by changing the characteristics of the pest species, or by modifying environmental conditions.

It is possible to change characteristics of the pest species by introduction of sterile males into wild population or by use of chemosterilants, or by introduction of harmful genes.

Attractants, repellants, misleading stimuli can be used to modify environmental conditions.

The most commonly methods used for modifying environmental conditions involve creation of shortages of space (cultural control) or the use of chemical or biological agents.

Agroecosystems vary widely, and can be manipulated in such way that other means of pest control may not be required. Intimate knowledge of the crop, the pest and its natural enemies, growing system can be altered more in favour of the crop and the naturally occurring enemies of the pest than in favour of the pest.

Cultural methods of pest management are many and include (a) *Crop rotation* to prevent build up of a pest with restricted host range, a number of non-susceptible hosts being grown in rotation with susceptible ones, (b) *crop hygeine* e.g. removal of trashes, crop residues, self sown tillers or plants, ratoons, (c) *adjustment of the planting time* so that the susceptible stages of the crops are tuned to coincide with the lowest or least damaging pest populations; (d) *adjustment of harvesting* to disrupt the requirements of pest regarding food, shelters of ovi position; (e) *host-free season*, this is similar to crop rotation with the difference that no alternative crop is grown, and the land is left bare or fallow; and (f) *habitat diversification*. Mono-culturing usually favour pests. Introduction of crop mixture, trap crops etc. can reduce the incidence of pest attack.

So far pesticides are concerned, there are two approaches, namely adoption of selective action pesticides and selective use of pesticides. Ideally selective action pesticides which have narrow specificity and a short life are desirable; but they are expensive. A broad spectrum pesticide can be used more effectively by adopting of suitable application techniques which include (a) *dosage*—application rates can be so adjusted as to kill the injurious pests without harming beneficial organisms. Over application leads to environmental pollution, development of resistance in target pests and an increase in the number of secondary pests; (b) *correct time of applications*: application of pesticide to a crop, should be restricted or made when target populations reach economy injury levels; and (c) *formulation and application*.

Most work on IPM has been carried in developed countries. These include: (a) apple pests in Eastern Canada. The control programme involves the conservation of natural enemies of the key pests; use of selective action pesticides, application of which are carefully timed, (b) cotton pests in USA Pesticide applications are limited and are based on population counts and economic injury levels, pests and natural enemy numbers are manipulated through cultural practices; (c) tobacco pest in North Carolina, USA. Cultural practices, improved crop hygiene and insecticide application on the basis of population counts have resulted in build of natural enemies and a reduction in major pest populations, and (d) sorghum pest in U.S.A. the green bug is controlled by a mixture of cultural practices and regulated application of pesticides.

QUESTIONS FOR DISCUSSION

1. How has the concept of integrated pest management originated ?
2. What do you understand by integrated pest management ? Why do different systems of pests control or management need to be integrated ?
3. What is the underlying concept of integrated pest management ?
4. What are the limitations of integrated pest management ?
5. Why is multidisciplinary approach necessary for integrated pest management ?
6. What are the different steps or phases in the development of integrated pest management ?
7. State the examples where integrated pest management has shown promising results.
8. Why is too much reliance on chemical methods not beneficial ?
9. What is the significance of integrated pest management in intensive crop production ?

will protract and review the conservation of natural enemies of major pests in a suitable, and in a suitable, application of which are applied. When this, there is not feasible application measures of and are based on population, injure and economic injury level, pests and natural enemy interaction control measures of the cultural, and control of natural pest or Rabb (mollusc) USA of and practices, conserved crop hygiene and insecticides application in a series of population bursts have resulted in faunal of natural enemies and the reduction in major pest. In the series of U.S.A. the green bug is controlled by a mixture of cultural practices and regulated application of pesticides.

CHAPTER 24

THE ASSESSMENT OF PESTS

In all aspects of pest management, assessment of pests and the effect they produce on host plants is important, as it constitutes a base for decision on pest control measures. In pest management, the major consideration is economic threshold, which is based on the relation between yield (Y) and infestation of (I*)

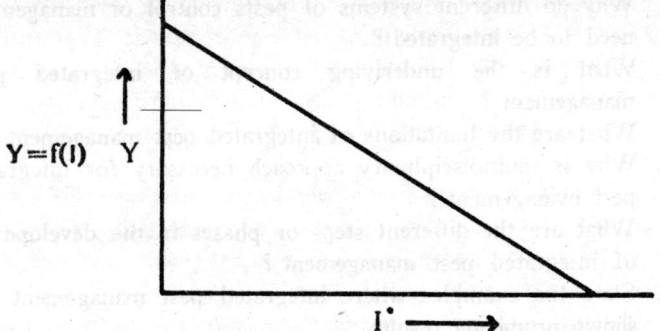

A number of methods are available for assessment of pests. These involve: (i) the study of biology of pests; their life cycles, ecology, migration etc.; (ii) survey of pests in place and time; (iii) assessment of the effects of pesticide trials; (iv) relation of crop losses to economic thresholds; and (v) prediction of pest outbreaks.

Methods can be divided into (i) direct counts (n) on plants or in the environment, or (ii) indirect counts of some effect caused by the pest, such as injury or damage. Both can be assessed as an index, score or on rating scale. It is however, necessary to find out relationship between the index of rating (X) and direct count (n).

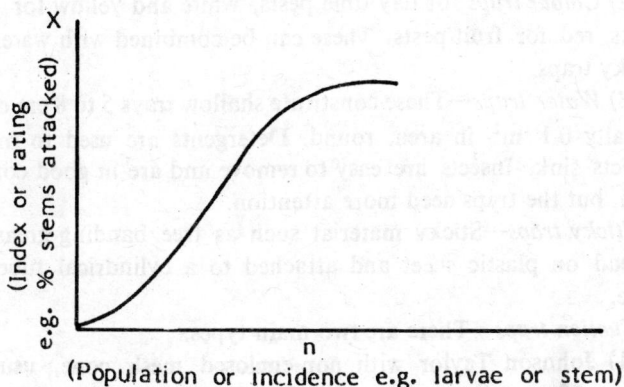

(Population or incidence e.g. larvae or stem)

There is however, difference between incidence, i.e. number of insects or plants attacked, and intensity which denotes degree or measure of infestation or attack.

Direct counts may be either absolute where total number of pests is counted or a sample or estimate can be made. The ratio of sample or estimate of population to total number of population is known as sampling efficiency. In case of sample, it has to be defined, such as size, position and number of leaves or tillers or stems, flower heads etc.

There are different ways in which counts of pest numbers can be taken. Some of these are : (a) cutting open affected parts as in case of stem borers, larvae in fruits and pods; (b) crushing or imprinting on glass paper or ninhydrin paper as in aphids, mites; (c) beating on to a sheet or inverted umbrella; (d) brushing by hand or machine on to a sticky plate or slide as in case of small insects or mites; (e) washing using if necessary a detergent or solvent in a known volume and pests are measured by volume of pests; (f) sweep net of different types; and (g) knock down in which tree or plant is sprayed with non-persistent chemicals e.g. Pyrethrum or dichlorovos and the treated tree or plant is shaken into a sheet.

Direct counts in the environment

(1) *Attractant traps* with chemicals or baits which attract several species (Kairomones) or sex attractants for one species

(Pheromones). These can be combined with water or sticky traps.

(2) *Colour traps* for day time pests, white and yellow for leaf pests, red for fruit pests. These can be combined with water or sticky traps.

(3) *Water traps*—These constitute shallow trays 5 to 8 cm deep usually 0.1 m² in area, round. Detergents are used to make insects sink. Insects are easy to remove and are in good condition, but the traps need more attention.

Sticky traps—Sticky material such as tree banding grease is spread on plastic sheet and attached to a cylindrical tube or pipe.

Suction traps—There are two main types.

(1) Johnson Taylor with non-enclosed mesh cone, usually about 23 cm diameter opening, may be 15 cm in shettered sites.

(2) Enclosed cone types usually 75 cm diameter and 2 m long for aerial work or 45 cm diameter in a metal cylinder.

Portable suction traps based on a vacuum cleaner have been used, sometimes for sampling in a walled area.

Light traps

There are five main types of light traps namely

(1) Base trap with overlapping glass plates on one side.

(2) Funnel trap (Helstand's) with a light at the centre of four vanes, a funnel, and collecting tube at the bottom.

(3) The Williams or Rothamsted type with glass walls and baffles and a killing bottle at the base.

(4) The Robinson with a light at the top of a metal drum.

(5) The Entech with mains or battery driven light in the top of a collapsible metal box with daylight operated switch.

Soil and debris sampling

Soil is usually sampled with a cutting borer or by digging a measured area and depth. Separation is by dry sieving by Berlese funnel or wet washing or flotation.

Emergence traps.—The sampling area of crop or soil is covered with a dark trap into which pests emerge to be collected in a tube by attraction to light.

Pitfall traps—Smooth-sided glass or plastic containers in the ground, their top level with the surface. They must be protected from rain, predatory birds, ants.

Indirect methods

Counts of injury, damage or other effects of pest attack or ratings or grades are often used as assessment of pest intensity. These are (a) damage to the whole plant,—the number of plants killed, wilted, dying back, in case of stem borers, aphids, soil larvae; (b) damage to roots—weight, volume, length, grade of root crop as in case of root weevils; (c) damage to stems—cut worm damage at soil level, dead hearts due to borer, exit holes, internodes attacked by borers; (d) damage to leaves; windows and shredding by stem borers, uniform holes by *Heliothis*, irregular areas eaten by locusts, *Epilachna* small lesions by sucking bugs. Leaf area damaged can be measured by photographic, airflow, scanning, measurement, weighing; (e) damage to fruits and seeds, cotton boll by boll worms, maize cobs, sorghum heads, coconuts by thrips, rice by borer (white heads).

Rating scales may be either descriptive namely nil, light, medium etc. or numerical rating namely 0, 1, 2, 3 etc. The scale may be arithmetic or logarithmic and scores can be added or transformed and analysed like counts.

Remote sensing

Photography from a tower, aeroplane or satellite may used to estimate pest populations or their effects and disease frequency and intensity. Special films or filters may show the effect more clearly.

Identification of pests and crop growth stages

Adequate descriptions of the stage, age and sex of pests to be sampled are needed as descriptive charts, diagrams or photographs. Similarly, description of the morphological and physiological stages of the host when they are vulnerable to attack of the pests are valuable.

Time of sampling of pests

Pests should be sampled when their effects on yield are maximum, i.e. at the most critical stages of crop growth.

Number and size of samples

Number and size of samples depend on the number of pests per sample, the variation between the samples and the degree of precision needed.

Distribution of pests

The frequency distribution or in other words, the number of sampling units containing different number of pests is important at the sample planning stage. If attack is not random, some sampling methods will not detect it; if the attack is systematic, systematic sampling may lead to error.

In analysis of data, if attack is not random, da   may be used incorrectly in calculating variance, sampling ε or, and the significance of means and regressions.

If the frequency distribution is not "normal",   ita should be transformed to normalize, for which Statistic n has to be consulted. The type of distribution and transfor ation can be found from Taylor's "Square Law" by plotting lε variance (S²) against logmean (X̄).

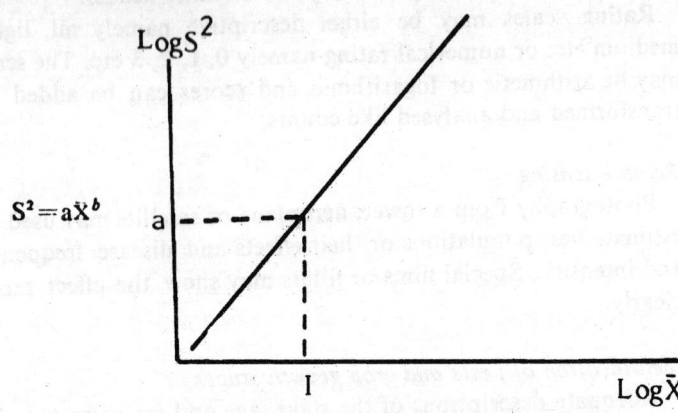

(a) is found as the intercept on the (S²) axis at X=I.

(b) the slope of the regression line is found from log S²=log a+b log x̄.

Value of P is obtained from P=1 —½b.

If P=0, log transformation of (X) is used.

If P=0.5, the square root of (X).

If P=0.5, the reciprocal square root.

If P=1.0, the reciprocal of (X̄).

Percentages are usually transformed by the angular or sine transform. If transformed, the results should de-transformed, for general understanding.

QUESTIONS FOR DISCUSSION

1. What are the methods for assessment of pests?
2. What are the methods of direct counts in the environment?
3. What are the indirect methods of assessment of pests?

CHAPTER 25

LOSSES CAUSED BY PESTS

Whatever may be the nature of damage direct or indirect, pests cause loss in yield, which may be from insignificant to enormous, depending on the intensity of the attack. Various attempts have been made from time to time to assess losses caused by plant pests, but fairly accurate estimates are difficult to obtain because the methodology adopted so far have not taken into account different factors. One of the obvious difficulties is that the yield that would have been obtained in the absence of the pests is not known. The second important aspect which is not taken care of is the deterioration in quality which fetches poorer price, e.g., shrivelled grains, fruits with blemishes, etc. Very often the loss is expressed in terms of yield or tonnage but is not clear whether the "yield signifies actual yield obtained or theoretical yield" that would have been obtained in the absence of the pest. In case of severe pest infestation or infection post harvest damages that may also result are not taken into consideration, as in case of potato tubers collected from fields affected with late blight, bacterial wilt, or tuber moth or fruits affected with borers, or blemishes in the fruit which may lead to deterioration in storage. Sometimes the loss is expressed in terms of money. As prices of agricultural commodities are often elastic following the laws of demand and supply, monetary losses may not convey the actual loss sustained. In cases of failure of crops, in which pests may also play an important part, prices may soar disproportionately high due to less availability, whereas in bumper years in which pest control may also contribute, prices may fall. Besides, in calculating monetary losses, the costs of pesticides and their application are not taken into account. They should be taken into account, because they add to the cost of cultivation to meet essential prerequisite, but for which pest damage would have been serious.

Ordish (1952), in his book "Untaken Harvest" reviewed the amount of losses due to pests in terms of additional acres that would be needed to meet the depredations. He estimated that an extra 88 million acres would be needed to meet the losses due to pests in the United States. Of course, the basis of this estimate has to be extremely local in relation to area and time. Loss in terms of acre cannot be same in different areas, as production type per acre varies in different areas as well as in different periods, for example loss in terms of acre before introduction of high yielding varieties will not be the same after introduction. Hence such estimate will be of local value.

Decker (1962), estimated monetary losses due to pests at about 25 per cent of the nation's annual production and that too in a country where pest control methods are used intensively.

Basing on the data in the Production Year Book of F.A.O. of 1965 on the prices paid to the growers in various countries and survey of world literature on the subject, Cramer (1967) attempted to determine the cost of pests in agriculture. An amount of seventy to ninety thousand million U.S. dollars per annum was estimated as the cost to be paid on account of pests. This estimate however does not include the cost of pesticides used to keep down the pest population and minimise the losses. Approximately one thousand million dollars are spent per annum for pesticides used in agriculture.

Cramer estimated that percentage loss due to pests to be 55 per cent of actual production and 35 per cent of potential production—13.8 per cent due to insects, 11.6 per cent due to diseases and 9.5 per cent on account of weeds.

Any attempt to estimate loss due to pests needs to be assessed over a period of years taking into consideration diverse factors affecting yield in the period and cost involved in the application of pesticides and other plant protection measures. It may not be possible to have accurate figures of losses in yield due to pests in terms of tonnage or money or deterioration in marketable quality. But it is well known that pests have created problems in agriculture by effecting serious reduction in yield which has resulted in famines. Such situations have been attended with social consequences with far reaching effect, which have, however not been adequately covered. While considering damages caused by pests, social and economic consequences need also to be taken into account.

In 1967, a symposium on crop losses was convened by the Food and Agriculture Organisation of the United Nations to emphasise the need for the development and use of experimental methods to estimate crop losses annually. Basing on the recommendations of the symposium the F.A.O. in collaboration with the Commonwealth Agriculture Bureaux, England has brought out a manual on crop loss' assessment methods.

Economics of pest control

Pest control measures are undertaken to increase the outturn of the crop. Adoption of such methods is advocated also for the stabilisation of yield as a guarantee against losses caused by pests and better quality of the products intended for market. Legislative measures quarantine, foreign or domestic, are enforced by the country or State and the cost : benefit ratio in such cases cannot be calculated on the basis of area or individuals. The use of quality seeds and certified planting materials is a guarantee against the possible loss in yield due to poor quality of seedlings or plants arising out of the materials infected with pests, or loss in stand due to seedling blight or damping off. The benefit derived from such measures may be assessed in comparison to the yield in plots using uncertified materials. The results are sometimes obvious particularly in cases of virus diseases, and many fungal diseases, e.g., red rot of sugar cane, loose smut of wheat. Biological suppression of pests has to be undertaken at the National and State levels. This calls for long-term investment in research and development of appropriate technology which may be profitably applied in relation to the pest in a specific area. Once a technology has proved successful, then recurring cost has to be incurred in mass rearing of insects and their release. It has been observed that once parasites and predators are established in an area, expenditure becomes much less. In such cases the economic aspect of control measure has to be assessed over the entire period of a project till the desired results have been achieved to find out whether achievements are economically justified. Similarly use of resistant variety is attended with beneficial results, at a very low cost at the level of farmers, because it involves the replacement of a variety. In the initial stage, when such varieties are introduced and seeds or planting materials are in short supply, increased prices may have to be paid. In such cases investment on research and cost of multiplication of nucleus seeds and programme of popularisation are not taken into account, but in assessment of economic gain, these should be considered. Normally initial investment costs are borne by research institutions where such work forms a part of normal activity hence these escape attention, because results of research are not often valued on cost/benefit rates.

Improved cultural practices may not cost much but benefits derived may be more in relation to the additional expenditure that a farmer may have to bear for the purpose.

An individual farmer may adopt direct plant protection measures by use of pesticides which are comparatively costly. Pesticide application will be justified if the damage is economic i.e., losses that will be incurred will amply justify the additional cost to be incurred for direct plant protection. It is, however, difficult to estimate beforehand whether the loss will be at an economic level or not, because—forecast of pest population at which economic injury may take place can only be made at a short interval. In many cases prophylactic application of pesticides has to be taken as a routine measure. It is particularly true of high yielding varieties of rice which are subject to attack by a number of pests which can be controlled by pesticides. Successful cultivation of these varieties demands regular application of pesticides. Use of pesticides in many cases particularly against diseases and many borer insects, is preventive in nature and is aimed as an insurance or guarantee against fluctuation in yield. Hence costs incurred forms a regular feature for cultivation, as for example, spraying of potato crop for protection against blight ; apple against scab ; coffee plants against rust and leaf blight ; rubber against leaf falls and mildews ; banana against sigatoka disease ; tea against blister blight ; cotton against bollworms, etc., has become a part of the entire operations and it is not possible to think of growing such crops without these measures.

Though the diseases and pests may cause serious damage resulting in appreciable loss of yield when they appear in a severe form, nevertheless their incidence may not be large in all years and damages caused by them may be quite insignificant. In such cases the economic aspect has to be determined over a period of years in areas with varying intensities of incidence of pests.

Damage due to pests needs to be assessed from two dimensions which may be termed qualitative or vertical and quantitative or horizontal. The former refers to the intensity of attack or degree of severity which is often categorised by numerical indices and number of plants affected. A weighted mean combining the two gives a picture of the extent of damage. Besides, as already pointed out the degree of infestation does not always depict commercial losses e.g., scab infected potato, apples infested with codling moth, etc.

From the equation of Van der Plank (1963) $Wt = Woe^{rt}$, it would be evident that greater the initial infestation, greater will be cost : benefit ratio. Losses may range from severe to slight, so also a farmer may take complete, partial or no control measures. Hence theoretically a number of ratios are expected. Basing on these assumptions Ordish (1952) has calculated potential cost : benefit ratio for a variety of crops in the U.S.A.

It is expected that control measures will be justified if the pest injury rises above the economic level and the benefit will be up to a point, correlated with the degree of infection or infestation. In this context, effect of uncertain weather elements on ultimate yield has to be borne in mind. This may upset the normal calculations and cost : benefit ratio. However degree of pest damage, as assessed by the usual methods may not be reflected in yield as for example root rot unless it reaches a critical level or takes place at critical period may not affect yield, because in many plants the root system is in excess of requirement or plants can easily regenerate root system. Foilar infections in the initial stages may not be reflected in the yield. From the experiments conducted by Justesen and Tammes, it may be concluded that the relationship between damage and initial population density of pests is exponential. Where a number of factors are influencing population density of pests. it is very difficult to predict the loss of yield a particular infection or infesta tion may cause. But the farmer as routine measure has to apply pesticides, hence the necessity of cost : benefit ratio arises. It is natural that the law of diminishing returns will also apply to the use of pesticides for the control of pests for higher yield. Hillebrandt (1960), found a distorted sigmoid shaped curve while computing relationship between pest damage and increase in yield. The log transformed data show a typical sigmoid dosage response curve. In cases of direct damage, greater response is likely to be found than in the cases of indirect damage.

Using micro-economic methods Headley (1972), has postula ted certain models on sophisticated lines. He has assumed that land and labour are fixed costs, while other inputs including pesticides are variable costs. Hence increase in cost will lead to increase in output up to a point then it will level off or may fall to a degree according to the response curve of law of diminishing return. Profit maximisation will occur at a place when the cost will be equal to the market price. But if there is pest attack and the pest is not controlled, then the cost remaining same, out-turn will fall appreciably and profit will be much less. Under such conditions greater profit in relation to cost (not greater yield) is expected if the cost is reduced because the margin of profit will not be appreciably scaled down.

In case of partial control of pests, an intermediate position would be attained, as anticipated and the cost of pesticides input will result in greater benefit at lower levels, with increasing costs, profit will be less. He has given mathematical expression to the theoretical cost : benefit ratios under varying conditions.

A farmer, however, does not work according to theoretical models, and he is interested in the maximisation of out-turn. In the absence of a complete biological data on pests and damages caused under varying conditions of weather and other relevant factors, the approach to a large extent will be empirical and the farmer will be the best judge in many cases. The matter is often complicated by a number of practical considerations namely lack of knowledge of the right type of pesticides to be used, multiplicity of pests necessitating use of different pesticides and efficiency of pesticide application equipments.

While assessing the cost : benefit ratio of pesticide application in terms of profit in an individual crop in a season, long-term effect of pesticide application in causing environmental pollution, resistance of pests to pesticides, toxicity hazards are not considered, but they are causing concerns. Social costs need also to be kept in view of potential dangers or hazards.

Decision in good pest management is usually based on the economic effects, in quantitative terms, caused by pests, diseases or weeds, as the case may be. Study of crop losses is important for (a) determining the economics of control by comparing the value of loss with the cost of control; (b) finding out the importance of different causes of loss and (c) planning for basic research on the cause of loss and method of prevention of the same.

Loss due to pests and diseases is usually expressed as the reduction in potential, maximum or pest-free yield (m) and a percentage of it, and is expressed as.

$$W = \frac{(m - y_1)}{m} \times 100$$

W = loss in yield.

m = potential, maximum or pest-free yield.

y_1 = actual yield.

Loss may be pre-harvest or post-harvest.

Reduction in yield is often attempted to be related with intensity of infestation.

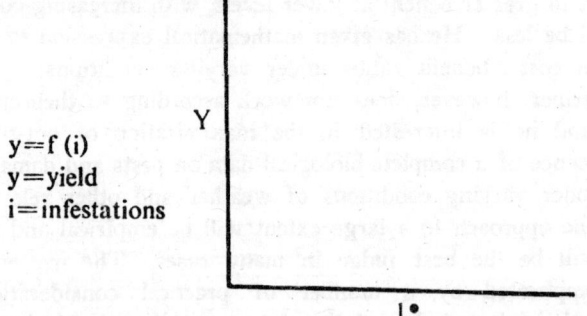

y = f (i)

y = yield

i = infestations

Yield for this purpose is usually taken as the economic product harvested which may be grains, fruit, tubers, or the usual plant part—leaves in tea, or berries in coffee etc. It may denote also loss in quantity namely in protein or oil content, or grade of fruit or vegetable.

The shape of curve of the value (V) of the yield plotted against infestation (1*) will influence the cost of control. It may indicate unresponsive to low levels of infestation as in (I),

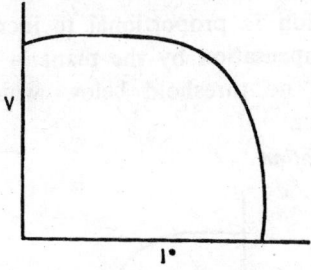

proportional to level of infestation as in (2).

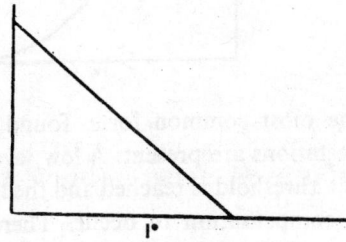

or very sensitive to low levels of infection.

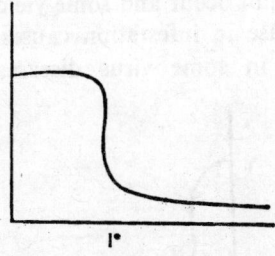

The relationship between yield and infestation as expressed graphically will give some idea the nature of reduction in yield. The following main types of relationship are usually obtained.

(a) *Straight line reduction*

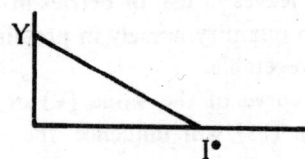

Yield reduction is proportional to increase in infestation. There is no compensation by the plant in increased yield due to attack and no threshold below which attack has no measurable effect.

(b) *Sigmoid relation*

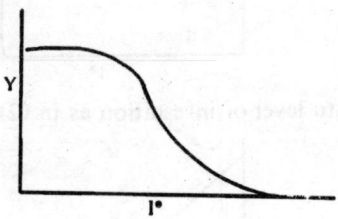

This is the most common form found, particularly if full range of infestations are present. A low level of attack has little effect unless a threshold is reached and the attack is light or early enough for compensation to occur. There is often a straight line portion and flattening at the end, because 100 per cent attack often does not occur and some yield is usually obtained.

(c) Small increase in infestation causes rapid loss in yield. This is expected in some virus diseases where vectors are involved.

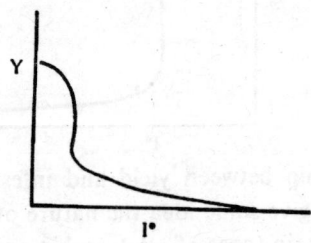

(d) Sometimes reduction is related to *logarithm of number of pests*, giving a parabolic curve typical of mobile or rapidly multiplying pests or one of their stages. There may be a thresh hold of attack, and compensation by the plant for attack.

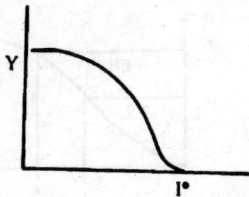

(e) *An increase in yield*

Sometimes yield may increase with increasing infestation, before falling perhaps in sigmoid form. This may be due to increased number of tillers or fruiting points after destruction of the apical dominance, increased maturity after pest attack etc.

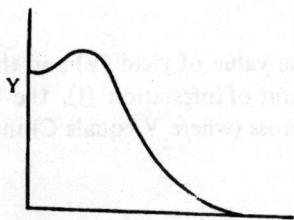

(f) *Little relation between yield and attack*

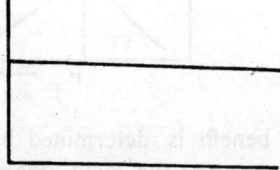

If the results are variable, plotting average yields and infestation may lead to the conclusion that no relation exists.

Economic Threshold (ET) and Economic Injury Level (EIL). Decisions on pest management are based on *economic thres-*

hold (ET), defined as the amount of infestation (I*) at which control must be applied to prevent the infestation (I*¹) rising to a level (I*²) causing economic loss (Y_2)—economic injury level (EIL).

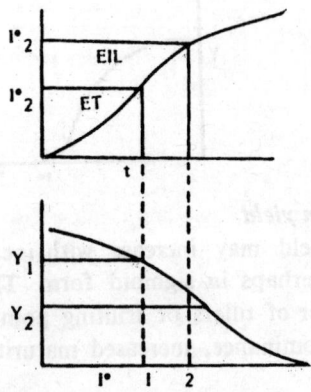

By plotting the value of yield (V) and the cost of control (C) against the amount of infestation (I), the level of (I) at which the two curves cross (where V equals C) indicates the ET.

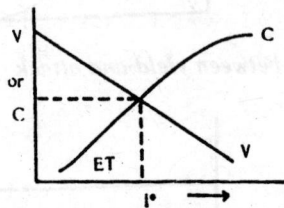

The level of benefit is determined by (a) the yield value: yield quantity ratio, (b) the yield quantity: infestation Value ratio and (c) the infestation level: control cost ratio.

For study of the relationship between the level of pest infestation and yield, a number of methods are available. These are described below:

1. *Using natural infections*

Plots or fields showing varying degrees of infestation are selected at random. The infestation is estimated and yields taken. A modification of this method may be to take attacked and unattacked plants in groups, measure the yield.

Disadvantages of this method are that variations due to soil and climate are often great, besides there may be a number of diseases and pests, the effects of which cannot be studied separately.

2. *Controlling the infestations with chemicals*

Different levels or degree of attack may be obtained by applying pesticides at different times, different concentrations or of different types in randomised blocks or randomly selected plots in fields. Yields of plots are taken and compared with those of different levels of pest attack.

Disadvantages are that the pesticides may have a stimulatory effect or phytotoxic effect in an indirect way by reducing yield or affecting other pests which affect yield.

3. *Caging the crop or plants to keep pests out*

Pest infested plants are kept inside the cages with wiremesh to prevent re-entry of pests. Pests inside are killed with non-persistent insecticides like pyrethrum preparations. Yields of cages with and without pests are compared.

Cages affect light and air movement hence there may be an effect on yield due to caging.

4. *Artificial removal of pests*

Eggs can be removed, larvae can be picked off or adults attracted by light traps. Thus artificially pest free populations in plots are created and yields may be compared with those with infestations. The main disadvantage is that this method is difficult to operate and test plants may be damaged due to removal of pests.

5. *Artificial infestations*

This is the most common method. Known number of pests

are put on the crop or plants which may be inside the cages and within enclosed walls. The natural infestation is removed and regulated levels of artificial infestations are introduced. Infestations can be done by putting infested material among test plants.

This method involves collection or breeding of pests. It can be laborious, expensive or difficult. The timing of an attack should be appropriate and is often critical.

6. *Use of resistant varieties*

Susceptible and resistant varieties can be found which have similar yields when not infested by pests or infected with diseases. Exposure to natural infections will result in different infestation rates and different yields. The major disadvantage of this method is that the different varieties yield differentially and are therefore difficult to compare.

7. *Causation of artificial damage*

The damages caused by different levels of pest infestation and disease attack are simulated or artificially produced and the effect of the damage on the yield is measured.

The major disadvantages are (a) simulating damaging effect is difficult. It may be easy to damage leaves, but damage to the growing point or roots is difficult and uncertain; (b) compensation for loss of plants or plant parts often occurs; (c) pest or disease damage cannot always be related to degree of infestation due to number of factors; and (d) amount of damage depends on the stage of attack or infestation and plants parts affected.

Loss of crop production due to pests or diseases over a large area, say district, region or national basis can be studied by (a) *actual survey of yields* in fields in cases of variable attacks and averaged to obtain estimate of yield, hence loss, and (b) *infestation surveys*—in case where relationship between level of infection or infestation and expected yield is known from experimental results or previous survey, then surveys of infestations can be made and the expected yields, hence losses can be estimated.

QUESTIONS FOR DISCUSSION

1. Why is it difficult to have fairly accurate estimates for losses ?
2. What are the different factors that have to be or are not normally taken into account in the assessment of loss ?
3. Can some examples be cited of the estimation of loss caused by diseases and pest in plants ?
4. Describe the different aspects of the economics of pest control.
5. What do you understand by economic loss ?
6. Why it is stated that the use of pesticides in many cases is termed as an insurance or guarantee against fluctuation in yield ? Can you cite cases where prophylactic pest control measures are considered more or less obligatory ?
7. What do you understand by the statement "damage needs to be assessed from two dimensions" ? What are the two dimensions ?
8. Under what conditions greater profit in relation to cost is expected in respect of pest control measures ? What do you understand by cost : benefit ratio ?
9. Why is it said that the farmer is the best judge in many cases ?
10. What are the different methods of assessment of losses.

CHAPTER 26

ORGANISATION OF PLANT PROTECTION IN INDIA

Acute food shortage following World War II and the Bengal famine of 1943, one of the major causes of which was the failure of rice crop due to the attack of brown spot incited by *Helminthosporium oryzae*, focussed attention on prevention of damage caused by pests to various crops in the field and during storage. The Woodhead Commission (1945) appointed by the Government of India to enquire into the causes of the Bengal famine of 1943 recommended the establishment of a comprehensive plant protection organisation to tackle effectively problems of pests and diseases of crop plants. A permanent Plant Protection and Quarantine Organisation as a Directorate of the Ministry of Agriculture, Government of India was set up in May, 1946. The Locust Control Organisation set up earlier was merged into this Directorate.

The main functions of the Directorate are as follows :

(1) Rendering advice and assistance to the Government of India in all matters pertaining to plant protection including its international obligations ;

(2) Collection, collation and dissemination of information about insect pests, plant diseases, noxious weeds and non-insect animal pests ; and methods of their control ;

(3) Locust survey intelligence, investigation and control of locusts in the desert areas of India ;

(4) Operation of a plant quarantine service for India and rendering advice to State Governments about domestic (inter-State) plant quarantine measures ;

(5) Organisation of advanced training programme in plant protection at the Central Plant Protection Training Institute,

Hyderabad and *ad hoc* training programme at the central Plant Protection Stations ;

(6) Surveillance of important pests and plant diseases with a view to develop a forecasting and warning service ;

(7) Implementation of Insecticides Act, 1968 and Insecticides (Amendments) Act, 1977 ;

(8) To help in formulating import policy on pesticides and plant protection equipment and in regulating and promoting, the production/formulations of indigenous pesticides and equipment ;

(9) To advise the State Governments and Union territories in formulating and coordinating their plan schemes on plant protection and assist in securing financial assistance for State Governments for programmes under plant protection, diseases and pests outbreaks ;

(10) To study pest infestation in storage and suggest remedial measures for prevention of loss in storage ;

(11) Establishment of parasites, predators and pathogens for the biological control of insect pests and weeds ;

(12) To conduct tests and trials of pesticides and plant protection machines including agricultural aircrafts ;

(13) Publication of a quarterly magazine "Plant Protection Bulletin", monthly magazine "Plant Protection Gleanings", fortnightly "Locust Bulletin" and "Pest Surveillance Bulletin" ; and

(14) To assist the Indian Standards Institution in formulating specifications for pesticides, plant protection equipment and related matters.

Locust Warning and Control Organisation

India forms the easternmost extremity of the 'locust belt' extending from Morocco in the West to Tanzania in the South, and Turkey in the North. Normally the regular outbreak of locust occur over an area of 2,58,000 sq km in the States of Gujarat, Rajasthan and Haryana. The strategy of locust control consists primarily of careful surveillance in the desert areas for timely detection of locust build-up and their destruction before the swarm formation takes place. For operational purposes, desert areas have been divided into four circles, nine zones with altogether thirty-four outposts with field headquarters at Jodhpur.

Network of information is maintained between outposts, zones, circles, field headquarters and headquarters through wireless system. To meet situations, number of outposts may be increased. Regular surveys are carried out throughout the year for which proper staff is provided at appropriate levels. Adequate control potential in the form of chemicals, equipments and transport is kept at various points.

Locusts spread from one country to another and for their effective control international cooperation is essential. Coordination of locust control is effected through the services of the Food and Agriculture Organisation of the United Nations. The entire locust invasion area is divided into several surveillance of locust movement and breeding. Iran, Afghanistan, Pakistan and India constitute the West-Asian Region. Locusts breed in summer in the Indo-Pak desert area, while winter and spring breeding takes place in Afghanistan, Iran and Pakistan. There is a cyclic movement between the two breeding areas. The F.A.O Regional Control Office at Tehran maintains liason between the member countries and the F.A.O.

India is subject to depredations of three species of locust of which the most important is desert locust (*Schistocerca gregaria*). The other two species are the Bombay locust [Patanga (*Cyrtacanthacris succincta*)] which has been a serious pest in the West Coast and peninsular areas of India and the Indian migratory locust (*Locusta migratoria*) which is commonly found in different parts of India in solitary phase, but can assume serious proportions under peculiar circumstances.

Hence apart from rigorous surveillance on desert locust, the population of Bombay locust and migratory locust is also watched.

There is a field station at Bikaner (Rajasthan) for investigation into various aspects of locust behaviour, biology and ecology under field conditions and to evolve and evaluate new methods of survey and control. Important lines of study pursued are correlation of meteorological factors and incidence of locusts ; the role of biotic factors—parasites and predators in control of locusts in periods of recession, evaluation and evolution of new methods of control including the application of new insecticides. A training course on locust biology and control is also organised in this station.

Central Plant Protection Stations

The main functions of these stations are to assist the State Governments and Union territories with technical personnel, plant protection machines, pesticides, vehicles, etc., for control of rather sudden outbreaks of pests and diseases, under which conditions resources at the disposal of the State Governments and Union territories may not be adequate. For this purpose each station maintains a buffer stock of different pesticides and plant protection machineries. Besides regional surveys are conducted and *ad hoc* training programme in plant protection are organised.

There are at present nine stations located at New Delhi, Pathankot (Punjab), Bilaspur (Madhya Pradesh), Bihar Sarif (Bihar), Cuttack (Orissa), Gauhati (Assam), Margao (Goa), Tiruchirapally (Tamil Nadu) and Ernakulam (Kerala). In addition four more stations one each in Sikkim (Gangtok), Nagaland (Dimapur), Mizoram (Aizwal), and Andaman and Nicobar Islands (Port Blair) have been established.

The main purposes of these stations is to provide guidance and assistance to the States and farmers as it had been repeatedly seen that in sudden emergencies, created by heavy and widespread infestations of crops by pests and diseases, the resources of the State Governments and Union Territories needed to be supplemented. These Stations are to assist State Governments and others with technical personnel, plant protection machines, pesticides and vehicles for the control of pests and diseases outbreaks. Organisation of regional rapid roving surveys on pests and diseases of major crops is also one of the functions of these stations.

Plant quarantine is essentially designed to prevent the entry or introduction of dangerous pests and diseases from one country into another (foreign quarantine) and of their spread within the country itself (Domestic quarantine). Measures of plant quarantine are enforced under the Government of India's Destructive Insects and Pests Act of 1914 and the various notifications issued under it from time to time.

Plant Quarantine and Fumigation Stations have been functioning at the sea ports of Bhavanagar, Bombay, Cochin, Calcutta, Madras, Rameshwaram, Tuticorin, Nagapattinam and Kandala. The stations are also functioning at the airports at Amritsar, Bombay, Calcutta, New Delhi, Madras, Tirichurapalli, Trivandrum, Patna, Varanasi, Hyderabad and Bangalore. Quarantine stations have also been established at the land frontiers of Attari Wagah Border, Attari Rail Road, Amritsar Rail Head (Indo-Pak Border), Gede, Bangaon, Agartala, Karimganj ferry ghat (Indo-Bangladesh Border) and Sukiapokhari, Kalimpong (Indo-Nepal Border).

Pest and Disease Surveillance Service

A project on surveillance of pests and diseases is being implemented by the Directorate of Plant Protection, Government of India. Seventeen Central Surveillance Stations have been set up in as many districts in fourteen States of India. They are located as follows : Eluru, Nizambad (A.P.) ; Malda (West Bengal) ; Kamrup (Assam) ; Baramulla (J & K) ; Palanpur (H.P.), Karnal (Haryana) ; Jullundur (Punjab) ; Indore, Bilaspur (M.P.) ; Deoria, Lucknow (U.P.) ; Tiruchirapalli (Tamilnadu) ; Shimoga (Karnataka) ; Samastipur (Bihar) ; Kota (Rajasthan) ; Cuttack (Orissa). Reporting centres have been established in 72 N.E.S. Blocks to record quarantine data on population of pest and disease-producing organisms in wheat, rice, sugar cane and potato.

Implementation of Insecticides Act

A separate cell has been created in the Directorate of Plant Protection for implementation of Insecticides Act. It is headed by the Secretary, Central Insecticides Board, who is assisted by supporting staff. The Central Insecticides Laboratory, as per provisions of the Act is to be set up to meet requirements under Insecticides Act.

Biological Control Stations

Biological Control Stations have been set up at Faridabad (Haryana), Srinagar (J & K), Khattsi (Ladakh), Gorakhpur (U.P.), Bangalore (Karnataka), and Hyderabad (A.P.) for mass rearing of dominant indigenous and exotic parasites and predators of cotton boll warms, sugar cane borers, rice stem borers, San Jose Scale, codling moth, *Heliothis* spp., potato tuber moth, and weeds-water hyacinth and *Lantana camara* for releasing them in the infested fields in specified doses, as an experimental measure for determining their efficacy under field conditons in multilocational trials.

Central Plant Protection Training Institute

The Institute located at Hyderabad offers courses on advanced plant protection ; pesticide formulation ; pesticide residue analysis ; pest surveillance, forecasting and forewarning ; plant quarantine ; integrated pest control ; and pest control machineries in the service training of subject matter specialists.

A setup of the Directorate of plant protection Quarantine and Storage, Ministry of Agriculture, Government of India is given herewith.

Organisation of plant protection work in the States/Union Territories

Apart from the Plant Protection Organisation at the Centre, in each State and Union Territory, there is a plant protection organisation. Agriculture being a State subject, it is the responsibility of the respective State Governments to take adequate steps for providing necessary assistance to the growers in control of pests, diseases and weeds. In case of plantation crops, namely, tea, coffee, rubber, there are separate organisations for each of the crop which normally looks after plant protection work in all aspects—technical, supply, service and advisory. Uniform pattern is not followed in respect of administrative setup, nevertheless the basic objectives as detailed below are essentially the same.

Extension work in respect of plant protection forms a major work of the organisation in the States. It involves : (a) preparation of plant protection schemes for control of pests, diseases and weeds often in collaboration with the Agriculture University of the State ; (b) Necessary arrangement of supply of pesticides. Though pesticides are marketed and sold through normal trade channels, nevertheless it is the responsibility of State Organisation to prepare annual requirement and quarterly indent to the appropriate authority of the Central Government for release of pesticides of different categories in time and to see that there is sufficient stock of pesticides ; (c) Suitable arrangement of supply of plant protection application machineries so that they are available to the growers in time, (d) Maintenance of a buffer stock of pesticides and plant protection application machineries to meet any emergent situation in the event of serious outbreaks of pests and diseases ; (e) To undertake specific campaigns against pests, diseases, weeds including rodents, as the case may be, as decided by the State Government; (f) Advisory work to guide the farmers in respect of plant protection work ; (g) Arrangement of training of the extension staff and farmers ; (h) Issue of forewarning wherever possible ; a: 1 (i) To conduct survey and surveillance for pest and disease s tuation as well as estimation of losses due to attack of pests and ciseases. This includes participation in the surveillance work organised by the Government of India.

Apart from extension work, Plant Protection Organisation at the State level has to carry out certain regulatory work. It has specific responsibilities in respect of enforcing the Insecticides Act. The State Government is to appoint a licensing officer, insecticide inspectors and insecticide analysts as well as establish an insecticide analysis laboratory for quality control. The State Government is required to notify the appeallate authority, terms and conditions of appeal in respect of matters arising out of cancellation of licence, etc. The Plant Protection Organisation of the State Government is actively involved in the process of implementation of provisions of the Act so far they relate to action on the part of the State Government.

The State Governments have been authorised to issue phytosanitary certificates for plants and plant commodities intended for export. Normally this assignment is carried out by the officers of the Plant Protection Organisation of the State. Besides it may be required to render assistance in the enforcement of regulations and notifications in respect of domestic or inter-State Quarantine.

SELECTED REFERENCES

Advani, M. L. (1968). 'Dusting powder formulations', *Pesticides*, **1,** 45.

Agricultural Research Council, U.K. (1970). *'Third Report of the Research Committee on Toxic Chemicals'*, HMSO, London.

Albert, A (1968). *'Selective Toxicity'*, Methuen, London.

Anon (1968). *'Proceedings of the joint· United States—Japan Seminar on Microbial Control of Insect Pests'*, Fukoka. US—Japan Comm. Sci. Coop.

Anon (1970). 'Sound as a rodent deterrant', *Bull. Grain Tech.*, **8,** 71.

Anon (1970). 'The selection and design of buildings for food storage', in *Food Storage Manual,* WFP., Rome.

Apple, J. L. (1972). 'Intensified pest Management needs of developing nations', *Bioscience,* **22,** 461-64.

Apple, J. L. and R. F. Smith (Eds.) (1976). *'Integrated Pest Management'*, Plenum Press, New York.

Ascher, K. R. S. and S. Nissim (1964). 'Organotin compounds and their potential use in insect control', *World Rev. Pest Control,* **3,** 188-211.

Asian Productivity Organisation, Production Unit (1974). *'Training Manual. Post-harvest Prevention of Waste and Loss of Food Grains'*, Philippines.

Askew, R. R. (1971). *'Parasitic Insects'*, American Elsevier, New York.

Audus, L. J. (Ed) (1976). *'Herbicides'* Vols. 1 & 2, Academic Press, New York.

Bailey, J. B. and J. E. Swift (1968). *'Pesticide Information and Safety Manual'* Univ. of California, Berkley, U.S.A.

Bailey, S. W. and H. J. Banks (1974). 'The use of controlled atmosphere for the storage of grain', *Proa 1st Int. Working Conf. Stored-Product Entomol.* Savannah, Georgia, U.S.A.

Baker, K. F. and W. C. Snyder (Eds.) (1965). *'Ecology of Soil-Borne Pathogens'*, Univ. of California Press, Berkley, U.S.A.

Baker, R. (1968). 'Mechanism of biological control of soil-borne pathogens', *Ann. Rev. Phytopathology*, **6**, 263-94.

Baker, T. W. (1963). 'The protection of stored products', in *Crop Protection* by G. J. Rose, Leonard Hill, London.

Barkovee, A. B. (1966). 'Insect chemosterilants', *Advances in Pest Control Research*, **7**, 1-140.

Bartels, W. and H. H. Cramer (1966). 'Side effects of plant diseases, plant pests and weeds on the health of man and animals on the quality of harvested products', *Pyanzenschutz. Nachrichten*, Bayer, **19**, 125-86.

'Bayer Pflanzenschults Compendum' (1964). Farben-fabriken Bayer, A. G., Leverkusen, W. Germany.

Beck, S. D. (1965). 'Resistance of plants to insects', *Ann. Rev. Entomol*, **10**, 207-32.

Behnke, John A. (Ed) (1972). *'Challenging Biological Problems, Directing towards their solution'*, Oxford University Press, New York.

Beirne, B. P. (1967). *'Pest Management'*, Leonard Hill, London.

Beirne, B. P. (1975). 'Biological control attempts by introductions against pest insects in the field in Canada', *Can. Entomologist*, **107**, 225-36.

Beye, F. (1978). 'Insecticides from the vegetable Kingdom' *Plant Research and Development*, Institute of Science Co-operation, Tubingen, W. Germany, **7**, 13-31.

Bhatnagar, A. P. and Harcharan Singh (1970). 'Wheat grain moisture requirement under air-tight storage conditions', *J. agric. Engg.* **7**, 46-59.

Biffen, R. H. (1906). 'Mendel's laws of inheritance and wheat breeding', *J. agric. Sci. Camb.* **1**, 4-48.

Bindra, O. S. and Harcharan Singh (1977). *'Pesticide Application Equipment'*, Oxford and IBH, New Delhi.

Bindra, O. S. and H. S. Toor (1972). *'The Common Harmful*

Birds of Punjab and their Control', Punjab Agricultural University Publ., India.

Birch, M. C. (Ed.) (1974). *'Pheromones'*, North Holland, London.

Blum, M. S. (1969). 'Alarm pheromones', *Ann Rev. Entomol.*, **14**, 57-80.

Boosalis, M. G. and R. Mankau (1965). 'Parasitism and predation of soil microorganisms', in *Ecology of Soil-Borne Pathogens*, K. F. Baker and W. C. Snyder (Eds.) Univ. of California Press, Berkley.

Borg, K., H. Wantrup, K. Erne and E. Hanks (1966). 'Mercury poisoning in Swedish wild life' (Summary) 'Pesticides in the environment and their effects on wild life', *J. Appl. Ecol.* **3** (Suppl.), 171-2.

Bosch, R. Vanden and P. S. Messenger (1973). *'Biological control'*, Intext Educational, New York.

Borlaug, N. E. (1968). 'The use of multilineal or composite varieties to control airborne epidemic diseases of self-pollinated crop plants'. *Proc. 1st. Int. Wheat Genet. Symp.*, Winnipeg, Canada.

Borlaug N. E. (1971). 'Mankind and Civilisation at another cross road', 1971 McDougal Memorial Lecture in Food and Agriculture Organisation Conference, Rome.

Bowers. W. S. (1971). 'Juvenile hormones', in *Naturally Occurring Insecticides* M. Jacofson and D. G. Crosby (Eds.), Marcel-Dekker, New York.

Brandes, G. A. (1971). 'Advances in fungicide utilization', *Ann. Rev. Phytopathology*, **9**, 363-386.

Brazelton, R. W., N. B. Akesson and W. E. Yates (1972). *'The safe application of agricultural chemicals—equipment and calibration'*, Univ. of California, Berkley.

Brian, P. W. (1954). 'The use of antibiotics for control of plant disease caused by bacteria and fungi', *Jour. Appl. Bact.*, **17**, 142-151.

Brian, P. W. (1957). 'The ecological significance of antibiotic production', in *Microbial Ecology*, R. E. O. Williams and C. C. Spicer (Eds.), Cambridge University Press, London.

Brian, P. W. (1960). "Griseofulvin", Presidential Address, *Brit. Mycol. Soc.,* **43,** 1-13.

Brian, P. W. (1972), 'The metabolic background for disease resistance', *Proc. 6th. Eucarpia Congress,* Cambridge.

Broadbent, L., P. H. Gregory and T. W. Tinsley (1950). 'Rogueing potato crops for virus diseases', *Ann. Appl. Biol.,* **37,** 640-650.

Brown, A. W. A. (1951). *'Insect Control by Chemicals',* John Wiley, New York.

Brown, A. W. A. and R. Pal (1971). *Insecticide Resistance in Arthropods',* WHO, Geneva.

Brown, W. (1936). 'The physiology of host-parasite relations', *Bot. Rev.,* **2,** 236-281.

Browning, C. B. (1972). 'Systems of the pest management and plant protection', *Report of the RICOP Committee on Plant Protection,* St. Louis, Missouri, U.S.A.

Browning, J. A. and K. J. Frey (1969). 'Multiline cultivators as a means of disease control', *Ann. Rev. Phytopathology,* **7,** 355-382.

Bruehl, G. W. (Ed.) (1975). *'Biology and Control of Soil-Borne Pathogens',* The American Phytopathological Society, St. Paul, Minnesota.

Burgess, H. D. and N. W. Hussey (1971). *'Microbial Control of Insects and Mites',* Academic Press, New York.

Busvine, J. R. (1966). *'Insects and Hygiene',* Methuen, London.

Busvine, J. R. (1971). 'Cross resistance in arthropods of public health importance', Parts I and II, *WHO/VBC/71.307.*

Butler, C. G. (1967). 'Insect pheromones', *Biol. Rev.,* **42,** 42-87.

Byass, J. S. (1963). 'Modern spraying techniques, its principles and practices', *World Crops,* **15,** 276-282.

Byass, J. B. (1969). 'Equipment and method of orchard spray application research', *J. agric. Engg. Res.,* **14,** 78-88.

Calderon, M. (1974), 'The possible role of aeration in the control of stored product insects in warm climates', *Proc. 1st Int. Working Conf. Stored-Product Entomol.,* Savannah, Georgia, U.S.A.

Cameron, E. A. (1973). *'Disparlure* : a potential tool for gypsy moth population manipulation', *Bull. Entomol. Soc. Am.,* **19,** 15-19.

Campion, D. G. (1972). 'Insect chemosterilants : a review', *Bull. Entomol. Res.*, **61**, 577-635.

Cantwell, G. S. (1974). *'Insect Diseases'*, Vols. I & II, Marcel Dekker, New York.

Carlson, G. A. (1970). 'The microeconomics of crop losses' in *Symposium on Economic Research on Pesticides for Policy-decision Making*, U. S. Dept. of Agriculture, Economic Research Service Washington, D.C.

Carson, R. (1963). *'Silent Spring*, Hamish Hamilton, London.

Cassida, J. E. (1963). 'Mode of action of carbamates', *Ann. Rev. Entomol.*, **8**, 39-58.

Cassida, J. E. (Ed. 1973). *'Pyrethrum, The Natural Insecticide'*, Academic Press, New York.

Chapman, R. F. (1969). *'The Insects : Structure and Function* : English Universities Press, London.

Charpentier, L. J. and R. Mathes (1969). 'Cultural practices in relation to stalk moth borer infestation in sugar cane', in *Pests of Sugar cane*, J. R. Williams, J. R. Metcalfe, R. W. Montgomery and R. Mathes (Eds.) Elsevier, New York.

Chatterji, S. (1953). 'Effect of humidity on some pests of stored cereals', *Indian J. Entomol.*, **15**, 327-339.

Cheng, T. H. (1963) 'Insect control in Mainland China', *Science*, **140**, 269-277.

Chensokov, P. G. (1962) *'Methods of Investigating Plant Resistance to Pests'*, (Translation from Russian), Israel Prog. Sci. Trans., Jerusalem.

Chester, K. S. (1947) *'Nature and Prevention of Plant Diseases'*, Blakiston, Philadelphia.

Chester, K. S. (1950) 'Plant disease losses, their apprisal and interpretation', *Pl. Dis Reptr. Suppl.*, **193**, 190-362.

Chester, K. S. (1955) 'Scientific and economic aspects of plant disease loss appraisal', *Ann. Appl. Biol.*, **42**, 335-343.

Chirappa, L. (1974) 'Possibility of supervised plant disease control in pest management systems', *FAO Plant Prot. Bull.*, **22**, 65-68.

Christensen, C. M., E. C. Stakman and J. J. Christensen (1947). 'variation in pathogenic fungi', *Ann, Rev, Microbiol.*, **1**, 61-81.

Christensen, C. M. (Ed.) (1974) *Storage of Cereal Grains and their Products*, Am. Assoc. Cereal Chemists, St. Paul, Minnesota, U.S.A.

Clayphon, G. E. and G. A. Mathews (1972). 'Care and maintenance of spraying equipment in the tropics', *Pans*, **19**, 13-23.

Commonwealth Agriculture Bureaux (1971) *'Crop Loss Assessment Methods'*, (FAO Mannual).

Cope, O. B. (1971) 'Interaction between pesticide and wild life', *Ann. Rev. Entomol.*, **16**, 325-364.

Coppel, H. C. and J. W. Mertins (1977) *'Biological Insect Pest Supression'*, Springer-Verlag, New York.

Corbett, J. R. (1974) *'The Biochemical Mode of Action of Pesticides'*, Academic Press, London.

Council of Environment Quality, (CEQ) (1972) *'Integrated Pest Management'*, U.S. Govt. Printing Office, Washington, D. C.

Council of Europe (1967) *'Agricultural Pesticides'*, Strasbourg, France.

Cornwell, P. B. (1966) *'The Entomology of Radiation Disinfection of Grain'*, Pergammon, Oxford.

Cotton, R. T. (1963) *'Pests of Stored Grain and Grain Products'*, Burgess, Minneapolis, U.S.A.

Crafts, A. S. and W. W. Robins (1962) *'Weed Control'*, McGraw-Hill, New York.

Crammer, H. H. (1967) 'Plant Protection and World Crop Production', *Pflanzenschutz-Nachrichten, Bayer*, **20**, 1-524.

Cruikshank, I. A. M. (1966) 'Defence Mechanisms in plants', *World Rev. Pest Control*, **5**, 161-175.

Damulji, S. F. (1962) 'Mercurial poisoning with the fungicide Granosan-M', *J. Facul. Med.*, Bagdad, **4**, 83.

Damulji, S. F. and S. Tkriti (1972) 'Mercury poisoning from wheat', *Br. Med. J.*, **1**, 804.

Davidson, A. and Richard B. Norgaard (1973) 'Economic aspects of pest control', *Europ. Plant Prot. Org. Bull.*, **3**, 63-75.

Davidson, G. (1974) *'Genetic Control of Insect Pests'*, Academic Press, London.

Day, P. R. (1968) 'Plant disease resistance', *Sci. Prog. Oxf.*, **56**, 357-370.

DeBach, P. (Ed) (1964) *'Biological control of Insect Pests and Weeds'*, Reinhold, New York.

DeBach, P. (1974) *'Biological Control of Natural Enemies'*, Cambridge University Press, London.

Dekker, J. (1971) 'Agricultural use of antibiotics', *World Rev. Pest Control.*, **10**, 9-23.

Deoras, P. J. (1968) 'Two rat traps for control and ecological studies', *Pesticides*, **1**, 38-44.

Department of Education and Science, U.K. (1971) *'Further Review of certain Organochlorine Pesticides used in Great Britain'*, HMSO, London.

Dickman, J. D. (1972) 'Use of insect hormones in pest control', *National Insect Pest Management Workshop*, Purdue University, Laffayette, Indiana, U.S.A.

Dimond, A. E. (1963) 'The selective control of plant pathogens', *World Rev. Pest Control*, **2**, 7-17.

Drummond, R. C. (1966) 'Recent developments in the control of commensal rodents', *Chem. and Industry* (1966), 1371-1375.

Duddington, C. L. (1962) 'Predacious fungi and control of eelworms', in *Viewpoints in Biology*, Vol. I, J. D. Carthy (Ed.), Butterworths, London.

Dutt, N. (1977) *'Rodent Pests and their Management'*, N.S.S. leaflet, Bidhan Chandra Krishi Viswa Vidyalaya, West Bengal, India.

Dykstera, T. P. (1961) 'Production of disease free seed', *Bot. Rev.*, **27**, 445-500.

Dykstera, W. W. (1966) 'The economic importance of commensal rodents', *WHO Seminar in Rodents and Rodent Ectoparasites*, Geneva.

Edwards, C. A. and G. W. Heath (1964) *'Principles of Agricultural Entomology'*, Chapman and Hall, London.

Edwards, C. A. (1969) 'Soil pollutants and soil animals', *Sci. Amer.*, **220**, 88-89.

Federico, V. R. (1967) *'Construction and Operation of Electric Rat Fence'*, The International Rice Research Institute. Philippines.

Fitzwater, W. D. and Iswar Prakash (1966) *'Handbook of Verte-*

brate Pest Control', Central Arid Zone Research Institute, Jodhpur, India.

Flemming, W. E. (1957) 'Soil management and insect control' in *Soil—The Yearbook of Agriculture*, U.S.D.A., Washington D.C.

Flor, H. H. (1971) 'Current status of gene for gene control', *Ann. Rev. Phytopathology*, **9**, 275-296.

Food and Agriculture Organization (FAO) (1966) *'Proceedings of the FAO Symposium on Integrated Pest Control'*, 1965, Vols. 1-3, Rome.

Food and Agriculture Organization (FAO) (1970) *'A Model Scheme for the Establishment of National Organizations for the Official Control of Pesticides'*, Rome.

Food and Agriculture Organization (FAO) (1970) *'Handling and Storage of Food Grains in Tropical and Subtropical Areas'*, Agriculture Development Paper No. **90,** Rome.

Food and Agriculture Organization (FAO) (1973) *'Report of the Fourth Session of the FAO Panel of Experts on Integrated Pest Control'*, December, 1972, Rome.

Food and Agriculture Organization (FAO) (1973) *'The Use of Viruses for the Control of Insect Pests and Disease Vectors'*, FAO Agri. Studies, **91,**.

Food and Agriculture Organization (FAO) (1974) 'The development and application of integrated pest control in agriculture', in *Report on the adhoc session of the FAO Panel of Experts on Integrated Pest Control*, October, 1974, Rome.

Food and Agriculture Organization (FAO) (1974) *'The Use of Mercury and Alternative Compounds as Seed Dressings'* FAO Agri. Studies, **95.**

Franz, J. M. (1964) 'Microorganism in the biological control of insects', in *Global Impacts of Applied Microbiology*, M. P. Starr (Ed.), John Wiley, New York.

Frazer, R. P. (1958) 'The fluid kinetics of application of pesticide chemicals', in *Advances in Pest Control Research*, Vol. II, 1-106, R. L. Metcalf (Ed.), Interscience, New York.

Frear, D. E. H. (1947) 'A Catalogue of Insecticides and Fungicides', Chronica Botania, Waltham, Mass., U.S.A.

Frings, H. and M. Frings (1971) 'Sound production and reception of stored product insect pests—a review of present knowledge', J. Stored Prod. Res., 7, 153-162.

Furmidge. C. G. L. (1972) 'General principles governing the behaviour of granular formulations', Pesticide Sci., 3, 745-751.

Fulton, R. H. (1965) 'Low volume spraying', Ann. Rev. Phytopathology, 3, 175-196.

Galley, R. A. E. (1971) 'The contribution of pesticides used in public health programme to the pollution of the environments', I. General and DDT, WHO/VBC/71, 326.

Garret, S. D. (1944) 'Root Disease Fungi', Chronica Botanica Waltham, Mass., U.S.A.

Garrett, S. D. (1965) 'Towards biological control of soilborne pathogens', in Ecology of Soil-Borne Plant Pathogens, K. F. Baker and W. C. Snyder (Eds.) Univ. of California Press, Berkley.

Georghiou, G. P. (1965) 'Genetic studies on insecticide resistance', Advances in Pest Control Research, 6. 171-230.

Gerott, T. P. (1969) 'Mode of entry of contact insecticides', J. Insect Physiol, 15, 563-580.

Gibbs, A. J. (Ed.) (1973) 'Viruses and invertebrates', North Holland, London.

Gohlich, H. and J. Zaske (1973) 'Research and application technique in crop protection'. Applied Sciences and Development, 2, 50-64, Institute for Science Cooperation, Tubingen, W. Germany.

Gould, R. F. (Ed.) 'New Approaches to Pest Control and Eradication', Advances in Chemistry Series, 41, American Chemical Society, Washington D.C.

Grossman, F. (1968) 'Conferred resistance in the host', World Rev. Pest Control, 7, 176-183.

Gunn, D. L. and J. G. R. Stevens (Eds.) (1976) 'Pesticides and Human Welfare', Oxford University Press, London.

Gupta, B. D. and D. P. Kulshrestha (1957) 'Control of Johnson grass (Sorghum haplense) an alternate host of sugar cane stalk borer, *Chilotraea quricilia* Dudg', *Newsl. Indian Inst, Sugar Res.*, Lucknow, **3,** 1-2.

Hall, D. W. (1955) 'Problems of food storage in tropical countries', *Ann. Appl. Biol.*, **42,** 85-97.

Hammons, J and R. Pal (1968) 'Practical implication of insecticide resistance in arthropods of medical and veterinary importance', *WHO/VBC/68.* **106.**

Hartley, G. S. and T. F. West (1969) *'Chemicals for Pest Control'*, Pergammon, Oxford.

Hatchett, J. H. and R. L. Gallun (1970) 'Genetics of the ability of Hessian fly, *Mayetiola destructor* Say to survive on wheats having different genes for resistance', *Ann. Entomol. Soc. Am.*, **63,** 1400-1407.

Headley, J. C. and J. N. Lewis (1967) 'The pesticide problem : an economic approach to public policy', in *Resources for the Future*, John Hopkins Press, New York.

Headley, J. C. (1972) 'Economics in agricultural pest control', *Ann. Rev. Entomol.*, **17,** 273-286.

Headley, J. C. (1972) 'Defining the economic threshhold', in *Pest Control Strategies for the Future*, National Academy of Science, Washington D.C.

Herford, G. V. B. (1961) 'Food lost in store by insect attack', *Span,* **4,** 40-42.

Higgins, D. J. (1961), *'Fungicides in Agriculture and Horticulture'*, S.C.I. (U.K.) Monograph, **15.**

Hillebrandt, P. M. (1960) 'The economic theory of the use of pesticides', Part I. *Am. J. Agric. Econ.*, **23,** 464-472 and Part II, *ibid,* **24,** 52-61.

Himmel, C. M. (1969) 'The fluorescent particle spray droplet tracer method', *J. Econ. Entomol.*, **62,** 912-916.

Himmel, C. M. (1969) 'The optimum size for insecticidal spray droplets', *J. Econ. Entomol.*, **62,** 919-925.

Holman, L. E. (1966) 'Aeration of grains in commercial storages', *Mktg. Res. Report*, U.S.D.A., **178.**

Holmes, P. M. (1971) 'Pyrotechnic smoke as a simple means of pest control', *Pans,* **17,** 520-521.

Hooker, A. L. (1962) 'Genetics and expression of resistance in plants to rusts of the genus *Puccinia*', *Ann. Rev. Phytopathology*, **5,** 163-182.

Horsfall, J. G. (1956) 'Principles of Fungicidal Action', Chronica Botanica, Waltham, Mass., U.S.A.

Hough, W. S. and A. F. Mason (1951) 'Spraying, Dusting and Fumigating Plants', Macmillan, New York.

Hueffaker, C. B. (1957) 'Fundamentals of biological control of weeds', Hilgardia, 27, 101-157.

Hueffaker, C. B. and C. A. Croft (1975) 'Integrated pest management in the U.S.A.—progress and promise', Environ. Health Pers.

Hyde, M. B. (1965) 'Principles of wet grain conservation', J. Proc. Instn. Agric. Engrs., 21, 75-82.

Hyde, M. B., A. A. Baker, A. C. Ross and C. O. Lopez (1974) 'Airtight grain storage', AGS Bull., 17, FAO, Rome.

'Indian Standards guide for handling cases of pesticide poisoning : symptoms, diagnosis and treatment', I.S. 4015 (Part II), Indian Standards Institution, New Delhi.

International Atomic Energy Agency (1974) 'Sterility Principle for Insect Control', IAEA, Vienna.

Irvine, D. E. H. and B. Knights (Eds.) (1974) 'Pollution and Use of Chemicals in Agriculture', Ann. Arbor. Sci., Michingan, U.S.A.

Isley, D. (1941) 'Methods of Insect Control', Part I, Burgess, Minneapolis, U.S.A.

Jacobson, M. (1965) 'Insect Sex Attractants', John Wiley, New York.

Jacobson, M. (1972) 'Insect Sex Pheromones'. Academic Press, New York.

Jalili, M. A. and A. H. Abbasi (1961) 'Poisoning by ethyl mercury toluene-sulphanilide', Br. J. Ind. Med., 18, 303.

Jones, F. G. W. and D. M. Parrott (1968) 'Potato production using resistant varieties on land infested with potato cyst eelworm Heterodera rostochiensis Woll', Outlook on Agriculture, 5, 215-222.

Justseen, S. H. and P. H. L. Tammes (1960) 'Studies of Yield losses, I : The self limiting effect of injurious or competitive organisms on crop yield', T. Pl. Ziekten, 66, 281-287.

Kearns, H. G. H. (1954) 'Hydraulic spraying machinery for fruit crops. The choice of power equipments', *Ann. Rep. Agric.* and *Hort. Res. Sta.,* Long Ashton, U.K.

Kendrick, J. B., Jr. (1975) 'Pesticide and work safety', *Calif. Agric.,* **29,** 2.

Khan, A. U., A. Amilhussain, J. R. Arboldea, A. S. Manalo and W. J. Chancellor (1974) 'Accelerated drying of rice using heat conduction media', *Trans. Am. Soc. Agric. Engrs.,* **17,** 549-555.

Khaney, M. R. E. (1962) 'Control of flying foxes, *Peteropus giganteus* Brunnich by trapping in Orissa', *Pl. Prot. Bull.,* New Delhi, **14,** 61.

Kilgore, W. W. and R. L. Doutt (Eds.) (1967) *'Pest Control : Biological, Physical and Selected Chemical Methods'* Academic Press, New York.

Klassen, W. (1975) 'Impressions of applied insect pathology in the U.S.S.R.', *U. S. Dept. Agr. Res. Serv.,* Hyattsville, U.S.A.

Klassen, W. (1975) 'Pest Management. Organisation and resources for implementation', in *Insecta, Science and Society,* D. Pimentel (Ed.), Academic Press, New York.

Knipling, E. F. (1960) 'The eradication of screw-worm fly', *Sci. Amer.,* **203,** 54-61.

Knipling, E. F. (1967) 'Sterility technique-principles involved in current application, limitations and future application', in *Genetics of Insect Vectors of Diseases,* J. W. Wright and R. Pal (Eds.) Elsevier, Amsterdam.

Knight, R. L. and F. H. Alston (1974) 'Pest resistance in fruit breeding', in *Biology in Pest and Disease control,* D. Price Jones and M. E. Solomon (Eds.) Blackwell, Oxford.

Krishnamurthy, K. (1965) 'Use of radiation for disinfesting food grains', *Bull. Grain Tech.,* **3,** 59-61.

Kuc, J. (1968) 'Biochemical control of disease resistance in plants', *World Rev. Pest Control,* **7,** 42-96.

Laudani, H. (1967) *'Technology of grain irradiation',* Market Quality Research Division, ARS, U.S.D.A.

Lawrence, W. J. C. (1968) *'Plant Breeding',* Edward Arnold, London.

Lichtenstein, E. P. (1966) 'Persistence and degradation of pesticides in the environment', in *U.S. National Academy of Science*, 1966, 221-9, Washington D.C.

Ling, Lee (1952) *'Digest of Plant Quarantine Regulations'*, FAO Development Paper, **23**, FAO, Rome.

Ling, Lee (1953) 'International Plant Protection Convention : its History, objectives and present status', *FAO Pl. Prot. Bull.*, **1**, 65-68.

Ling, Lee (1954) *'Digest of Plant Quarantine Regulations, Supplement'*, FAO Development Paper, **24**, FAO, Rome.

Link, K. P. and J. C. Walker (1933) 'The isolation of catcheol from pigmented onion scales and its significance in relation to disease resistance', *J. Biol. Chem.*, **100**, 379-384.

Lipa, J. J. (1975) *'An Outline of Insect Pathology'*, (Transl. from Polish), Foreign Sci. Publ. Dep. Nat. Center Sci. Tech. Econ. Inform. Warsaw.

Luginbill, P. (1969) 'Developing resistant plants—the ideal method of controlling insects', *U.S. Dep. Agric. Prod. Res. Rept.*, **111.**

Lupton, F. G. H. (1967) 'The use of resistant varieties in crop protection', *World Rev. Pest Control*, **6**, 47-58.

Lupton, F. G. H. and R. Johnson (1970) 'Breeding for mature plant resistance to yellow rust in wheat', *Ann. Appl. Biol.*, **66**, 137-143.

Lupton, F. G. H. (1974) 'Plant breeding and disease resistance', in *Biology in Pest and Disease Control*, D. Price Jones and M. E. Solomon (Eds.), Blackwell, Oxford.

Maas, W. (1971) *'ULV Application and Formulation Techniques'*, N. V. Philips—Duphar Crop Protection Division, Amsterdam.

Macdonald, D. and J. A.'brook (1963) 'Growth of *Aspergillus yaws* and production of aflatoxin in groundnuts', *Trop. Sci.* **5**, 208-14.

Mani, M. S. (1973) *'General Entomology'*, Oxford and I.B.H., New Delhi.

Maramarosch, K. (Ed.) (1968) *'Insect Viruses'*, Springer, New York.

Marsh, R. W. (Ed.) (1977) *'Systemic Fungicides'*, Longman, New York.

Martin, H. (1964) *'The Scientific Principles of Crop Protection'*, Edward Arnold, London.

Martin, H. (Ed.) (1969) *'Insecticide and Fungicide Handbook'*, Blackwell, Oxford.

Martin, H. and C. R. Worthing (Eds.) (1977) *'Pesticide Manual'*, Brit. Crop. Prot. Council.

Mathews, G. A. and J. E. Clayphon (1973) 'Safety precautions for pesticide application in the tropics', *Pans*, **19**, 1-12.

Mathews, G. A. (1975) 'Determination of droplet size', *Pans*. **21**, 213-225.

Maxwell, F. G., J. N. Jenkins and W. L. Parrott (1972) 'Resistance of plants to insects', *Advan. Agron.*, **24**, 187-265.

Maxwell, F. G. and F. A. Harris (Eds.) (1974) *'Proceedings of the Summer Institute on Biological Control of Plant Insects and Diseases'*, Univ. Mississippi, Jackson, U.S.A.

Mayer, P. (1959) *'4500 Jahre Pflanzenschutz'*, (4500 years of Plant Protection), Ulmer, Stuttgart, W. Germany.

McCubbin, W. A. (1946) 'Preventing plant disease introduction', *Bot. Rev.*, **12**, 101-139.

McCubbin, W. A. (1954) *'The Plant Quarantine Problem'*, Munksgaard, Copenhagen.

Mcfarlane, N. K. (Ed.) (1977) *'Crop Protection Agents—Their Biological Evaluation'*, Academic Press, New York.

Mellanby, K. (1967) *'Pesticides and Pollution'*, Collins, London.

Menn, J. J. and M. Beroza (1972) *'Insect Juvenile Hormones— Chemistry and Action'*, Academic Press, New York.

Metcalf, R. L. (1955) *Organic Insecticides'*, Inter Science, New York.

Metcalf, R. L. and J. J. McKelvey, Jr. (1976) *'The Future of Insecticides* : *Needs and Prospects'*, Wiley—Inter Science, New York.

Milani, R. (1963) 'Genetical aspects of insecticide resistance', *Bull. World. Hlth. Org.*, **29**, 77-87.

Miller, P. R. and M. B. Linn (1954) 'The efficacy of fungicides in the control of certain genera of plant pathogenic fungi, a literature review', *Pl. Dis. Reptr. Suppl.*, **226**, 54-71.

Milthorpe, F. L. (1945) 'The compatibility of protectant seed dusts with root nodule bacteria', *J. Austr. Inst. Agric. Sci.*. **11**. 89-92.

Ministry of Agriculture, Fisheries and Food, U.K. (1971) 'Pesticide Safety Precautions scheme agreed between Government Departments and Industry', HMSO, London.

Moody, D. E. M. and D. P. Moody (1963) 'Toxic products in groundnuts', Nature, London, 198, 1062-1063.

Moore, N. W. (1967) 'A synopsis of pesticide problem', Adv. Ecolog. Res., 4, 75-130.

Morgan, N. C. (1972) 'Spray application in the plantation crops', Pans, 18, 316-326.

Mundkur, B. B. (1967) 'Fungi and Plant Diseases', Macmillan, Calcutta.

Munro, J. W. (1966) 'Centrifugal force', in Pests of Stored Product', Hutchinson, London.

Murton, R. K. and E. W. Wright (Eds.) (1968) 'The Problems of Birds as Pests', Institute of Biology/Academic Press, New York.

Nagarkatti, S. and K. Ramchandran Nair (1973) 'The influence of wild and cultivated Graminae and Cyperaceae on populations of sugar cane borers and their parasites in North India', Entomophaga, 18, 419-430.

National Council of Applied Economic Research, (1967) 'Pesticides in Indian Agriculture', NCAER, New Delhi

Nash, R. G. and T. H. Cheng (1965) 'Research and development of food resources in Communist China', Part I, Bioscience, 15, 643-656.

Nation, H. J. (1970) Separation of the application factors which determine sprayer performance', Proc. Brit. Crop. Protect. Counc. Sympos. on Pesticide Application, London.

National Academy of Sciences (1970) 'Principles of Plant and Animal Pest Control' Vol. 5, Vertebrate pests; Problems and Control', Washington, D.C.

National Academy of Sciences (1972) 'Pest Control Strategies for the Future', Washington D.C.

Neergaard, P. (1972) 'International and national cooperation in seed health testing and certification', Proc. Int. Seed Test. Ass., 37, 117-138.

Neff, J. A. and R. T. Mitchell (1955) 'The rope fire cracker : a device to protect crops from bird damage', U.S. Fish and Wildlife Service, Wildlife leaflet, 365, 1955.

Negherbon, W. O. (1956-59) 'Hand-book of Toxicology', W. B. Saunders, Philadelphia.

Nelson, S. O. (1974) 'Radiofrequency, infrared, and ultra-violet radiation for control of stored product insects : prospects and limitations', Proc. 1st. Int. Working Conf. Stored-Product Entomol. Savannah, Georgia, U.S.A.

Nene, Y. L. (1971) 'Fungicides in Plant Diseases Control', Oxford and IBH, New Delhi.

Newhall, A. G. (1955) 'Disinfestation of soil by heat, flooding and fumigation', Bot. Rev., 21, 189-250.

Noorgaard Richard B. (1976) 'Integrating economics and pest management', in Integrated Pest Management, J. L. Apple and R. F. Smith (eds.) Plenum. Press, New York.

O'Brien, R. D. (1967) 'Insecticides : Action and Metabolism', Academic Press, New York.

O'Brien, R. D. and J. Yamamoto (1970) 'Biochemi al Toxicology of Insects', Academic Press, New York.

Oppenoorth, F. J. (1965) 'Biochemical genetics of insecticide resistance', Ann. Rev. Entomol., 10, 185-206.

Ordish, G. (1952) 'Untaken Harvest', Constable, London.

Ordish, G. (1967) 'Biological Methods in Crop Pest Control', Constable, London.

Pal, R. and M. J. Whitten (Eds.) (1974) 'The Use of Genetics in Insect Control', American Elsevier, New York.

Pathak, M. D. (1970) 'Genetics of plant in pest management', in Concepts of Pest Management, R. L. Rabband, and F. E. Gutherie (Eds.) State University, Raleigh, North Carolina, U.S.A.

Pearson, A. J. A. (1977) 'Fungicides—the new generation', Span, 20, 129-132.

Pederson, P. N. (1960) 'Methods of testing pseudo-resistance of barley to infection by loose smut (Ustilago nuda Gens.) (Rostr.)', Acta. Agric. Scand., 10, 312-322.

Pingale, S. V., K. Krishnamurthy and T. Ramsivan (1967) 'Rats, Food grain Technologists' Research Association, Hapur, U.P., India.

'Pesticides in the modern world' (1972), A symposium prepared by members of the Co-operative Programme of Agro-allied Industries with FAO and other United Nations Organizations.

Phillips, R. W. (1955) 'Plant Protection activities of FAO designed to strengthen national services and develop international cooperation', *FAO Pl. Prot. Bull.*, **3**, 128-138.

Poinar, G. O., Jr. (1975) *'Entomogenous Nematodes'*, E. J. Brill, Leiden.

Polon, J. A. (1973) 'Formulation of pesticidal dusts, wettable powders and granules', in *Pesticide Formulation*, W. V. Valkenburg (Ed.), Marcel-Dekker, New York.

Potter, M. C. (1908) 'On a method of checking parasitic disease in plants', *J. agric. Sci.*, **3**, 102-107.

Pradhan, S. (1968) 'Analysis of grain storage problems in India *Indian J. Entomol.*, **30**, 94-103.

Pramer, D. (1955) 'Antibiotics against plant diseases', *Sci. Amer.*, **192**, 82-88.

Price Jones, D. and M. E. Solomon (1974) *'Biology in Pest and Disease Control'*, Blackwell, Oxford.

Propic, Z. (1953) 'Quarantine service in plant protection', *Agron. Glasnik*, **3**, 522-527.

Pruthi, H. S. and Mohan Singh (1950) *'Pests of Stored Grain'*, Indian Counc. Agric. Res., New Delhi.

Pruthi, H. S. (1969) *'Textbook on Agricultural Entomology'*, Indian Counc. Agric. Res., New Delhi.

Pyenson, L. L. (1951) *'Elements of Plant Protection'*, John Wiley, New York.

Rabb, R. L., R. E. Stimer. G. A. Carlson (1974) 'Ecological Principles as a basis for pest management in agro-ecosystem', in *Proceedings of the Summer Institute on Biological Control of Plant Insects and Diseases*, F. G. Maxwell and F. A. Harris (Eds.) Univ. Mississippi. Jackson, U.S.A.

Rao, V. P., M. A. Ghani, T. Sankaran and K. C. Mathur (1971) 'A review of the biological control of insects and other pests in South-East Asia and Pacific regions', *Commonwealth, Inst. Biol. Contr. Tech. Commun.*, **6**.

'Rat Control in Rice Fields', (1976) Joint recommendations of the Bureau of Plant Industry, the College of Agriculture,

University of Philippines at Los Banos, the Rodent Research Centre and the Phillipine-German Crop Protection Programme.

Reddy, D. B. (1956) 'Insecticides, fungicides and weedicides and application equipment', Pl. Prot., Vols. 2 & 3, Hyderabad Farmers' Union, India.

Reddy, D. B. (1967) 'The importance of pesticides in Indian food production', Proc. Roy. Soc. London B, 167, 145-154.

Reddy, D. B. (1968) 'Plant Protection in India', Allied Publishers, Calcutta.

Reinwater, H. L. and C. A. Smith (1966) 'Quarantine—first line of defence', in Protecting Our Food, Year Book of Agriculture, 1966, USDA, Washington D.C.

Renfro, B. L. (1969) 'The concept of systemic control of fungal diseases', Indian Phytopathology, 22, 157.

Renjen, P. L., M. V. Venkatesh and N. C. Joshi (1960) 'Plant quarantine in India', Sci. and Cult., 28, 215-218.

'Report of the National Commission on Agriculture, Part X, Inputs, Ministry of Agriculture and Irrigation, Government of India, New Delhi, 1976.

Rice, B. (1970) 'Ground-crop sprayer testing ; A review of procedures and techniques, Proc. Brit. Crop Prot. Counc. Sympos. on Pesticide Application, London.

Richards, O. W. (1961) 'The theoretical and practical study of natural insect populations', Ann. Rev. Entomol. 6, 147-162.

Ripper, W. E. (1955) 'Application method for crop protection chemical', Ann. Appl. Biol., 42, 288-324.

Robb, R. L. and F. E. Gutherie (Eds.) (1970) 'Concepts of Pest Management', State University, North Carolina, Raleigh, U.S.A.

Roberts, D. E. and D. B. Broker (1975) 'Grain drying with a recirculator', Trans. Am. Soc. Agric. Engrs., 18, 181-184.

Robinson, J. (1970) 'Birds and pest control chemicals', Bird Study, 17, 195.

Rose, G. J. (1963) 'Crop Protection', Leonard Hill, London.

Rudd, R. L. (1964) *'Pesticides and Living Landscape'*, Faber and Faber, London.

Sanford, G. B. (1926) 'Some factors affecting the pathogenecity of *Actinomyces scabies'*, *Phytopathology*, **16**, 525-547.

Schrader, G. (1963) *'Die Entwicklung neuer insektiz der Phosphorosaure-Ester'*, Verlag Chemie, G.M.B.H. Weinheim/ Bergstr. W. Germany.

Schumutter, H. (1969) *'Pests of crops of North-east and Central Africa'*, Gustav-Fischer, Stuttgart, W. Germany.

Schutmann, W. (1968) 'Chronic liver disease after occupational exposure to dichloro diphenyl trichloroethane (DDT) and hexachloro cyclo hexane (HCH)' (in German), *Int. Arch Gewerbepath*, **24**, 193-210, WHO/BVC/Toxc./ 69.3

Scott, W. P., E. P. Lloyd, J. O. Bryson and T. B. Davich (1974) 'Trap plots for suppression of low density overcounted populations of boll weevils', *J. Econ. Entomol.*, **67**, 281-283.

Sharvalle, E. G. (1961) *'The Nature and Uses of Modern Fungicides'*, Burgess, Minneapolis, U.S.A.

Shaver, T. N. (1974) 'Biochemical basis of resistance of plants to insects : antibiotic factors other than phytohormones', in *Proceedings of the Summer Institute on Biological Control of Plant Insects and Diseases*, F. G. Maxwell and F. A. Harris (Eds.), Univ. Mississippi, Jackson, U.S.A.

Shephard, H. H. (1939) *'The Chemistry and Toxicity of Insecticides'*, Burgess, Minneapolis, U.S.A.

Shephard, H. H. (1951) *'The Chemistry and Action of Insecticides'*, McGraw-Hill, New York.

Siama, K., M. Romanuk and F. Sorm (1974) *'Insect Hormones and Bioanalogues'*, Springer, New York.

Siminoff, P. and D. Gottileb (1951) 'The production and role of antibiotics in the soil', *Phytopathology*, 41, 420-430.

Sisler, H. D. (1963) 'Fungitoxic mechanism', in *Perspectives of Biological Plant Pathology*, S. Rich (Ed.) *Conn. Agric. Exp. Sta. Bull.*, **663**.

Slater, Sir W. (Ed.) (1967) 'A discussion on pesticides, benefits and dangers', *Proc. Roy. Soc. London B.*, **167**, 88-163.

Smith, C. N. (Ed.) (1966) 'Insect Colonization and Mass Production', Academic Press, New York.

Smith, L. B. (1974) 'The role of low temperature to control stored food pests', Proc. 1st. Int. Conf. Stored-Product-Entomol., Savannah, Georgia, U.S.A.

Smith, R. F. and R. Vanden Bosch (1967) 'Integrated control in pest control', in Biological, Physical and Selected Chemical Methods, W. W. Kilgore and R. L. Doutt (Eds.), Academic Press, New York.

Smith, R. F. (1972) 'The impact of green revolution on plant protection in tropical and subtropical countries', Bull. Entomol. Soc. Am., 8, 7-14.

Snyder, W. C. (1960) 'Antagonism as a plant disease control principle', in Biological and Chemical control of Plant and Animal Pests, L. P. Reitz (Ed.) Am. Assoc. Advan. Sc. Publ., 61.

Society of Chemical Industry, England (1970) 'Technological Economics of Crop Protection and Pest Control', Monograph no. 36.

Sondheimer, E. and J. B. Simenone (1970) 'Chemical Ecology', Academic Press, New York.

Stakman, E. C. and J. G. Harrar (1957) 'Principles of Plant Pathology', Ronald Press, New York.

Staniland, L. N. (1969) 'The principles of hotwater treatment of plants', in Plant Nematology, J. F. Southey (Ed.) Min. Agri. Fish and Food, U.K., Tech. Bull. 7, HMSO, London.

Steinhaus, E. A. (Ed.) (1963) 'Insect Pathology : An Advanced Treatise', Vols. I & II, Academic Press, New York.

Stevens, R. B. (1960) 'Cultural practices in disease control', in Plant Pathology—An Advanced Treatise J. G. Horsfall and A. E. Dimond (Eds.), Academic Press, New York.

Strayer, J. (1972) 'The pest management concept, the extension entomologist's view', Proc. Tall Timbers Conf. Ecol. Animal Contr. Habitat Managem., 3, 21-22.

Strobel, A. (1975) 'A mechanism of disease resistance in plants', Sci. Amer., 232, 80-88.

Thakur, C. (1977) 'Weed Science', Metropolitan Book Co., New Delhi.

Thirumalachar, M. J., P. W. Rahalkar, R. S. Sukapure and K. S. Gopalkrishnan (1964) 'Aureofungin' a new heptene antibiotic. I. "Microbiological Studies" ', *Hindusthan Antibiot. Bull.* **6,** 108-111.

Thorn, G. D. and R. A. Ludwig (1962) *'The dithiocarbamates and related compounds',* Elsevier, Amsterdam.

Tilton, E. W. and J. H. Barrower (1973) 'Status of USDA research on irradiation disinfestation of grain and grainproducts', *Radiation Preservation Food Conf. Proc.,* Bombay, India. IAEA.

Torgensen, D. C. (Ed.) (1967) *'Fungicides : An Advanced Treatise',* Vols. I & II, Academic Press, New York.

Tothill, J. D. (1958) 'Some reflections on causes of insect outbreaks', *Proc. 10th. Int. Congr. Entomol.,* Montreal, 1956.

U. S. Dept. of Agriculture Symposium (1966) 'Pest Control by Chemical, biological, genetical and physical means', ARS, 33-110.

Vanider Laan, P. A. (Ed.) (1967) *'Insect Pathology and Microbial Control',* Proc Int. Colloquium on Insect Pathology and Microbial Control, Wageningen, North Holland, Amsterdam.

Vanider Plank, J. E. (1963) *'Plant Diseases : Epidemics and Control',* Academic Press, New York.

Von Schmeling, B. and M. Kulka (1966) 'Systemic fungicidal activity of 1,4-Oxathin derivatives', *Science.* **152,** 659-660.

Wainman, H. E., B. Chakrabarti and A. H. Harris (1974) 'Fumigation of grain in butyl rubber silos', *Int. Pest Control,* **16,** 5-9.

Watt, K. E. F. (1973) *'Principles of Environmental Science',* McGraw-Hill, New York.

Walters, F. L. (1972) 'Control of storage insects by physical means', *Trop. Stored Prod. Inf.,* **23,** 13-28.

Walker, J. C. (1953) 'Disease resistance in vegetable crops' *Bot. Rev.,* **19,** 606-643.

Walker, J. C. (1957) *'Principles of Plant Pathology',* McGraw-Hill, New York.

Watson, T. F, K. K. Barnes, J. E. Slosser and D. G. Fulerton

(1974) 'Influence of plowdown dates and cultural practices on spring moth emergence of pink bollworm', *J. Econ. Entomol.*, **67**, 207-210.

Wegler, R. (Ed.) (1970) *'The chemistry of Plant Protection and Pest Control Agents'*, Springer-Verlag, Berlin, New York.

Wheeler, B. E. J. (1976) *'Diseases in Crops'*, Oxford and IBH, New Delhi.

Weindling, R. (1934) 'Studies on a lethal principle effective in parasitic action of *Trichoderma lignorum* on *Rhizoctonia solani* and other soil fungi', *Phytopathology*, **24**, 1154-1179.

Whitney, W. K., S. O. Nelson and H. H. Walkden (1961) 'Effects of high frequency electric fields on certain species of stored-grain insects', MRR No. **455**, ARS, USDA.

WHO (1962) *'Principles governing consumer safety in relation to pesti.ide residues'*, Tech. Report Series No. **240**.

WHO (1974) *'Equipments for Vector Control'*, Geneva.

Wilkin, D. R. and A. H. Green (1970) 'Polythene sacks for the control of insects in bagged grain', *J. Stored Prod. Res.* **6**, 97-101.

Williams, C. M. (1967) 'Third generation pesticides', *Sci. Amer.*, **217**, 13-17.

Wingard, S. A. (1953) 'The nature of resistance to disease', in *Plant Disease* Yearbook, USDA. U. S. Govt. Printing Press, Washington, D.C.

Wood, R. K. S. (1967) *'Physiological Plant Pathology'*, Blackwell, Oxford.

Woods, A. (1974) *'Pest Control : A Survey'*, McGraw-Hill, Maidenhead, England.

Woodwell, G. M. (1967) 'Toxic substances and ecological cycle', *Sci. Amer.*, **216**, 24-31.

APPENDIX

Some important provisions of
the Insecticides Act, 1968, Govt. of India
No. 46 of 1968
and
the Insecticides (Amendment) Act, 1977,
Govt. of India No. 24 of 1977
and
the Insecticide Rules, 1971 and its amendment
in 1977, Government of India

Purpose

An Act to regulate the import, manufacture, sale, transport, distribution and use of insecticides with a view to prevent risk to human beings or animals and matters connected therewith.

Definitions

Insecticide means
(a) Any substance specified in the schedule,
(b) Any other substances (including fungicides and weedicides) as the Central Government may, after consultation with the Board, include in the schedule,
(c) Any preparation containing one or more of such substances.

Label means any printed, written or graphic matter on the immediate package and on every other covering in which the package is placed or packed or includes any written, printed or graphic matter accompanying the insecticide.

Manufacture in relation to any insecticide includes :
(a) Any process or part of process for making, altering, finishing, packing, labelling, breaking up or otherwise treating or adopting any insecticide with a view to its sale distribution or use, but does not include the packing or breaking up of any insecticide in the ordinary course of retail business, and
(b) Any process by which a preparation containing an insecticide is formulated.

Misbranded, an insecticide shall be deemed to be misbranded

(a) If its label contains any statement, design, or graphic representation relating thereto which is false, or misleading in any material particular or if its package is otherwise deceptive in respect of its contents, or

(b) If it is an imitation of, or is sold under name of another insecticide, or

(c) If its label does not contain a warning or caution which may be necessary and sufficient if complied with, to prevent risk to human beings or animals, or

(d) If any word, statement or other information required by or under this Act to appear on the label is not displayed thereon in such conspicuous manner as the other words, statements, designs or graphic matter have been displayed on the label and in such terms to render it likely to be read and understood by any ordinary individual under customary conditions of purchase and use ; or

(e) If it is not packed or labelled as required by or under this Act ;

(f) If it is not registered in the manner required by or under this Act ; or

(g) If the label contains any reference to registration other than the registration number ; or

(h) If the insecticide has a toxicity which is higher than the level prescribed or is mixed or packed with any substance so as to alter its nature or quality or contains any substance which is not included in the registration ;

Package means a box, bottle, casket, tin, barrel, case, receptacle, sack, bag, wrapper or any other thing in which the insecticide is placed or packed.

Premises means any land, shop, stall or place where any insecticide is sold or manufactured or stored or used and includes any vehicle carrying insecticides.

"Sale" means sale of any insecticide whether for cosh, or on credit, and whether by wholesale or retail and includes an

agreement for sale, an offer for sale. the exposing for sale, and having in possession for sale of any insecticide and includes also an attempt to sell any such insecticides.

CENTRAL INSECTICIDES BOARD

(1) Central Insecticides Board is constituted to advise the Central Government and State Governments on all technical matters arising out of the administration of this Act and to carry out other functions assigned to the Board by or under the Act.

(2) The matters on which the Board may advise shall include matters relating to :

(a) The risk of human beings or animals involved in the use of insecticides and safety measures necessary to prevent such risks ;

(b) The manufacture, sale, storage, transport and distribution of insecticides with a view to ensure safety to human beings or animals.

(3) The Board shall, in addition to the functions assigned to it by the Act, carry out the following function :

(a) Advise the Central Government on the manufacture of insecticides under the Industries (Development and Regulation) Act, 1951 (65 of 1951) ;

(b) Specify the uses of classification of insecticides on the basis of their toxicity as well as their being suitable for aerial application ;

(c) Advise tolerance limits for insecticides residue and on establishment of minimum intervals. between the application of insecticides and harvest in respect of various commodities ;

(d) Specify the shelf-life or insecticides ;

(e) Suggest colourisation including colouring matter which may be mixed with concentration of insecticides, particularly those of highly toxic nature ;

(f) Carry out such other function as are supplemental, incidental or consequential to any of the functions conferred by this Act or these rules.

(4) The Board shall consist of following members :

(a) The Director General of Health Services, ex officio, who shall be the Chairman ;

(b) The Drug Controller, India, ex officio ;

(c) The Plant Protection Adviser to the Government of India, ex officio ;

(d) The Director of Storage and Inspection, Ministry of Food, Agriculture, Community Development and Cooperation (Department of Food) ex officio ;

(e) The Chief Adviser of Factories, ex officio ;

(f) The Director, National Institute of Communicable Diseases, ex officio ;

(g) The Director General, Indian Council of Agricultural Research, ex officio ;

(h) The Director General, Indian Council of Medical Research, ex officio ;

(i) The Director, Zoological Survey of India, ex officio ;

(j) The Director General, Indian Standards Institution, ex officio ;

(k) The Director General of Shipping, or in his absence, the Deputy Director General of Shipping, Ministry of Transport and Shipping, ex officio ;

(l) The Joint Director, Traffic (General) Ministry of Railways, (Railway Board) ex officio ;

(m) The Secretary, Central Committee of Food Standards, ex officio ;

(n) The Animal Husbandry Commissioner, Department of Agriculture, ex officio ;

(o) The Joint Commissioner (Fisheries), Department of Agriculture, ex officio ;

(p) The Deputy Inspector General of Forests (Wild Life), Department of Agriculture, ex officio ;

(q) The Industrial Adviser (Chemicals) Directorate General of Technical Development, ex officio ;

(r) One person to represent the Ministry of Petroleum and Chemicals to be nominated by the Central Government ;

(s) One pharmacologist to be nominated by the Central Government ;

(t) One medical toxicologist to be nominated by the Central Government ;

(u) One person who shall be in charge of the department dealing with public health in a State to be nominated by the Central Government ;

(v) Two persons who shall be the Director of Agriculture in States to be nominated by the Central Government ;

(w) Four persons, one of whom shall be an expert in industrial health and occupational hazards, to be nominated by the Central Government.

REGISTRATION COMMITTEE

(1) The Central Government shall constitute a Registration Committee consisting of a chairman, and not more than five persons who shall be members of the Board (including the Drugs Controller, India and the Plant Protection Adviser to the Government of India) :

(a) To register insecticides after scrutinising their formulations and verifying claims made by the importer or the manufacturer, as the case may be, as regards their efficacy and safety to human beings and animals, and

(b) To perform such other functions as are assigned to it by or under this Act.

(2) The Registration Committee shall in addition to the functions assigned to it by the Act perform the following functions :

(a) Specify the precautions to be taken against poisoning through the use of handling of insecticides,

(b) Carry out such other incidental or consequential matters necessary for carrying out the functions assigned to it under the Act or the rules made thereunder.

Secretary

The Central Government shall :

(a) Appoint a person to be the Secretary of the Board who shall also function as Secretary to the Registration Committee, and

(b) Provide the Board and the Registration Committee with such technical and other staff as the Central Government considers necessary.

Registration of insecticides

(1) Any person desiring to import or manufacture insecticide may apply to the Registration Committee for the registration of such insecticide and there shall be a separate application for each insecticide.

(2) Every application in subsection (1) shall be made in such form and contain such particulars as may be prescribed.

(3a) On receipt of any such application for the registration of an insecticide, the Committee may, after such enquiry as it deems fit and after satisfying itself that the insecticide to which the application relates conforms to the claims made by the importer or by the manufacturer as the case may be, as regards the efficacy of the insecticide and its safety to human beings and animals, register on such conditions as may be specified by it and on payment of such fees as may be prescribed, the insecticide, allot a registration number thereto and issue a certificate of registration in token thereof within a period of twelve months from the date of receipt of the application.

Provided that the Committee may, if it is unable within the said period to arrive at a decision on the basis of the materials placed before it, extend the period by a further period not exceeding six months.

Provided further that if the Committee is of opinion that the precautions claimed by the applicant as being sufficient to ensure safety to human beings or animals are not such as can be easily observed or that notwithstanding the observance of such precautions the use of insecticide involves serious risk to human beings and animals, it may refuse to register the insecticide.

(3b) Where the Registration Committee is of the opinion that the insecticide is being introduced for the first time in India, it may, pending any enquiry, register it provisionally for two years on such conditions as may be specified by it.

(3c) The Registration Committee may, having regard to the efficacy of an insecticide and its effect on human beings and animals, vary the conditions subject to which a certificate of registration has been granted and may for that purpose require the certificate-holder by notice in writing to deliver up the certificate within such time as may be specified in the notice.

(4) Notwithstanding anything contained in section, where

an insecticide has been registered on the application of any person, any other person desiring to import or manufacture the insecticide or engaged in the business of, import or manufacture thereof shall on application and on payment of prescribed fee be allotted a registration number and granted a certificate of registration in respect thereof on the same condition on which the insecticide was originally registered.

Appeal against non-registration or cancellation

Any person aggrieved by a decision of the Registration Committee, may within a period of thirty days from the date on which the decision is communicated to him, appeal in the prescribed manner and on payment of prescribed fee to the Central Government whose decision thereon shall be final.

Provided that the Central Government may entertain an appeal after the expiry of the said period, if it is satisfied that the appellant was prevented by sufficient cause from filing the appeal in time.

Powers of revision of Central Government

The Central Government may, at any time, call for the records relating to any case in which the Registration Committee has given a decision for the purpose of satisfying itself as to the legality or propriety of any such decision and may pass any such order in relation thereto as it thinks fit.

Provided that no such order shall be passed after the expiry of one year of the decision.

Provided further that the Central Government shall not pass any order prejudicing any person unless that person has had a reasonable opportunity of showing cause against the proposed order.

Licensing Officers

The State Government may, by notification in the Official Gazette, appoint such persons as it thinks fit to be licensing officers for the purpose of this Act and define the areas in respect of which they shall exercise jurisdiction.

Grant of Licence

(1) Any person desiring to manufacture or to sell, stock or exhibit for sale or distribute any insecticide or undertake commercial pest control with the use of any insecticide may make an application to the licensing officer for grant of a licence ;

(2) Every application under subsection (1) shall be made in such form and contain such particulars as may be prescribed ;

(3) On receipt of any such application for the grant of a licence, the licensing officer may grant a licence in such form on such conditions and on payment of such fees as may be prescribed ;

(4) A licence granted under this section shall be valid for the period specified therein and may be renewed from time to time for such period and on payment of such fees as may be prescribed ;

Provided that where a licence has been granted to a person who has made application under the appropriate provision of the law that the licence shall be deemed to be cancelled in relation to any insecticide, the application for registration whereof has been refused or the registration whereof has been cancelled, under this Act with effect from the date on which refusal or cancellation is notified in the Official Gazette.

Revocation, suspension and amendment of licences

(1) If the licensing officer is satisfied either on a reference made to him in this behalf or otherwise, that :

(a) the licence granted under Section 13 (grant of licence) of the Act has been granted because of misrepresentation as to an essential fact ; or

(b) the holder of a licence has failed to comply with the conditions subject to which the licence was granted or has contravened any of the provisions of this Act or the rules made thereunder, then without any prejudice to any other penalty to which the holder of the licence may be liable under this Act, the licensing officer may after giving the holder of the licence an opportunity of showing cause, revoke or suspend the licence.

(2) Subject to any rules that may be made in this behalf,

the licensing officer may also vary or amend a licence granted under section 13 (grant of licence) of the Act.

Appeal against the decision of Licensing Officer

(1) Any person aggrieved by a decison of a licensing officer under Section 13 or 14 (revocation, suspension and amendment of licence) may within a period of thirty days from the date on which the decision is communicated to him, appeal to such authority in such manner and on payment of such fee as may be prescribed :

Provided that the appellate authority may entertain an appeal after the expiry of the said period if it is satisfied that the applicant was prevented by sufficient cause from filing the appeal in time.

(2) On receipt of an appeal, under subsection (1) the appellate authority shall, after giving the appellant an opportunity of showing cause, dispose of the appeal ordinarily within a period of six months and the decision of the appellate authority shall be final.

Central Insecticides Laboratory

The Central Government may by notification in the Official Gazette, establish a Central Insecticides Laboratory under the control of a Director to be appointed by the Central Government to carry out the functions entrusted to, by or under this Act.

Provided that if the Central Government so directs by a notification in the Official Gazette, the functions of the Central Insecticides Laboratory shall, to such extent as may be specified in the notification, be carried out at any such institution as may be specified therein and thereupon the functions of the Director of the Central Insecticides Laboratory shall to the extent so specified be exercised by the head of that institution.

The functions of the Laboratory shall be as follows :

(a) To analyse such samples of insecticides sent to it under the Act by any Officer or authority authorised by the Central or State Governments and submission of analyses to the concerned authority ;

(b) To analyse samples of materials for insecticide residues under the provisions of the Act ;

    (c) To carry out such investigations as may be necessary for the purpose of ensuring the conditions of Registration of Insecticides ;

    (d) To determine the efficacy and toxicity of insecticides ;

    (e) To carry out such other functions as may be entrusted to it by the Central Government or by a State Government with the permission of the Central Government and after consultation with Insecticide Board.

Prohibition of import and manufacture of certain insecticides

No personal shall by himself or by any person on his behalf import or manufacture :

    (a) Any misbranded insecticide ;

    (b) Any insecticide the sale, distribution or use of which is prohibited under appropriate section of the Act ;

    (c) Any insecticide except in accordance with the conditions on which it is registered ;

    (d) Any insecticide in any other provision of this Act or of any rule made thereunder.

Provided that any person who has applied for registration of an insecticide under proviso to subsection (1) of Section 9 of the Act may continue to import or manufacture any such insecticide and such insecticide shall not be deemed to be a "misbranded" insecticide until he has been informed by the Registration Committee of its decision to refuse to register the said insecticide.

No person shall, himself or by any person on his behalf, manufacture any insecticide except under, and in accordance with the conditions of a licence issued for such purposes under this Act.

Prohibition of sale, etc. of certain insecticides

(1) No person shall himself or by any person on his behalf, sell, stock or exhibit for sale, distribute, transport, use, or cause to be used by any worker :

    (a) Any insecticide which is not registered under this Act :

(b) Any insecticide, the sale, distribution or use of which is prohibited for the time being under Section 27 (Prohibition of sale, etc. of insecticides for reasons of public safety) of the Act.

(c) Any insecticide in contravention of any other provision of this Act or of any rule made thereunder.

(2) No person shall, himself or by any person on his behalf sell, stock or exhibit for sale, or distribute, or use for commercial pest control operations any insecticide except under, and in accordance with the conditions of a licence issued for such purpose under this Act.

For the purpose of this section, an insecticide in respect of which any person has applied for a certificate of registration under proviso to subsection (1) of Section 7 of the Act shall be deemed to be registered till the date on which the refusal to register such insecticide is notified in the Official Gazette.

Insecticide Analysts

The Central Government or a State Government may, by notification in the Official Gazette, appoint persons in such number as it thinks fit and possessing such technical and other qualifications as may be prescribed to be Insecticide Analysts for such areas and in respect of such insecticides or class of insecticide may be specified in the notification :

Provided that no person who has any financial interest in the manufacture or sale of any insecticide, shall be so appointed.

The Insecticide Analyst shall have the power to call for such information or particulars or do anything as may be necessary for the proper examination of the samples sent to him either from the Insecticide Inspector or the person from whom the samples were obtained.

The Insecticide Analyst shall analyse or cause to be analysed or test or cause to be tested such samples of insecticides as may be sent to him by the Insecticide Inspector under the provisions of the Act and shall furnish reports or results of such tests or analysis.

An Insecticide Analyst shall, from time to time, forward to the State Government reports giving the result of analytical work and investigation with a view to their publication at the discretion of the Government.

Insecticide Inspectors

(1) The Central Government or a State Government may, by notification, in the Official Gazette, appoint persons in such number as it thinks fit and possessing such technical and other qualifications as may be prescribed to be Insecticide Inspectors for such areas as may be specified in the notification :

Provided that any person who does not possess the required qualifications may be so appointed for the specified purposes (mentioned in clauses (a) and (d) of subsection (1) of Section 21 (Powers of Insecticide Inspector).

Provided that no person who has any financial interest in the manufacture, import or sale of any insecticide shall be so appointed.

(2) Every Insecticide Inspector shall be deemed to be a public servant within the meaning of Section 21 of the Indian Penal Code and shall be officially subordinate to such authority as the Government appointing him may specify in this behalf.

(3) The Insecticide Inspector shall have the following duties :

(a) To inspect not less than three times in a year all establishment selling insecticides within the area of his jurisdiction ;

(b) To satisfy himself that the conditions of licence are being complied with ;

(c) To procure and send for test and analysis, samples of insecticide which he has reason to suspect are being sold, stocked or accepted for sale in contraventions of the provisions of the Act or rules made thereunder ;

(d) To investigate any complaint in writing which may be made to him ;

(e) To institute prosecutions in respect of breaches of the Act and the rules made thereunder ;

(f) To maintain a record of all inspections made and action taken by him in the performance of his duties including the taking of samples and the seizure of ⁓ and to submit copies of such record to the Li, fficer ;

(g) To make such enquiries and inspections as may be necessary to detect the sale and use of insecticides in contravention of the Act.

(4) Inspectors specially authorised to inspect manufacture of insecticides shall have the following duties;

(a) To inspect not less than twice a year all premises licensed for the manufacture of insecticides within the area of his jurisdiction and to satisfy himself that the conditions of the licence and the provisions of the Act or the rule made thereunder are being observed ;

(b) To send forthwith to the Licensing Officer after each inspection, a detailed report indicating the conditions of the licence and the provisions of the Act and the rules made thereunder which are being observed and the conditions and provisions, if any, which are not being observed ;

(c) To draw samples of insecticides manufactured on the premises and send them for test or analysis in accordance with the rules made under the Act ;

(d) To report to the Government all occurrences of poisoning.

Powers of Insecticide Inspectors

An Insecticide Inspector shall have power to.

(a) Enter, and search at all reasonable times and with such assistance, if any, as he considers necessary, any premises in which he has reason to believe that an offence under this Act or the rules made thereunder has been or is being or is about to be committed, or for the purpose of satisfying himself that the provisions of this Act or the rules made thereunder or the conditions of any certificate of registration or licence issued thereunder are being complied with ;

(b) Require the production of, and, to inspect, examine and make copies of, or take extracts from, registers, records or other documents kept by a manufacturer, distributer, carrier, dealer or any other person in pursuance of the provisions of this Act or the rules made thereunder and seize the same, if he has reason to believe that all or any of them may furnish evidence of the commission of an offence punishable under this Act or the rules made thereunder ;

(c) Make such examination and enquiry as he thinks fit in order to ascertain whether the provisions of this Act or the rules made thereunder are being complied with and for that purpose stop any vehicle ;

(d) Stop the distribution, sale or use of an insecticide which he has reason to believe is being distributed, sold or used in contraventions of provisions of this Act or the rules made thereunder, for a specified period of not exceeding twenty days, or unless the alleged contravention is such that the defect may be removed by the possessor of the insecticide, seize the stock of such insecticide ;

(e) Take samples of any insecticide and send such samples for analysis to the Insecticide Analyst for test in the prescribed manner ; and

(f) Exercise such other powers as may be necessary for carrying out the purposes of the Act or the rules made thereunder.

(1) The provisions of the Code of Criminal Procedure, 1973 shall, so far as may be, apply to any search or seizure under this Act, as they apply to any search or seizure made under the authority of a warrant issued under section 94 of the said Code.

(2) An Insecticide Inspector may exercise the powers of a police officer under Section 42 of the Code of Criminal Procedure, 1973 for the purpose of ascertaining the true name and residence of the person from whom a sample is taken or an insecticide is seized.

Procedure to be followed by Insecticide Inspector

(1) Where an Insecticide Inspector seizes any record, register or document under appropriate provisions of the Act, he shall, as soon as may be; inform a Magistrate and take his orders as to the custody thereof.

(2) Where an Insecticide Inspector takes any action under clause 1(d) of the Powers of Insecticide Inspectors ;

(a) He shall use all despatch in ascertaining whether or not the insecticide or its sale, distribution or use contravenes any of the provisions of "Prohibitions of sale, etc. of certain insecticides" and if it is ascertained that the insecticide or its sale, distribution or use does not so contravene, forthwith revoke the order passed under the said clause or, as the case may be, take such action as may be necessary for the return of the stock seized ;

(b) If he seizes the stock of the insecticide, he shall, as soon as may be, inform a Magistrate and take his orders as to the custody thereof ;

(c) Without prejudice to the institution of any prosecution if the alleged contravention be such that the defect may be remedied by the possessor of insecticide, he shall, on being satisfied that the defect has been so remedied, forthwith revoke his order and in case where the Insecticide Inspector has seized the stock of the insecticide, he shall, as soon as may be, inform a Magistrate and obtain his orders as to the release thereof.

(3) Where an Insecticide Inspector takes any sample of an insecticide, he shall tender the fair price thereof and may require a written acknowledgement thereof.

(4) Where the price tendered under subsection (3) is refused or where the Insecticide Inspector seizes the stock of any insecticide under clause (d) of subsection (1) of Section 21 of the Act, therefor in the prescribed form.

(5) Where an Insecticide Inspector takes a sample of an insecticide for the purpose of test or analysis, he shall intimate such purpose in writing in the prescribed form to the person from whom he takes it and, in the presence of such person unless he willingly absents himself, shall divide the sample into three portions and effectively seal and suitably mark the same and permit person to add his own seal and mark to all or any of the portions so sealed and marked :

Provided that where the insecticide is made up in containers of small volume, instead of dividing a sample as aforesaid, the Insecticide Inspector may, and if the insecticide be such that it is likely to deteriorate or otherwise damaged by exposure, shall, take three of the said containers after suitably marking the same and, where necessary, sealing them.

(6) The Insecticide Inspector shall restore one portion of a sample so divided or one container, as the case may be, to the person from whom he takes it and shall retain the remainder and dispose of the same as follows :

(i) One portion of the container he shall forthwith send to the Insecticide Analyst for test and analysis ; and

(ii) The second, he shall produce to the court before which proceedings, if any, are instituted in respect of the insecticide.

Persons bound to disclose the place where insecticides are manufactured or kept

Every person for the time being in charge of any premises where an insecticide is being manufactured or is kept for sale or distribution shall, on being required by an Insecticide Inspector so to do, be legally bound to disclose to the Insecticide Inspector the place where the insecticide is being manufactured or is kept, as the case may be.

Report of Insecticide Analyst

(1) The Insecticide Analyst to whom a sample of any insecticide has been submitted for test or analysis under subsection (6) of Section 22 of the Act, shall within a period of sixty days, deliver to the Insecticide Inspector submitting a signed report in duplicate in the prescribed form.

(2) The Insecticide Inspector on receipt thereof shall deliver one copy of the report to the person from whom the sample was taken and shall retain the other copy for use in any prosecution in respect of the sample.

(3) Any document purporting to be a report signed by an Insecticide Analyst shall be evidence of the facts stated therein, and such evidence shall be conclusive unless the person from whom the sample was taken has within twentyeight days of the receipt of a copy of the report notify in writing to the Insecticide Inspector or the Court before which proceedings in respect of the sample are pending that he intends to adduce evidence in controversion of the report.

(4) Unless the sample has already been tested or analysed

in the Central Insecticide Laboratory, where a person has
notified his intention of adducing evidence in controversion of
the Insecticide Analysts' report, the court may, of its own
motion or in its discretion at the request either of the
complainant or of the accused, cause the sample of the insecticide
produced before the Magistrate under subsection (6) of Section
22 of the Act to be sent for test or analysis to the said laboratory, which shall make the test or analysis and report in writing
signed by, or under the authority of the Director of the Central
Insecticide Laboratory the result thereof, and such report shall
be conclusive evidence of the facts stated therein,

Confiscation

(1) Where any person has been convicted under this Act
for contravening any of the provisions of this Act or any of
the rules made thereunder, the stock of the insecticide in respect
of which the contravention, has been made shall be liable to
confiscation.

(2) Without prejudice to the provisions contained in
(1) where the Court is satisfied on the application of an Insecticide Inspector or otherwise and after such enquiry as may be
necessary that the insecticide is a misbranded insecticide, such
insecticide shall be liable to confiscation.

Notification of poisoning

The State Government may, by notification in the Official
Gazette, require any person or class of persons specified therein
to report all occurrences of poisoning (through the use or
handling of an insecticide) coming within his or their cognizance
to such officers as may be specified in the said notification.

Prohibition of sale, etc., of insecticides for reasons of public safety

(1) If on reciept of a report under Section 26 (notification
of poisoning) of the Act or otherwise, the Central Government
or the State Government is of opinion, for reasons to be recorded
in writing that the use of any insecticide specified in sub clause
(iii) of clause (e) of section 3 (any preparation containing one
or more of insecticides) or any specific batch thereof is likely

to involve such risk to human beings or animals as to render it expedient or necessary to take immmediate action, then that Government may, by notification in the Official Gazette, prohibit the sale, distribution or use of the insecticide or batch in such area, to such an extent and such period (not exceeding sixty days) as may be specified in the notification pending investigation into the matter.

Provided that where the investigation is not completed within the said period, the Central Government or State Government, as the case may be, may extend it by such further period or periods not exceeding thirty days in the aggregate as it may specify in a like manner.

(2) If as a result of its own investigation or on receipt of report from the State Government, and after consultation with the Registration Committee, the Central Government, is satisfied that the use of the said insecticide or batch is or is not likely to cause any such risk, it may pass such order (including an order refusing to register the insecticide or cancelling the certificate of registration, if any, granted in respect thereof), as it deems fit, depending on the circumstances of the case.

Notification of cancellation of registration etc.

A refusal to register any insecticide or cancellation of the certificate of registration of any insecticide shall be notified in the Official Gazette and in such manner as may be prescribed.

Offences and Punishment

(1) Whoever :

(a) Imports, manufactures, sells, stocks or exhibits for sale or distributes any insecticides deemed to be misbranded under subclause (i) (if its label contains any statement, design or graphic representation relating thereto which is false or misleading in any material particular, or if its package is otherwise deceptive in respect of its contents) or subclause (iii) (if its label does not contain a warning or caution which may be necessary and sufficient, if complied with, to prevent risk to human beings or animals) or subclause (viii) (if the insecticide has toxicity which is higher than the level prescribed or is mixed or packed with any substance so as to alter its nature or quality or contains any substance which is not included in the registration), of clause (k) (definition of "misbranded") of Section 3 of the Act, or

(b) Imports or manufactures any insecticide without a certificate of registration ; or

(c) Manufactures, sells, stocks or exhibits for sale or distributes an insecticide without a licence ; or

(d) Sells or distributes an insecticide, in contravention of Section 27 of the Act (prohibition of sale etc. of insecticides for reasons of public safety) ; or

(e) Causes an insecticide, the use of which has been prohibited under Section 27 of the Act, to be used by any worker ; or

(f) Obstructs an Insecticide Inspector in the exercise of his powers or discharge of his duties under this Act or the rules made thereunder, shall be punishable :

(i) for the first offence, with imprisonment of a term which may extend to two years or with a fine which may extend to two thousand rupees or with both ;

(ii) for the second and a subsequent offence, with imprisonment for a term which may extend to three years, or with fine or with both.

(2) Whoever uses an insecticide in contravention of any provision of this Act or any rule made thereunder shall be punishable with fine which may extend to five hundred rupees.

(3) Whoever contravenes any of the other provisions of this Act or any rule made thereunder or any condition of a certificate of registration or licence granted thereunder shall be punishable—

(a) For the first offence, with imprisonment for a term which may extend to six months, or with fine, or with both ;

(b) For the second and a subsequent offence, with imprisonment for a term which may extend to one year, or with fine, or with both.

(4) If any person convicted of an offence under this Act commits a like offence afterwards it shall be lawful for the court before which the second or subsequent conviction takes

place to cause the offender's name and place of residence, the offence and the penalty imposed to be published in such newspapers or in such manner as the court may direct.

Defences which may or may not be allowed in prosecutions under this Act.

(1) Save in hereafter provided in this Section, it shall be no defence in a prosecution under this Act to prove merely that the accused was ignorant of the nature and quality of the insecticide in respect of which the offence was committed or of the risk involved in the manufacture, sale or use of such insecticide or of the circumstances of its manufacture or import.

(2) For the purpose of Section 17 (prohibition of import and manufacture of certain insecticides) of the Act, an insecticide shall not be deemed to be misbranded only by reason of the fact that :

(a) There has been added thereto some inocuous substance or ingredient because the same is required for the manufacture or the preparation of an insecticide as an article of commerce in a state fit for carriage or consumption, and not to increase the bulk weight or measure of the insecticide or to conceal its inferior quality or other defect ; or

(b) In the process of manufacture, preparation or conveyance some extraneous substance has unavoidably become intermixed with it.

(3) A person not being an importer, or a manufacturer of any insecticide or his agent for the distribution thereof, shall not be liable for a contravention of any provision of this Act, if he proves :

(a) That he acquired the insecticide from an importer or a duly licensed manufacturer, distributer, dealer thereof ;

(b) That he did not know and could not with reasonable diligence, have ascertained that the insecticide contravened any provision of this Act ; and

(c) That the insecticide, while in his possession, was properly stored and remained in the same state as when he acquired it.

Power of Central Government to make rules

The Central Government may, after consultation with the Central Insecticides Board, make rules for the purpose of giving effect to the provisions of this Act.

Power of State Government to make rules

(1) The State Government may, after consultation with the Central Insecticides Board, make rules for giving effect to the provisions of this Act and not inconsistent with the rules, if any, made by the Central Government.

(2) In particular and without prejudice to the generality of the foregoing power, such rules may provide for :

(a) The authority to which, the manner in which and the fee on payment of which, an appeal may be filed under Section 15 (appeal against the decision of a licensing officer) of the Act and the procedure to be followed by the appeallate authority in disposing of the appeal ;

(b) The delegation of any of the powers and functions conferred by this Act on the State Government to any officer or authority specified by that Government.

Exemption

(1) Nothing in the Insecticide Act shall apply to :

(a) The use of any insecticide by any person for his own household purpose or for kitchen garden or in respect of any land under his cultivation ;

(b) Any substance specified or included in the Schedule or any preparation containing one or more such substances, if such substance or preparation is intended for purposes other than preventing, destroying, repelling or mitigating any insects, rodents, fungi, weeds or other forms of plant or animal life not useful to human beings.

(2) The Central Government may, by notification in the official Gazette and subject to such conditions, if any, as it may specify therein, exempt from all or any of the provisions of this Act or the rules made thereunder, any educational,

scientific or research organisation engaged in carrying out experiments with insecticides.

Rules relating to packing and labelling

(1) No person shall stock or exhibit for sale or distribute any insecticide unless it is packed and labelled in accordance with the provision of these rules.

(2) *Packing of insecticides.* Every package containing an insecticide, shall be of a type approved by the Registration Committee and a sample container in which the insecticide is proposed to be packed, shall either accompany the application for registration or shall be supplied to the Registration Committee separately.

(3) *Leaflet to be contained in a package.* The packing of every insecticide shall include a leaflet containing the following details :

    (a) The plant disease, insects or noxious animals or weeds for which the insecticide is to be applied, the adequate direction concerning the manner in which the insecticide is to be used at the time of application ;

    (b) Particulars regarding chemicals harmful to human being, animals and wild lives, warning and cautionary statements including the symptoms of poisoning, suitable and adequate safety measures and emergency first aid treatment where necessary ;

    (c) Caution regarding storage and application of insecticides with suitable warnings relating to inflammable, explosive or other substances harmful to the skin ;

    (d) Instructions concerning the contamination or safe disposal of used containers ;

    (e) A statement showing the antidote for the poison shall be included in the leaflet and the label ;

    (f) If the insecticide is irritating to the skin, nose, throat or eyes, a statement shall be included to that effect.

Manner of labelling

(1) The following particulars shall be either printed or written in indelible ink in the label of the innermost container

of any insecticide and on the outermost covering in which the container is packed :

(a) Name of the manufacturer (if the manufacturer is not the person in whose name the insecticide is registered under the Act, the relationship between the person in whose name the insecticide has been registered and the person who manufactures, packs or distributes or sells shall be stated) ;

(b) Name of the insecticide (brand name or trade mark under which the insecticide is sold) ;

(c) Registration number of the insecticide ;

(d) Kind and name of active ingredients and percentage of each (common name accepted by the International Standards Organisation or the Indian Standards Institution of each of the ingredients shall be given and if no common name exists, the correct chemical name which conforms most closely with the generally accepted rules of chemical nomenclature shall be given) ;

(e) Net content of volume (the net content shall be exclusive of wrapper or other material. The correct statement of the net content in terms of weight, measure, number of units of activity, as the case may be shall be given. The weight and volume shall be expressed in the metric system) ;

(f) Batch number ;

(g) Expiry date, i.e., up to the date the insecticide shall retain its efficiency and safety ;

(h) Antidote statement. .

(2) The label shall be so affixed to the container that it cannot be ordinarily removed.

(3) The label shall contain in a prominent place and occupying not less than one-sixteenth of the total area of the face of the label, a square, set at an angle of 45 degrees (diamond shape). The dimension of the said square shall depend on the size of the package on which the label is to be affixed. The said square will be divided into two equal triangles, the upper portion

shall contain the symbol and signal word and the lower portion shall contain the colour specified.

(4) The upper portion of the square shall contain the following symbols and warning statements ;

(a) Insecticides belonging to Category I (extremely toxic) shall contain the symbol of skull and crossbones and the word "POISON" printed in red ;

The following warning statements shall also appear on the label at appropriate place, outside the triangle.

(i) "KEEP OUT OF THE REACH OF CHILDREN" ;

(ii) "IF SWALLOWED, OR IF SYMPTOMS OF POISONING OCCUR, CALL PHYSICIAN IMMEDIATELY" ;

(b) Insecticides in Category II (highly toxic) will contain the word "POISON" printed in red and the statement "KEEP OUT OF THE REACH OF CHILDREN", shall also appear on the label at appropriate place, outside the triangle ;

(c) Insecticides in Category III (moderately toxic) shall bear the word "DANGER" and the statement "KEEP OUT OF THE REACH OF CHILDREN" shall also appear on the label at suitable place outside the triangle ;

(d) Insecticides in Category IV (slightly toxic) shall bear the word "CAUTION".

(5) The lower portion of the square should contain the colour depending on the classification of the insecticides specified in the corresponding entry in column (1) of the following table :

(6) In addition to the precautions to be undertaken as specified, the label to be affixed in the package containing insecticides which are highly inflammable shall indicate that it is inflammable or that the insecticide should be kept away from heat or open flame and the like.

(7) The label and leaflets to be affixed or attached to the packages containing insecticides shall be printed in Hindi, English and in one or two regional languages of the areas where the said packages are likely to be stocked, sold or distributed.

(8) Labelling of insecticides must not bear any unwarranted claims for the safety of the product or its ingredients. This includes statements such as "SAFE", 'NON-POISONOUS', 'NON-

INJURIOUS' or 'HARMLESS' with or without such qualified phrase
as "when used as directed".

TABLE

Classification of the insecticide	Medium lethal dermal route (dermal toxicity) LD 50 mg/kg body weight of test animals	Medium lethal dose by the oral route (accute toxicity) LD 50 mg/kg body weight of test animals	Colour of identification band on the label
1	2	3	4
1. Extremely toxic	1-50	1-200	Bright red
2. Highly toxic	51-500	201-2000	Bright yellow
3. Moderately toxic	501-5000	2001-20000	Bright blue
4. Slightly toxic	More than 5000	More than 20000	Bright green

Prohibition against altering inscriptions etc. on containers, labels or wrappers of insecticides.

No person shall alter, obliterate or deface any inscription
or mark made or recorded by the manufacturer on the container,
label or wrapper of any insecticide :

Provided that nothing in this rule shall apply to any
alteration of any inscription or mark made on the container,
label, or wrapper of any insecticide, at the instance, direction
or permission of the Registration Committee.

Transport and storage of insecticides by rail, road or water.

Manner of packing, storage while in transit :

(1) Packages containing insecticides, offered for transport
by rail, shall be packed in accordance with the conditions
specified in the Red Tariff issued by the Ministry of Railways.

(2) No insecticide shall be transported or stored in such

a way as to come into direct contact with foodstuffs or animal feeds.

(3) No foodstuffs or animal feeds which got mixed up with insecticides as a result of damage to the package containing insecticides during transport or storage shall be released to the consignee unless it has been examined for possible contamination by competent authorities as may be notified by the State Government.

(4) If any insecticide is found to have leaked out in transport or storage, it shall be the responsibility of the transport agency or storage owner to take such measures urgently to prevent poisoning and pollution of soil or water, if any.

Conditions to be specified for storage of insecticides.

(1) The packages containing insecticides shall be stored in separate rooms or premises away from the rooms or premises used for storing other articles or shall be kept in separate almirahs under lock and key depending upon the quantity and nature of insecticides.

(2) The rooms or premises meant for storing insecticides shall be well built, dry, well lit and ventilated and of sufficient dimension.

Medical examination

(1) The persons who will be engaged in the work of handling insecticides during its manufacture, formulation, transport, distribution or application, shall be examined medically before their employment and periodically while in service by a competent physician who is aware of the risks to which such workers will be exposed.

(2) Any person showing symptoms of poisoning shall be immediately examined and given proper treatment.

First aid measures

In all cases of poisoning, first-aid treatment shall always be given before the physician is called. The Indian Standards Guide for handling pesticides poisoning Part I, First-Aid Measures [I.S. 4015 (Part II) 1967] should be consulted for such first-aid treatment in addition to other books on the subject. The workers also should be educated regarding the effects of poisoning and the first-aid treatment to be given.

Protective clothing

(1) Persons handling insecticides during its manufacture, formulation, transport, distribution, or application shall be adequately protected with appropriate clothing.

(2) The protective clothing should be used wherever necessary in conjunction with appropriate respiratory devices.

(3) The protective clothing shall be made of materials which prevent or resist the penetration of any form of insecticide formulations. The materials shall also be washable so that the toxic elements may be removed after each use.

(4) A complete suit of protective clothing shall consist of the following items :

    (a) protective outer garment/overalls/hood/hat ;

    (b) rubber gloves or such other protective gloves extending half way up to the forearm, made of materials impermeable to liquids ;

    (c) dust-proof goggles ;

    (d) boots.

Respiratory devices

For preventing inhalation of toxic dusts, vapours or gases, the workers shall use any of the following types of respirators or gas masks suitable for the purpose :

    (a) chemical-cartridge respirator ;

    (b) supplied air respirator ;

    (c) demand flow type respirator ;

    (d) full-face or half-face gas masks with canister.

In no case shall the concentrates of insecticides in the air where the insecticides are mixed exceed maximum permissible values.

Provisions regarding protective clothing, equipment and other facilities for workers during manufacture, etc. of insecticides.

(1) It shall be the duty of manufacturers, formulators of

insecticides and operators to dispose packages or surplus materials and washings in a safe manner so as to prevent environmental or water pollution.

(2) The used packages shall not be left outside so as to prevent their re-use.

(3) The packages shall be broken and buried away from habitation.

Aerial spraying Operations

The aerial application of pesticides shall be subject to the following provisions :

(a) Marking of the area shall be the responsibility of the operators ;

(b) The operators shall use only approved insecticides and their formulations at approved concentrations and height ;

(c) Washing, de-contamination and first-aid facilities shall be provided by the operators ;

(d) All aerial operations shall be notified to the public not less than twenty-four hours in advance through competent authorities ;

(e) Animals and persons not connected with the operations shall be prevented from entering such areas for a specific period ; and

(f) The pilots shall undergo specialised training including clinical effects of insecticides.

Places at which the insecticides may be imported

No insecticide shall be imported into India except through one of the following places :

Ferozepur Cantontment and Amritsar railway stations in respect of insecticides imported by rail across the frontier with Pakistan.

Madras, Calcutta, Bombay and Cochin in respect of insecticides imported by sea into India.

Madras, Calcutta, Bombay, Delhi and Ahmedabad in respect of insecticides imported by air into India.

INSECTICIDES

Approved by the Registration Committee

1. Aldrin
2. Aluminium Phosphide
3. Allethrin
4. Alachlor
5. Alpha Napthyl Acetic Acid
6. Atrazine
7. Aureofungin
8. Aldicarb*
9. B.H.C.+
10. Butachlor
11. Benomyl
12. Barium Carbanate
13. Benthiocarb
14. BPMC
15. Copper Oxychloride
16. Cuprous Oxide
17. Copper Sulphate
18. Carbaryl*
19. Chlordane
20. Captan
21. Calcium Cyanide
22. Chlorebenzilate
23. Chlorothalonil
24. Coumachlor
25. Chlormaquat Chloride
26. Carbendazim (Bavistin)
27. Carbofuran
28. Carboxin
29. Chlorfenvinphos
30. Chlorpyriphos
31. Calixin
32. Copper Acceto Arsente (Paris Green)
33. Cypermethrin
34. D.D.T.+
35. Dimethoate
36. Diazinon
37. Dalapon
38. Dichlorvos
39. D.D. Mixture
40. Diuron
41. Dicofol
42. Difenphos
43. Difolaton
44. Dinocap
45. Decamethrin
46. Dithianon
47. Dicamba (for export)
48. EMC
49. Ethylene Dibremide+
50. EDCT
51. Endosulphan*
52. Ethion
53. Ediphenphos (Hinosan)
54. Ethepon
55. Ferbam
56. Fenthion
57. Fenitrothion
58. Fluchloralin (Basalin)
59. Fermothion
60. Fenvalerate
61. Gibberllic Acid
62. Glyphosate
63. Heptachlor
64. Isoproturon
65. Kitazin
66. Lindane (Gamma B.H.C.)+
77. Lime Sulphur
68. Malathion
69. MEMC
70. Methyl Bromide*

71. MCPA
72. Maleic Hydrazide+
73. Metaldehyde
74. Monocrotophos*
75. Mancozeb
76. MSMA
77. Methyl Parathion+
78. Methabenzthiazuron (Tribunil)
79. Metoxuron
80. Menazon
81. Nicotine Sulphate
82. Nickel Chloride
83. Nitrofen £
84. Oxydemeton
85. PMA
86. Pyrethrum
87. PCNB*
88. Paradichlorobenzene
89. Pentachlorophenon (PCP)
90. Pendimethalin
91. Phosphamidon
92. Phorate*
93. Phenthoate
94. Phosalone
95. Paraquat Dichloride*
96. Propoxur
97. Propanil

98. Paraquat Dimethyl Sulphate
99. Pirimiphos Methyl
100. Quinalphos
101. Sodium Cyanide
102. Sulphur
103. Sirmate
104. Simazine
105. Streptocycline
106. Thiram
107. Thiometon
108. Trichlorophon (Dipterex)
109. Triforine
110. Tetradifon
121. Trichloro Acetic Acid (TCA)
112. Triallate
113. Thiophanate Methyl
114. Warfarin
115. Ziben
116. Ziram
117. Zinc Phosphide
118. 2, 4-D*
119. Toxaphene £
120. Dibromochloropropane (DBCP)£
121. Acephate
122. Bromodioloric
123. Metalaxyl

*Restricted in some other countries of Asia
●Restricted in some countries of Asia including India
+Banned in some other countries of Asia
£Banned in some countries of Asia including India
Source: Directorate of Plant Protection, Quarantine and Storage, Faridabad

INSECTICIDES NOT APPROVED FOR REGISTRATION

1. Calcium Arsenate
2. Lead Arsenate
3. Carbophenthion (Trithion)
4. Azinophos Methyl (Gusathion)
5. EPN
6. Mevinphos (Phosdrin)
7. 2, 4, 5 — T
8. Vamidothion
9. Mephosfolan
10. Azinphos Ethyl
11. Binapacryl
12. Dicrotophos
13. Thiodemeton/Disulphoton
14. Fentin Acetate
15. Fentin Hydroxide
16. Chinomethonate (Morestan)
17. Ammonium Sulphanate
18. Leptophos (Phosvel)

INSECTICIDES APPROVED—IMPORTS NOT PERMITTED

1. Toxaphene
2. Dibromochloropropane (DBCP)

INSECTICIDES FOR USE BY DIRECTORATE OF PLANT PROTECTION, QUARANTINE AND STORAGE IN LOCUST CONTROL

1. Dieldrin

INSECTICIDES PHASED OUT OF USE

1. Endrin
2. Ethyl Parathion

RULES FOR REGULATING THE IMPORT OF PLANTS, ETC. INTO INDIA

(Corrected up to March, 1967).

The following notification of Government of India for the purpose of prohibiting, regulating and restricting the import into India of the articles herein after specified is in vogue.

(1) In this order—

(i) *Official certificate* means a certificate granted by the proper officer or authority in the country of origin; and the officer, and authorities named in the third column of the first Schedule appended hereto are the proper officers and authorities to grant, in the countries named in the second column, the certificates required by the provisions referred to in the first column thereof;

(ii) *Plant* means a living plant or part thereof but does not include seeds; and

(iii) *Prescribed Port* means any of the following ports, namely, Bombay, Calcutta, Cochin, Rameshwaram, Madras, Nagapatam, Visakhapatnam and Tuticorin;

(iv) All provisions referring to plants or seeds shall apply also to all packing material used in packing or wrapping such plants or seeds.

(2) No plant shall be imported into India by means of the letter or sample post; provided that sugar cane for planting intended to be grown under the personal supervision of the Government Sugar Cane Expert, Coimbatore, may be imported by him by such post.

(3) No plant shall be imported into India by Air.

Provided that plants which are infested with living insects and are intended for the introduction of such living insects may be so imported if they are accompanied by a special certificate from the Head of Division of Entomology, Indian Agricultural Research Institute, New Delhi or Plant Protection Adviser to the Government of India, New Delhi, or the Forest Entomologist, Forest Research Institute, Dehra Dun, certifying such plants are imported for the purpose of introducing such insects;

Provided further that plants may be imported by air subject to the following conditions :

(i) Plants other than those species whose importation is totally prohibited or specifically restricted by the Notification of the Government of India in the late Department of Education, Health and Land, No. 320/35-A dated the 20th July, 1936, may be imported into India if accompanied by special permit in the form set forth in the "Schedule II" to this order.

(ii) All applications for certificate to import plants by air shall be sent to the Plant Protection Advisor ; Directorate of Plant Protection, Quarantine and Storage, Faridabad (Haryana) in advance in the form specified in the "Schedule I" to this order. The issue of certificate may be withheld without assigning any reasons therefor.

(iii) No plants shall be imported by air except through the airport of Santa Cruz at Bombay, Meenabakkam at Madras, Dum Dum at Calcutta, Palam or Safdarjang at New Delhi and Tiruchirapally.

(iv) All plants imported by air shall be accompanied by Green and Orange coloured tag, as shown in the "Schedule III" to this order issued by the Plant Protection Adviser to the Government of India and shall be used according to the instructions specified on the reverse thereof.

(v) All plants imported by air shall be inspected and if necessary, fumigated or otherwise disinfected at the port of entry by the Plant Protection Adviser to the Government of India or any person duly empowered by him in which case the importer shall pay to such officer a fumigation fee of Rs 5 per consignment-if it exceeds 1.5 cu m but does not exceed 3 cu m in volume and an equal amount for every additional 1.5 cu m or portion thereof.

(vi) The importer shall make arrangements himself or through his agents to take delivery of the consignment from the Collector of Customs at the port of entry, after payment of dues, if any.

Provided also that the imports of plant by any research institution or organisation under the control of the Central Government or the State Governments shall be subjected to all the conditions specified in the previous proviso, except the condition relating to the special permit to be issued by the Plant Protection Adviser or the payment of the prescribed fumigation fee.

(3) Provided also that the import by air from Afghanistan of fruits and vegetables intended for consumption may be permitted after fumigation at an airport at Amritsar, New Delhi (Safdarjang and Palam) or Bombay (Santa Cruz) on condition that the importer pays a fee of Rs 2 (Rupees two) only for every 50 kilogrammes of part thereof to the Plant Protection Adviser to the Government of India to meet the cost of fumigation, and also acts in accordance with such instructions as may from time to time be issued by that officer ; consignments of fruits not exceeding two kilogrammes each in eight imported as accompanied baggage by passengers shall, however, be fumigated free.

    (i) The import into India by land from Afghanistan of fruits or vegetables intended for consumptions may be permitted after inspection and also if necessary, after fumigation at Khalra in Amritsar District and Hussianiwala in Ferozepur District by the Plant Protection Adviser to the Government of India or any person duly empowered by him in this behalf.

    (ii) Whereafter inspection it is found necessary to fumigate the consignments, the importer shall pay a fee of Rs 2 (Rupees two only) for every 50 kilogrammes of the fruits or vegetables or part thereof to the said Plant Protection Adviser to meet the cost of fumigation , and act in accordance with such instructions as may from time to time be issued by him

Provided that, consignments of fruits or vegetables not exceeding two kilogrammes in weight imported as accompanied baggage by passengers shall, however, be fumigated free.

(4) No plants other than fruits and vegetables intended for consumption, potatoes, sugar cane and unmanufactured tobacco either raw or cured, shall be imported into India by sea except

after fumigation with hydrocyanic acid gas, methyl bromide or ethylene dibromide at a prescribed port.

Provided that plant, which are infected with living fungi and cultures of living fungi and are imported for the introduction of such fungi or for similar experiments, may be imported without fumigation if they are accompanied by a special certificate from the Forest Mycologist, Forest Research Institute, Dehra Dun, that such plants are imported for the purpose mentioned above.

Provided that plants which are infested with living parasitised insects and are intended for the introduction of such parasites may be imported without such fumigation if they are accompanied by special certificate from the Head of Division of Entomology, Indian Agricultural Research Institute, and Forest Entomologist, Dehra Dun that such plants are imported for the purpose of introducing such parasites.

Provided further that in the case of plant breeding materials imported by the Silviculturist, Forest Research Institute, Dehra Dun, such fumigation at the prescribed Ports shall be dispensed with on the condition that he makes himself personally responsible for the effective disinfection and disinfestation under the supervision of the Forest Entomologist and Forest Mycologist of the Forest Research Institute, Dehra Dun and before release the plants are certified as free from living insects and fungi by the said officers. All such plants shall be packed in such containers as will not permit the insects reaching or leaving the plants and that such containers shall not be opened in any part of India except at Dehra Dun.

(5i) No plants other than fruits and vegetables intended for consumption and potatoes shall be imported into India by sea unless accompanied by an official certificate that they are free from injurious insects and diseases.

(ii) The certificate shall be in the form prescribed in the Third Schedule or in a form as near thereto as may be and supplying all the information called for in that form.

(6) The import of potato plant (*Solonum tuberosum*) including the tubers, is prohibited.

Provided that potato, tubers for purposes of research and experimentation may be imported into India by sea and air only by scientific institutions under the Central or the State Governments, with a permit in the form prescribed in Schedule II to this Notification, through the seaports of Bombay, Calcutta, Cochin, Madras or Visakhapatnam and the airports of Santa Cruz (Bombay) Dum Dum (Calcutta) Meenambakkam (Madras) or Palam or Safdarjang (New Delhi) or Tiruchirapally under the following conditions namely :

A(i) A certificate from the consignor, stating the country and the district of such country in which the potato tubers were grown, shall accompany the consignment of potato tubers.

(ii) An official certificate stating the following shall accompany the consignment of potato tubers :

(a) That the said potato tubers were grown in areas free from the Wart Disease (*Synchytrium endobioticum*), Bacterial Ring Rot (*Corynebacterium sepedonicum*), Golden Nemtode (*Heterodera rostochiensis*) and Colorado Potato Beetle (*Leptinotarsa decemlineata*) ;

(b) That there was no occurrence of Bacterial Ring Rot and Colorado Potato Beetle during the last twelve months, immediately preceding the time of lifting the said potato tubers from the fields, and of Wart Disease and Golden Nematode at any time, in any stage of development and within a radius of eight kilometres of the field wherein the potato tubers included in the consignments were grown ;

(c) That the crop, from which the potato tubers were derived was inspected in the field at least 16 days before harvest and was found to be healthy and free from virus diseases ;

(d) That the potato tubers immediately prior to export, were examined and found to be free from insects, diseases and soil ;

(e) That the potato tubers included in the consignment were placed in a new, clean and unused packing ; and

(f) That the potato tubers included in the consignment are free from the Wart Fungus, Bacterial Ring Rot, Golden Nematode and Colorado Potato Beetle in any of their stages :

or

that the Wart disease of potato, Bacterial Ring Rot, Golden Nematode and Colorado Potato Beetle do not occur in the country of origin of the consignment.

(B) The quality of potato tubers imported at any one time shall not exceed the quantity to be specified in the permit prescribed in Schedule II to the said Notification.

(C) All requests for a permit to import potato tubers shall be made to the Plant Protection Adviser to the Government of India in the form prescribed in Schedule I to this notification at least two months in advance of the expected date of arrival of the consignments. A request for permit may be rejected without assigning any, reasons therefor.

(D) All potato tuber consignments shall be inspected and if necessary, treated at the Plant Quarantine stations at the port of entry by the Plant Protection Adviser to the Government of India or any person duly empowered by him in this behalf.

(E) If, in the opinion of the Plant Protection Adviser to the Government of India, or any officer duly authorised by him in his behalf, the consignment is required to be grown under quarantine, it shall be done so far at least one vegetative generation and only the healthy progeny of such tubers may be released to the consignee to the extent of the original consignment.

(F) The consignee shall make arrangement himself or through his agents to take delivery of the consignment from the Collector of Customs at the port of entry.

(G) Provided further that potatoes, grown in the countries of the South-East Asia and Pacific Region, as defined in the Plant Protection Agreement for the South-East Asia and Pacific Region of the Food and Agriculture Organisation of the United Nations of 1956, may be imported by sea or by air through the ports of Calcutta or Madras, subject to conditions (A), (C), (D), (E) and (F) specified in the preceeding proviso.

7. Rubber plants shall not be imported into India by sea unless, in addition to the general certificate required under paragraph 5, they are accompanied by an official certificate that the State from which the plants have originated or the individual plants are free from *Fomes lignosus, Sphaerostilbe repens,*

Dothidella ulei (*Melanopsamposis ulei*, *Fusicladium macrasporum*) and *Oidium heveae.*

8A. No Lemon plants, Lime plants, Orange plants, Grape fruit plants or other citrus plants and not cuttings of such plants shall be imported into India unless, in addition to the general certificate required under paragraph 5 they are accompanied by an official certificate that they are free from the *Mal Secco* caused by *Deuterophoma tracheiphila* or that the disease does not exist in the country in which they are grown.

8B.(i) Unmanufactured tobacco shall not be imported into India, except through the seaports of Bombay, Calcutta, Cochin, Madras, Trivandrum and Visakhapatnam, or the airports of Santa Cruz (Bombay), Dum Dum (Calcutta), Meenambakkam (Madras) and Palam and Safdarjung (New Delhi), and unless accompanied by an official phytosanitary certificate required under paragraph 5 with the additional declaration that it is free from any stage of the tobacco moth. *Fohostia elutella* or that the pest does not occur in country of origin.

(ii) Unmanufactured tobacco shall not be imported by letter or sample post. It may be imported by air. Provided that consignments of unmanufactured tobacco imported by air shall not exceed 10 kilogramme in gross weight and shall be packed in a manner which will not allow insects to enter into or escape from the package.

(iii) The consignment shall be plainly and clearly marked to show the general nature and the quantity of the contents, the locality and the country of origin of the contents and the name and address of the consignor and the consignee.

(iv) Unmanufactured tobacco shall be inspected and if necessary, fumigated or otherwise treated on arrival by the Plant Protection Adviser to the Government of India or any person duly authorised by him in this behalf on payment of a fumigation or treatment fee, as prescribed in subparagraph (v).

(v) A fumigation or treatment fee of Rs 6 per consignment of 1.5 cubic metres or less in volume and Rs 2 for every additional 1.5 cubic metres or part thereof shall be payable by the importer to the Collector of Customs concerned : provided that a consignment not exceeding 10 kilogrammes in gross weight shall be fumigated or otherwise treated without the payment of fee.

(vi) It shall be the responsibility of the importer to bring the consignment to the ₁Plant Quarantine Station concerned or to the place of inspection, fumigation or treatment and to open, repack, load into or unload from the fumigation chamber, seal the consignment, remove the consignment from the premises of the Plant Quarantine Station or the place of inspection, fumigation or treatment as directed by the Plant Protection Adviser to the Government of India or any person duly authorised by him in this behalf.

(vii) The Plant Protection Adviser to the Government of India or any officer duly authorised by him in this behalf, may draw such samples from the consignment as he may deem necessary for inspection and tests.

(viii) In the case of imports of unmanufactured tobacco by sea, the importers shall inform the officer-in-charge of the Plant Quarantine Station and Collector of Customs concerned, the country of origin of the consignment, the number and size of packages or cases, the nature of packages and the probable date of the arrival of the carrier, at least fourteen days in advance of the arrival of the carrier at the concerned seaport, provided that where the ordinary length of voyage from the country of export is less than 14 days, it shall be sufficient to furnish the information not less than seven days before the arrival of the carrier.

Provided that the above condition shall not apply to the case of sample consignment not exceeding 10 kilogrammes gross weight.

(ix) Unmanufactured tobacco intended for other countries shall be allowed transit through India or transhipment at any of the seaports in India mentioned in sub-paragraph (i), if some consignments are landed in India. Consignments which are landed in India for the above purposes shall be governed by the provisions of this paragraph :

Provided that the sample consignments of unmanufactured tobacco by air intended for other countries shall be allowed transit through India or transhipment at any of the airports in India mentioned in sub-paragraph (i) if such consignments are packed in a container in such a way as not to allow insects to enter into or escape from it and the container is not to be opened in any part of India.

Explanation

For the purposes of this paragraph,

(a) a *consignment* means one or more packages of unmanufactured tobacco of any one description consigned to one consignee by one consignor at any one time.

(b) *unmanufactured tobacco* means unmanufactured tobacco, either raw or cured.

8c. Tobacco seeds shall not be imported into India except for experimental purposes by the Director, Central Tobacco Research Institute, Rajahmundry, who shall ensure that the consignment of seeds so imported is accompanied by an official certificate stating that such seeds originated from a crop free from the disease known as "Blue mould" (*Peronospora tabacina*) or that such disease does not occur in the country where such seeds originated.

9(i) The importation of sugar cane into India by sea from the Fiji Islands, New Guinea, Australia or the Philippines Islands is prohibited absolutely.

(ii) The importation of sugar cane into India by sea from any other country is prohibited unless, in addition to the general certificate required under paragraph 5, it is accompanied by an official certificate that it has been examined and found free from cane borers, scale insects, white flies, root disease (any form), pineapple disease (*Ceratostomella paradoxa*) (*Thielaviopsis paradoxa*), sereh and cane gummosis, that was obtained from a crop which was free from mosaic disease and that the Fiji disease of sugar cane does not occur in the country of export ;

Provided that in the case of canes for planting imported direct by the Government Sugar Cane Expert, Coimbatore and intended to be grown under his personal supervision, such certificates shall be required only in respect of the freedom of the country of export from the Fiji disease of sugar cane.

10(i) Hevea rubber plants and hevea rubber seeds shall not be imported into India from America or from the West Indies except by the Director of Agriculture, Madras.

(ii) Rubber seeds from other countries may be imported into India only after fumigation and disinfection at the port of entry, namely, Madras or Bombay, as the case may be.

11(i) Seeds of flax and berseem shall not be imported by letter or sample post, or otherwise than by sea.

(ii) The importation of 'Mexican Jumping Beans' (*Sebastiana palmeri* of the family Euphorobiaceae) is absolutely prohibited.

(12) Coffee plants, coffee seeds and coffee beans shall not be imported into India except for experimental planting purposes only by the Director of Research, India Coffee Board, Bangalore, who shall take measures necessary to' ensure that such coffee plants, beans or seeds as are imported by him are free from plant diseases and injurious insects. Provided that nothing in this paragraph shall apply to roasted or ground coffee.

(13) Flax seeds and berseem (Egyptian clover) seeds shall not be imported into India by sea, unless the consignee produces before the Collector of Customs a licence from a Department of Agriculture in India in that behalf.

14(i) Unginned cotton other than cotton from a port of Saurasthra which has been produced in India shall not be imported by sea or by air.

(ii) Cotton seed shall not be imported save for experimental purposes by one of the officers named in the second schedule appended hereto and shall not be so imported by such officer save at the sea ports of Bombay, Bhavnagar, Calcutta, Cochin, and Madras or at the Airports of Bombay (Santra Cruz), Calcutta (Dum Dum), New Delhi (Palam/Safdarjung) and Madras (Menambakkam) and in quantities not exceeding one hundred weight in any one consignment and on condition that it will be fumigated on importation with carbon disulphide or methyl bromide.

Provided that, if the cotton seed is accompanied by a certificate from a Government entomologist of the country of origin to the effect that the seed and its container have been treated in such a way as to destroy all insect life, the seed shall be examined on importation by such officer as the Central Government may appoint and shall not be required to be fumigated unless such examination shows that refumigation is necessary.

(14a) Sunflower seed shall not be imported into India from Argentina and Peru by means of letter or sample post or as passenger's accompanied baggage or by any other means.

(14b) No bulbs or plants or onion (*Allium cepa*), garlic (*Allium sativim*), shallot (*Allium asclonicum*) leak (*Allium porrum*), chive (*Allium schoednoorasum*) shall be imported into India unless they are accompanied under paragraph 5, by an official phytosanitary certificate guaranteeing freedom from the fungus disease *Urocystis cepulae.*

14c(i) The importation of cocoa plants (*Theobroma cocoa* and other species of *Theobroma*) including seeds (in the unmanufactured state) from Africa, West Indies and Ceylon is prohibited.

(ii) Cocoa plants shall not be imported into India from any other country. Such import is however, allowed where it is for purposes of research and propagation by an institution or organisation under the control of the Central Government or State Government, subject to the condition that the cocoa plants are accompanied by an official certificate, as required under paragraph 5 and are inspected and if necessary, fumigated and disinfected by the Plant Protection Adviser to the Government of India or any person duly empowered by him in this behalf at Bombay or Madras, and are also accompanied by :

(a) a certificate from the consignor stating fully in what country and in what district cocoa plants were grown, and

(b) an official certificate stating that the plants have been examined and found to be free from 'Pod rot' (*Monilia roveri*), 'Mealy pod' (*Trachysphaera fructigena*), 'Witches broom' (*Cripipellia perniciosus ; Marasmius pernicious*) and that the 'Swollen Shoot' and other virus diseases of cocoa do not occur in the country of origin.

(14d) The importation of seedlings and seeds of groundnut from South America, North America, West Indies, Continental China and Soviet Russia are prohibited with a view to prevent the introduction of the disease known as groundnut rust (*Puccinia arachidis*) and *Sphaceloma arachidis* :

Provided that groundnut seedlings including decorticated seeds required for scientific purposes may be imported into India by sea and air only by scientific institutions under the Central

Government or any State Government subject to the condition that in addition to the general certificate required under paragraph 5 they are accompanied by an official certificate stating that the diseases *Puccinia arachidis* and *Sphacceloma arachidis* are not prevalent in the concerned importing country and that the seeds are disinfected with an appropriate fungicide before export from such country.

(14e) The importation of coconut plant and plant materials including seeds from Carribean area, Jamaica, Haiti, Florida, Ghana, Togoland, Philippines, British Guiana, West Indies and Guam are prohibited with a view to prevent the introduction of Red Ring (*Aphelenchoides cocophilus*), Lethal yellowing, Kaincope disease, Cadangcadang, Bronze leaf wilt and Guam coconut disease :

Provided that unsprouted nuts, from which the perianth has been removed, may be imported into India by sea and air only for scientific purposes by the Central Coconut Research Station subject to the condition that in addition to the general certificate required under paragraph 5, they are accompanied by an official certificate stating that the seeds come from trees showing no signs of any such disease :

Provided further that upon arrival in India such nut shall be inspected and fumigated or treated by any other method considered appropriate by the Plant Protection Adviser to the Government of India or any other officer authorised by him ;

Provided also that the imported seednuts shall be planted in ndividual containers in isolated quarantine for a period of one year and the diseased seedlings together with the containers and the planting medium, shall be destroyed by burning under supervision of an authorised officer :

(14f) The importation of forest plants, namely *Castanea*, *Ulnus* and *Pinus*, including seeds is prohibited as a safeguard against the introduction of destructive strains of disease pathogens *Endothia parasitica*, *Ceratocystis ulni* and *Cronartium ribicola*, respectively

Provided that the import of the above mentioned forest plants and seeds is allowed where it is for purposes of research and propagation by an institution or organisation under the control of the Central Government or State Government, subject to the following conditions, namely :

(a) the forest plants and seeds are accompanied by an official certificate, as required under paragraph 5 and are inspected and, if necessary, fumigated and disinfected by the Plant Protection Adviser to the Government of India or any person duly empowered by him in this behalf.

(b) the exporting country certifies that the forest plants were inspected in the field and were found to be free from any diseases and insects.

(c) the forest plants raised from the prohibited genera named above are subject to post entry inspection at regular intervals to ensure their freedom from diseases and insects.

(15) Nothing in these paragraphs shall be deemed to apply to :

(i) The bringing by sea or by air from one port or place in India to another such port or place.

(ii) the transit of plants through India by air or their transhipment, if they are accompanied by official certificates as presribed in clause (i) of paragraph I and are packed in such containers as will not permit the insects reaching or leaving the plant material and as are not to be opened in part of India.

(16) Where any plant which is imported into India is not accompanied by an official certificate or any other certificate required under any of the aforesaid paragraphs, the Plant Protection Adviser to the Government of India or any other plant if, after inspection or fumigation, the said Plant Protection Adviser or such other officer is satisfied that the plant is free from injurious pests and diseases :

Provided that before releasing any plant under this paragraph the said Plant Protection Adviser or such other officer shall record in writing the reason therefor.

(17) (1) The high quality ornamental plants and plant materials which are not covered by the proper import permit and phytosanitary certificate may continue to remain the property of the consignee, but can be held in quarantine by the National Botanical Garden at Trombay, Bombay or the Sibpur Botanical Garden, Calcutta or the Indian Agricultural Research Institute,

List of some diseases and pests which need careful attention for prevention of introduction into India

Insect Common Name	Scientific Name	Hosts	Economic Importance	Geographic Distribution	Quarantine Risk
1. Striped cucumber beetle	Acalymma vittata (Diabrotica fabricus)	Cucurbits including Squash, pumpkin, rock-melon, water-melon, gourd.	Very destructive in U. S. A. and carrier of bacterial wilt organism (Erwinia tracheiphila)	U.S.A., Canada	Slight risk of larvae carried in the soil ; over-wintering adults may be transferred as hitch hickers, risks generally minimal.
2. Bronze Birch borer	Agrilus anxius	Ornamental-birch, poplar, willow	Very destructive. Following browning of tips of upper branches the entire tree may die.	U.S.A.	Poplar or willow rounds, if imported, could be a means of gaining entry.
3. Leek moth	Acrolepia assectella	Onion, leek, garlic	Severe damage to leeks, onions, and related species is caused in Europe by mining and feeding within the foliage and bulb with subsequent collapse of plants and rotting of bulbs. Larvae feed on the seed stalk preventing seed formation.	Europe including northern portion of Mediterranean area U. K., Spain, Italy, Greece, Pacific-Hawai.	Introduction as larvae in onions & related plants a definite risk.

Insect Common Name	Scientific Name	Hosts	Economic Importance	Geographic Distribution	Quarantine Risk
4. Grape flea beetle	*Altica chalybea*	Grape, plum, apple, quince, birch, elm, Virginia creeper.	Buds of grapes are eaten off.	U.S.A.	Hitchhicking adults may be the means of entry and should not be overlooked.
5. Fruit tree leaf roller	*Archips argyrospila*	Nearly all deciduous fruit trees, many forest and herbaceous plants.	In early stages of fruit development, insect can ruin 80-90% apple crop by causing malformation of fruit as a result of feeding on them.	U.S.A.	Importation in the nursery stocks as over-wintering eggs—hence nursery stocks must be fumigated.
6. Squash bug	*Anasa tristis*	Cucurbits, specially squashes and pumpkins.	Extremely persistent and make it impossible to grow cucurbits.	U.S.A., Canada	Can be easily introduced as hitchhicking insects sheltering as overwintering adults.
7. Pigmy man-gold beetle	*Atomaria pinearis*	mangold, sugar-beet, spinach, raddish.	Very troublesome for sugar-beet cultivation.	Europe	There is a risk of moving this insect as hibernating adults in soil.

Insect Common Pests Name	Scientific Name	Hosts	Economic Importance	Geographic Distribution	Quarantine Risk
8. Buffalo treehopper	*Ceresa bubalus*	apple, pear, quince, cherry, cottonwood, elm, maize, grasses, legumes.	Double rows of curved slits are formed in the barks of small branches and twigs.	U.S.A., Canada	may be introduced into this country in infested dormant twigs and branches.
9. Strawberry root weevil	*Brachyrhinus ovatus*	Strawberry and related small fruit plants.	Attacked strawberry plants stunted with bunched and dying leaves.	U.S.A., Canada	Adults can gain entry with strawberry plants, while larvae and pupae may be carried over in soil.

Diseases	Common Name	Scientific Name	Hosts	Economic Importance	Geographic Distribution	Quarantine Risk
1.	Pod rot of cacao	*Monilia roveri*	Cocoa	Causes severe losses to cacao leaves.	Pathogen-Colombia, Eucador, Peru, Venezuela, Panama.	Disease of great importance to India. Disease carried through infected seeds. Importation of diseasefree certified seed material essential to check introduction.
2.	Downymildew of onion	*Peronospera destructor*	Onion, leek, garlic, shallot etc.	Very serious disease, affects bulb, reduces yield and seed production.	Fungus, Kenya, Libya Mauritius, South Africa, Southern Rhodesia, Tanganiyaka, China, Taiwan, Iraq, Israel, Japan, Australia, Newzealand, almost all European countries. West Indies, Bermuda, Costa Rica, Panama, Canada, U.S.A., Mexico, Argentina, Brazil, Peru, Bolivia, Chile, Uruguay, Venezuela.	Great quarantine risk as these often come in accompanied baggage, besides ships or aircrafts have these materials in stores.

Diseases	Common Name	Scientific Name	Hosts	Economic Importance	Geographic Distribution	Quarantine Risk
3.	Mealy pod of Cacao, fruit rot of Coffee, fruit rot of Banana	*Trachysphaera fructigema*	Cocoa, Coffee Banana, *Mimusops elengi*.	Of minor importance in coffee and cocoa, but in banana a serious problem causing both pre-harvest and post-harvest damage.	Zaire Republic, Ghana, Ivory Coast, Sierra-Leone	Great quarantine risk, as banana fruits have unrestricted import as accompanied baggages & airfreights.
4.	Groundnut rust	*Puccinia arachidis*	Groundnut (*Arachis hypogula*), *A. marginata*, *A. prostrata*	Causes severe damage to the crops if infected early in the wet season.	Kenya, Malwai, Mauritius, Rhodesia, Mozumbique, Zambia, Tanzania, Brunei, Japan, U.S.SR, Taiwan Thailand, China, Korea, Australia, Papua, New Guinea, Solomon Islands, U.S.A., most Central and Latin American countries.	Seed materials are always imported in small lots and they always carry the quarantine risk.

Diseases	Common Name	Scientific Name	Hosts	Economic Importance	Geographic Distribution	Quarantine Risk
5.	Neckrot of onion	Botrytis alli	Onion, garlic; shallot.	Very serious storage disease and one of the major bulb destroying diseases.	Canary Islands, South Africa, Kenya, China, Cyprus, Japan, Australia, Tasmania, New-zealand, Canada, U.S.A., Brazil, Austria, Bulgaria, Denmark, Finland, France, U.K., U.S.S.R., West Germany, Holland, Poland. Yugoslavia.	Fungus overwinters as sclerotia in the affected bulbs on scales, onion bulbs are frequently carried in ships and aircraft stores. Passengers bring them as accompanied baggages—great quarantine risk.

New Delhi or the Agricultural College and Research Institute at Coimbatore for a period of six months :

Provided that it may be returned to the consignee after payment, if necessary, of such quarantine charges as may be determined by the Head of the Institution under whose charge the consignment is held.

(2) The consignment shall after a lapse of the period of six months referred to sub-paragraph (1) be deemed to be the property of the Institution which held it in quarantine.

INTERNATIONAL PLANT PROTECTION CONVENTION

PREAMBLE

The contracting Governments, recognizing the usefulness of international co-operation in controlling pests and diseases of plants and plant products and in preventing their introduction and spread across national boundaries, and desiring to ensure close co-ordination of measures directed to these ends, have agreed as follows :

ARTICLE I

PURPOSE AND RESPONSIBILITY

(1) With the purpose of securing common and effective action to prevent the introduction and spread of pests and diseases of plants and plant products and to promote measures for their control, the contracting Governments undertake to adopt the legislative, technical and administrative measures specified in this Convention and in supplementary agreements pursuant to Article III.

(2) Each contracting Government shall assume responsibility for the fulfilment within its territories of all requirements under his Convention.

ARTICLE II

SCOPE

(1) For the purposes of this Convention the term "plants" shall comprise living plants and parts thereof, including seeds in so far as the supervision of their importation under Article VI of the Convention or the issue of phytosanitary certificates in respect of them under Articles IV (1), (a), (iv) and V of this Convention may be deemed necessary by

contracting Governments ; and the term "plant products" shall comprise unmanufactured and milled material of plant origin, including seeds in so far as they are not included in the term "plants".

(2) The provisions of this Convention may be deemed by contracting Governments to extend to storage places, containers, conveyances, packing material and accompanying media of all sorts including soil involved in the international transportation of plants and plant products.

(3) This Convention shall have particular reference to pests and diseases of importance to international trade.

ARTICLE III

SUPPLEMENTARY AGREEMENTS

(1) Supplementary agreements applicable to specific regions, to specific pests or diseases, to specific plants and plant products, to specific methods of international transportation of plants and plant products, or otherwise supplementing the provisions of this Convention, may be proposed by the Food and Agriculture Organisation of the United Nations (hereinafter referred to as FAO) on the recommendation of a contracting Government or on its own initiative, to meet special problems of plant protection which need particular attention or action.

(2) Any such supplementary agreements shall come into force for each contracting Government after acceptance in accordance with the provisions of the FAO Constitution and Rules of Procedure.

ARTICLE IV

NATIONAL ORGANISATION FOR PLANT PROTECTION

(1) Each contracting Government shall make provision, as soon as possible and to the best of its ability, for

(a) An official plant protection organisation, with the following main functions :

(i) The inspection of growing plants, of areas under cultivation (including fields, plantations, nurseries, gardens and greenhouses), and of plants and plant products in storage and in transportation, particularly with the object of reporting the existence, outbreak and spread of plant diseases and pests and of controlling those pests and diseases;

(ii) The inspection of consignments of plants and plant products moving in international traffic, and, as far as practicable, the inspection of consignments of other articles or commodities moving in international traffic under conditions where they may act incidentally as carriers of pests and diseases of plants and plant products, and the inspection and supervision of storage and transportation facilities of all kinds involved in international traffic whether of plants and plant products or of other commodities (particularly with the object of preventing the dissemination across national boundaries of pests and diseases of plants and plant products);

(iii) The disinfestation or disinfection of consignments of plants and plant products moving in international traffic, and their containers, storage places, or transportation facilities of all kinds employed :

(iv) The issue of certificates relating to phytosanitary condition and origin of consignments of plants and plant products (hereinafter referred to as "phytosanitary certificates") :

(b) The distribution of information within the country regarding the pests and diseases of plants and plant products and the means of their prevention and control :

(c) Research and investigation in the field of plant protection.

(2) Each contracting Government shall submit a description of the scope of its national organization for plant protection and of changes in such organisation to the Director-General of FAO, who shall circulate such information to all contracting Governments.

ARTICLE V

PHYTOSANITARY CERTIFICATES

(1) Each contracting Government shall make arrangements for the issue of phytosanitary certificates to accord with the plant protection regulations of other contracting Governments, and in conformity with the following provisions :

(a) Inspection shall be carried out and certificates issued only by or under the authority of technically qualified and duly authorized officers and in such knowledge and information available to those officers that the authorities of importing countries may accept such certificates with confidence as dependable documents.

(b) Each certificate covering material intended for planting or propagation shall be as worked in the Annex to this Convention and shall include such additional declarations as may be required by the importing country. The model certificate may also be used for other plants or plant products where appropriate and not inconsistent with the requirements of the importing country.

(c) The certificates shall bear no alterations or erasures.

(2) Each contracting Government undertakes not to require consignments of plants intended for planting or propagation imported into its territories to be accompanied by phytosanitary certificates inconsistent with the model set out in the Annex to this Convention.

ARTICLE VI

REQUIREMENTS IN RELATION TO IMPORT

(1) With the aim of preventing the introduction of diseases and pests of plants into their territories, contracting Governments shall have full authority to regulate the entry of plants and plant products, and to this end, may :

(a) Prescribe restrictions or requirements concerning the importation of plants or plant products ;

(b) Prohibit the importation of particular plants or plant products, or of particular consignments of plants or plant products ;

(c) Treat, destroy or refuse entry to particular consignments of plants or plant products, or require such consignments to be treated or destroyed.

(2) In order to minimise interference with international trade, each contracting Government undertakes to carry out the provisions referred to in paragraph 1 of this Article in conformity with the following :

(a) Contracting Governments shall not, under their plant protection legislation, take any of the measures specified in paragraph 1 of this Article unless such measures are made necessary by phytosanitary considerations.

(b) If a contracting Government prescribes any restrictions or requirements concerning the importation of plants and plant products into its territories, it shall publish the restrictions or requirements and communicate them immediately to the plant protection services of other contracting Governments and to FAO.

(c) If a contracting Government prohibits, under the provisions of its plant protection legislation, the importation of any plants or plant products, it shall publish its decision with reasons and shall immediately inform the plant protection services of other contracting Governments and FAO.

(d) If a contracting Government requires consignments of particular plants or plant products to be imported only through specified points of entry, such points shall be so selected as not unnecessarily to impede international commerce. The contracting Government shall publish a list of such points of entry and communicate it to the plant protection services of other contracting Governments and to FAO. Such restrictions on points of entry shall not be made unless the plants or plant products concerned are required to be accompanied by phytosanitary certificates or to be submitted to inspection or treatment.

(e) Any inspection by the plant protection service of a contracting Government of consignments of plants offered for importation shall take place as promptly as possible with due regard to the perishability of the plants concerned. If any consignment is found not to conform to the requirements of the plant protection legislation of the importing country the plant protection service of the exporting country shall be informed. If the consignment is destroyed, in whole or in part, an official report shall be forwarded immediately to the plant protection service of the exporting country.

(f) Contracting Governments shall make provisions which, without endangering their own plant production, will reduce to a minimum the number of cases in which a phytosanitary certificate is required on the entry of plants or plant products not intended for planting, such as cereals, fruits, vegetables and cut flowers.

(g) Contracting Governments may make provision for the importation for purposes of scientific research of plants and plant products and of specimens of plant pests and disease-causing organisms under conditions affording ample precaution against the risk of spreading plant diseases and pests.

(3) The measures specified in this Article shall not be applied to goods in transit throughout the territories of contract-ing Governments unless such measures are necessary for the protection of their own plants.

ARTICLE VII

INTERNATIONAL CO-OPERATION

The contracting Governments shall co-operate with one another to the fullest practicable extent in achieving the aims of this Convention, in particular as follows :

(a) Each contracting Government agrees to co-operate with FAO in the establishment of a world reporting service on plant diseases and pests, making full use of the facilities and services of existing organisations for this purpose, and when this is established, to furnish to FAO periodically the following information :

(i) Reports on the occurrence, outbreak and spread of economically important pests and diseases of plants and plant products which may be of immediate or potential danger ;

(ii) Information on means found to be effective in controlling the pests and diseases of plants and plant products.

(b) Each contracting Government shall, as far as is practicable, participate in any special campaigns for combating particular destructive pests or diseases which may seriously threaten production and need international action to meet the emergencies.

ARTICLE VIII

REGIONAL PLANT PROTECTION ORGANISATION

(1) The contracting Governments undertake to co-operate with one another in establishing regional plant protection organisations in appropriate areas.

(2) The regional plant protection organisations shall function as the co-ordinating bodies in the areas covered and shall participate in various activities to achieve the objectives of this Convention.

ARTICLE IX

SETTLEMENT OF DISPUTES

(1) If there is any dispute regarding the interpretation or application of this Convention, or if a contracting Government considers that any action by another contracting Government is in conflict with the obligations of the latter under Article V and VI of this Convention, especially regarding the basis of prohibiting or restricting the imports of plants or plant products coming from its territories, the Government or Governments concerned may request the Director-General of FAO to appoint a committee to consider the question in dispute.

(2) The Director-General of FAO shall thereupon, after consultation with the Governments concerned, appoint a committee of experts which shall include representatives of those

Governments. This committee shall consider the question in dispute, taking into account all documents and other forms of evidence submitted by the Governments concerned. This committee shall submit a report to the Director-General of FAO who shall transmit it to the Governments concerned, and to other contracting Governments.

(3) The contracting Governments agree that the recommendations of such a committee, while not binding in character, will become the basis for renewed consideration by the Governments concerned of the matter out of which the disagreement arose.

(4) The Governments concerned shall share equally the expenses of the experts.

ARTICLE X

SUBSTITUTION OF PRIOR AGREEMENTS

This Convention shall terminate and replace, between contracting Governments, the International Convention respecting measures to be taken against the *Phylloxera vastatrix* of 3 November, 1881, the additional Convention signed at Berne on 15 April, 1889 and the International Convention for the Protection of Plants signed at Rome on 16 April, 1929.

ARTICLE XI

TERRITORIAL APPLICATION

(1) Any Government may at the time of ratification or adherence or at any time thereafter communicate to the Director-General of FAO a declaration that this Convention shall extend to all or any of the territories for the international relations of which it is responsible, and this Convention shall be applicable to all territories specified in the declaration as from the thirtieth day after the receipt of the declaration by the Director-General.

(2) Any Government which has communicated to the Director-General of FAO a declaration in accordance with paragraph 1 of this Article may at any time communicate a

further declaration modifying the scope of any former declaration or terminating the application of the provisions of the present Convention in respect of any territory. Such modification or termination shall take effect as from the thirtieth day after the receipt of the declaration by the Director-General.

(3) The Director-General of FAO shall inform all signatory and adhering Governments of any declaration received under this Article.

ARTICLE XII

RATIFICATION AND ADHERENCE

(1) This Convention shall be open for signature by all Governments until 1 May, 1952 and shall be ratified at the earliest possible date. The instruments of ratification shall be deposited with the Director-General of FAO, who shall give notice of the date of deposit to each of the signatory Governments.

(2) As soon as this Convention has come into force in accordance with Article XIV, it shall be open for adherence by non-signatory Governments. Adherence shall be effected by the deposit of an instrument of adherence with the Director-General of FAO, who shall notify all signatory and adhering Governments.

ARTICLE XIII

AMENDMENT

(1) Any proposal by a contracting Government for the amendment of this Convention shall be communicated to the Director-General of FAO.

(2) Any proposed amendment of this Convention received by the Director-General of FAO from a contracting Government shall be presented to a regular or special session of the Conference of FAO for approval and, if the amendment involves important technical changes or imposes additional obligations on the contracting Governments, it shall be considered by an Advisory committee of specialists convened by FAO prior to the Conference.

(3) Notice of any proposed amendment of this Convention shall be transmitted to the contracting Governments by the Director-General of FAO not later than the time when the agenda of the session of the Conference at which the matter is to be considered is dispatched.

(4) Any such proposed amendment of this Convention shall require the approval of the Conference of FAO and shall come into force as from the thirtieth day after acceptance by two-thirds of the contracting Governments. Amendments involving new obligations for contracting Governments, however, shall come into force in respect of each contracting Government only on acceptance by it and as from the thirtieth day after such acceptance.

(5) The instruments of acceptance of amendments involving new obligations shall be deposited with the Director-General of FAO, who shall inform all contracting Governments of the receipt of acceptances and the entry into force of amendments.

ARTICLE XIV

ENTRY INTO FORCE

As soon as this Convention has been ratified by three signatory Governments it shall come into force between them. It shall come into force for each Government ratifying or adhering thereafter from the date of deposit of its instrument of ratification or adherence.

ARTICLE XV

DENUNCIATION

(1) Any contracting Government may at any time give notice of denunciation of this Convention by notification addressed to the Director-General of FAO. The Director General shall at once inform all signatory and adhering Governments.

(2) Denunciation shall take effect one year from the date of receipt of the notification by the Director-General of FAO.

Names of some important Plant Protection Chemicals

Common Name		Chemical Name	Other Name
I. Insecticides			
A. Chlorinated hydrocarbons			
1. Aldrin	....	1,2,3,4, 10-10-Hexachloro-1, 4-4a, 5,8,8a hexahydro-1, 4-endo-exo-5, 8-dimethano-napthalene.	*Aldrex*, Aldrosol, Drinox, Octalene, Seedrin.
2. BHC	....	1,2,3,4,5,6, Hexachlorocyclohexane or Benzene hexachloride.	Benzex, Benzichlor, Benzahex, Dol, Dolmix, *Gammexane*, HCCH, Hexachlor, Hexyclan Micosane, Soprocide.
3. Chlordane	....	1,2,4,5,6,7,8, 8-Octachlor-2,3,3a, 4,7,7a-hexahydro-4, 7-methanoindane.	*Chlordam*, Corodane, Chlorkil, Kypchlor, Ocatchlor, Ortho-Klor, Synklor, Topichlor 20.
4. DDT	....	Dichlorodiphenyl Trichloroethane or 1,1, 1-Trichloro-2, 2-bis (p-chlorophenyl) ethane.	Anofex, Dedelo, Didimac, Genitox, *Guesarol*, Gesapon, Gesarol, Gyron, Ixodex, Kopsol, Neocid, Penthachlorin, Rukseam, Zerdane.
5. Dieldrin	....	1,2,3,4,10, 10-Hexachloro-6, 7-epoxy-1,4,4a, 5,6,7,8, 8a-Octahydro-exo-1, 4-endo-5,8-dimethanonapthalene.	Alvit, *Dieldrex*, Octalox, Micodiedlrin, Panoram D-31.

Common Name		Chemical Name	Other Names
1. Insecticides (*contd.*)			
6. Endosulfan	....	6,7,8,9,10, 10-Hexachloro-1,5,5a,6,9,9a-hydro-6, 9-methano-2, 4, 3-benzo(e)-dioxathiepin-3 oxide,	Chlorthiepin, Cyclodan, Hexasulfan, Insectophene, Kop-Thiodan, Malix, Thifor, *Thiodan*, Thimul, Thionex.
7. Endrin	....	1,2,3,4,10, 10-hexachloro-6, 7-epoxy-1, 4,4a, 5,6,7,8, 8a-octahydro-exo-1, 4-exo-5, 8-dimethanonapthalene.	*Endrex*, Enzex, Hyxadrin, **Mendrin**, Tafadrin.
8. Heptachlor	....	1,4,5,6,7, 8-Heptachloro-3a. 7. 7a-tetrahydro-4, 7-methanoindene.	Drinox H-34, *Heptamul*.
9. Lindane	....	Gamma isomer of 1,2,3,4,5,6-Hexachlorocyclohexane.	Gamma-BHC, *Gammexane*, Gamaphex, Gammaline, Gammex, Lindafor, Lindagam, Lintox, Novigam, Silvanol.
10. Toxaphene	....	Chlorinated Camphene.	Alltox, clor Chem, Th-590, polychlorocamphene, Phenancide, Phenatox, Strobane-T, Toxakil.
B. Organophosphates			
11. (a) Azinphos (Ethyl)	...	O, O-Diethyl-S (4-oxo-3H-1, 2, 3-benzotriazine-3 yl methyl-dithio phosphate.	*Gusathion A.*
(b) Azinphos (Methyl)	...	O,O-Dimethyl-S [c-Oxo-1, 2, 3-benzotriaz-	*Guthion*, Carfene, Gusathion **M.**

Common Name		Chemical Name	Other Names
I. Insecticides (contd.)			
12. Chlorfenvinphos	....	2-chloro-1 (2, 4-dichlorophenyl)-vinyl die-thyl phosphate.	*Birlane*, Sapecron, Supona
13. Cytrolane	....	2-(diethoxyphosphinylimino)-4 methyl-1, 3 dithiolane.	....
14. Dicrotopnos	....	3-Hydroxy-N, N-dimethyl-cis-crotonamide dimethylphosphate.	*Bidrin*, Carbicron, Ektafos.
15. DDVP	....	2, 2 dichloro vinyl, 0, 0-dimethyl phosphate.	*Dichlorvos*, Dichlorphos, Dedevap, Divipan, Merkol, Mafu, Mervex, Nogos, No-Pest, *Nuvan*, Oko, Phosvit, Vapona.
16. Dimethoate	....	0,0-Dimethyl-(N-methylcarbamoyl methyl) phosphorothioate.	Cygon, Daphene, De-Fend, Fostion MM, *Rogor*, Roxion, Perfektion, Trimetion.
17. Diazinon	....	0,0-Diethyl 0-(2-isopropyl-4-methyl-6-pyri-midinyl) phosphorothioate.	*Basudin*, Diazajet, diazide, Diazol, Dazzel, Gardentox, Spectracide.
18. EPN	....	0-Ethyl-0-p-nitrophenylphenyl phospho-nothioate.	....
19. Fenitrothion	....	0,0-Dimethyl 0-(1-nitro-m- tolyl) phos-phorothioate.	Accothion, Folithion, Novathion, Nuvanol, *Sumithion*.

Common Name		Chemical Name	Other Names
I. Insecticides (Contd.)			
20. Fenthion	...	0, On-Dimethyl 0-[3-methyl-4-(Methyl-thio) phosphorothioate.	*Baviex*, Entex, Lebaycid, Tiguvon.
21. Formothion	...	0, 0-Dimethyl-S-(N-methyl-N-tormyl-carbomoylmethyl) phosphorodithioate.	Aflix, *Anthio*.
22. Malathion	...	0,0-Dimethyl phosphorodithioate of diethyl-mercapto succinate.	Carbofos, *Cythion*, Emmatos, Fyfanon, Karbofos, Kop-Thion, Kypfos, Malaspray, Malamer-Mercaptothion, Zithiol.
23. Metasystox	...	0,0-Dimethyl-0(and S) [2-(ethylthio) ethyl] phosphorothioates I and II.	*Demeton-methyl*.
24. Monocrotoplos	...	3-hydroxy-N-methyl-*cis*-crotonamide-dimethyl phosphate.	Azodrin, *Nuvacron*.
25. Menazon	...	S-(4, 6, Diamino-S-triazin-2-ylmethyl) 0, 0-dimethyl phosphorodithioate.	Azidithion, Saphi-Col, Saphizon, **Saphos, Sayfos, Sayphos**.
26. (a) Parathion (Ethyl)	...	0, 0-Diethyl-0-p-nitrophenyl phosphoro-thioate.	Alleron, Alkron, Bladan, Corothion, Etilon, *Ekatox*, Foliidol, Fospono, 50, Niran, Ortho-phos, Panthion, Paramar, *Parathion*, Porawet, Plioskil, Rhodiatox, Soprathion, Stathion, Tehiophos.

Common Name		Chemical Name	Other Names
Insecticides (Contd.) (b) Parathion (Methyl)	...	O,O-Dimethyl-O-p-nitrophenyl phosphoro-thioate.	*Dalf*, *Folidol-M*, **Metaphos**, **Metacide-50**, **Metron, Nitrox-80, Partron-M, Takwaisa.**
27. Phenthoate	...	O,O-Dimethyl-S. (a-ethoxy-carbonylbenzyl) phosphorodithioate.	Cidial, Dimephenthoate, *Elsan*, **Papthion.**
28. Phorate	...	O,O-Diethyl-S-(ethylthio)-methyl phos-phorodithioate.	*Thimet*, Timet.
29. Phosalone	...	O,O-Diethyl-S [C-Chloro-2. oxobenzoxa-zolin-3yl) methyl] phosphorodithioate.	*Zolone*.
30. Phosphamidon	...	2-Chloro-N, N-diethyl-3 (—dimethoxy-phosphinyloxy) crotonamide.	*Dimecron*.
31. Phosvel	...	O-(4-Bromo-2, 5-dichlorophenyl) O-methyl-phenyl-phosphonothioate.	Abar, Leptophos.
32. Quinalphos	...	O, O-Diethyl O-(2-chinoxalyl)-phosphoro-thioate	Bayrusil, *Ekalux*.
33. Thiodemeton	...	O, O-Diethyl S-(2-ethylthio) ethyl phos-phodithioate.	Disyston, Disulfoton, Dithiosystox. **Frumin, Al-Solvirex.**
34. Thiometon	...	O,O-Dimethyl-S(2-(ethylthio) ethyl phosh-porodithioate.	*Ekatin*.

Common Name		Chemical Name	Other Names
I. Insecticides (Contd.)			
35. Trichlorphon	....	Dimethyl (2, 2, 2-trichloro-1-hydroxyethyl) phosphonate.	*Dipterex*, Dylox.
36. Vamidothion	....	0-Dimethyl-S-(2-cl-methylcarbomyl-ethyl) thio) ethyl phosporodithioate.	*Kilval, Vamidothion.*
C. Carbamates and their Thio salts			
37. Carbaryl	....	1-Naphthyl N-methylcarbamate.	Hexavin, Karbaspray, Ravyon, Sepkene, *Sevin*, Tricarnam.
38. Carbofuran	....	2, 3-Dihydro-2, 2-dimethyl-7-benzofuranyl-methyl carbamate.	*Furadan.*
39. Methomyl	....	S-Methyl-N-(methylcarbamoyl) oxy) thio-acetimidate.	*Lannate.*
D. Others			
40. Aldicarb	....	2-Methyl-2-(methylthio) Propionaldehyde-0-(methylcarbamoyl)-oxime.	Ambush, *Temik.*
41. Nicotine	....	(S),-3-(1-Methyl-2 pyrrolioyl) pyridine.	....
42. Padan	....	1,3-Bis (Carbamoylthio)-2-(N,N-dime-thylamino) propane.	*Cartap.*

Common Name		Chemical Name	Other Names
I. Insecticides (Contd.)			
43. Telodrin	....	Minimum of 82% isobornylthio cyanoacetate+18% (Maximum) of active terpene.	*Isobenzan.*
44. Thanite	....	Isobornyl thiocyanoocetate, 82% ; other related terpenes, 18%.	....
II. Fungicides			
A. Metallic salts			
45. Copper Oxychloride	....	Basic cupric chloride.	Basic Copper Chloride, Coprantol, *Blitox,* Blue Copper 50, Colloidex, Coxysan, Cupramar, Cupric chloride, Cupravit, Cuprox, *Fytolan.*
46. Copper Sulphate	....	Cupric Sulphate Pentahydrate	Blue Stone, *Blue Vitrol,* Blue Copperas.
47. Cuprous Copper Oxide	....	Cupric oxide (CuO), Cuprous oxide Cu₂O.	Copper-Sandoz, Coppesan, Fungimar, Fytomix, *Perenox,* Yellow Cuprocid
48. Nickel Chloride	....	NiCl₂	....
B. Non-metallic Salts			
49. Sulphur dust	....	....	Brim stone, *Cosan,* Cofril.
50. Sulphur colloida	....	....	Sulkol.

Common Name	Chemical Name	Other Names
II. Fungicides (*contd.*)		
51. Sulphur wettable	....	Cosan, Hexasul, Thiovit, Sulkol, Sultof, Solbar, *Spersul*, Spitox, Sulfex.
C. Organomercurials		
52. Ethylmercury chloride	C_2H_5HgCl	*Ceresan*, Granosan.
53. Methoxy ethyl mercury-Chloride	$C_2H_5OC_2H_4HgCl$. About 2 to 6% mercury equivalent.	Aretan, Agallol (3% mercury equivalent), Agalloiforte (6% mercury equivalent), Baytan, *Ceresanwet* Ceresan-Universal-Nasbeize (2.5% to 3.5% mercury-equivalent) Tayssato (2.2% dust, 1.9% liquid).
54. Phenyl mercury chloride ...	C_6H_5HgCl.	Stopspot.
55. Phenyl mercury acetate	$C_6H_5H_8OOCCH_3$	Agrosan G. N., Ceresan, Gallotox, Hongnien, Liquiphene, Mersolite, *PMA*, Phenmad, Phix, Shimmerex
56. Phenyl mercury salicylate	$C_6H_4OHCOOH_8C_6H_5$	*Mercuime*, Mercusol.
D. Carbamate		
57. Benlate	Methyl-I (butyl carbamoyl)-2-benzimidazole-carbamate.	*Benomyl.*

Common Name		Chemical Name	Other Names
II. Fungicides (*Contd.*)			
E. Thiocarbamates			
58. Maneb	...	Manganese ethylene-1, 2-bisdithiocarbamate.	Chloroble M, *Dithane M-22*, Kypman 80, Maneb, Manesan, Manzate, Manzate D, Sopranebe, Trimangol.
59. Mancozeb	...	Coordinate product of Zinc ion and Manganese ethylene-bisdithio carbamate.	*Dithane M-45*, Fore
60. Nabam	...	Disodium ethylene-1, 2 bisdithio carbamate.	Chem Bam, *Dithane D-14*, DSE, Parzate, Spring-Bak.
61. Zineb	...	Zinc ethylene bisdithio carbamate.	Aspor, Chem Zineb, *Dithane Z-78*, Hexathane, Kypzin, Lonacol, Parzate C, Polyram Z, Tiezene, Tritoftorol, Zebtox, Zidan, Zinosan.
F. Dithiocarbamates			
62. Ferbam	...	Ferric dimethyl dithiocarbamate.	Fermate, Ferberk, Hexaferb, Trifungol.
63. Thiram	...	Bis(dimethylthio Carbamoyl)disulfide ; or tetramethylthiuram-disulfide.	Arasan, Fernide–850, Hexathir, Mercuram, Nomersam, Polyram-ultra, Pomarsol, fote, Spotrcte, Tersan, Thimer, *Thiride*, Thiotex, Thiramad, Thirasan, Thylate, Tirampa, TMTDS, Trametan, Tripomol, Tuads, Vancide, TM-95.
64. Ziram	...	Zinc dimethyldithiocarbamate.	*Cuman*, Fuklasin, Hexazir, Mezene, Pomarsol. Z forte. Tricarbamix Z. Triscabol,

Common Name		Chemical Name	Other Names
II. Fungicides (*Contd.*)			
G. Others			
65. Fentin Acetate	....	Triphenyltin Acetate	*Brestan*, Phentin acetate.
66. Fentin Hydroxide	....	Triphenyltin hydroxide	TPTH, *Du-Ter*.
67. Captan	....	cis-N-[(Trichloromethyl) thio)]-4-Cyclohexene-1,2 dicarboximide.	Merpan, *Orthocide*.
68. Fentiazon	....	3-Benzyldeneamino-4-phenylthiazoine-2-thione.	*Celdion*.
69. Hinosan	....	0-Ethyl-S, S-diphenyl-dithiophosphate.	Edifemphos.
70. Karathane	....	2-(1-Methyl-n-heptyl)-4-6-dimethyl crotonate with its isomer 4-(1-methyl nyheptyl) 2, 6-nitrophenyl crotonate.	Arathane, *Dinocap*, Iscothane, Mildex.
71. PCNB	....	Pentachloronitrobenzene	Avicol, Botrilex, *Brassicol*, Folosan, **Tritisan**, Terrachlor, Tilcarex.
72. Plantvax	....	Dioxide of Vitavax	Oxycarboxin.
73. Vitavax	....	5, 6-Dihydro-2-methyl-1, 4-Oxathlin-3-carboxanilide.	Carboxin.

Common Name		Chemical Name	Other Names
III. Weedicides			
A. Acetic Acid Derivatives			
74. 2, 4-D	....	2, 4-Dichlorophenoxy acetic acid.	Amoxone, Chloroxone, Crop Rider, Daca-mine, Ded-Weed, Esteron, Pennamine-D, Salvo, Tributon, Weedone.
75. 2, 4, 5-T	....	2, 4, 5-Trichlorophenoxyacetic acid.	Ded-Weed, Brush Killer, Esteron 245 Con-centrate, Fence Rider, Inverton 245, Line Rider, Reddon.
B. Amide Derivatives			
76. Alachlor	....	2-Chloro-2', 6', diethyl-N-(methoxy-methyl)-acetanilide.	*Lasso*, Lazo.
77. Butachlor	....	2-Chloro-2'6', diethyl-N-(Butoxymethyl)-acetanilide.	*Machete.*
78. Propachlor	....	2-Chloro-N-Isopropylacetanilide.	*Ramrod.*
79. Propanil	....	3-4-Dichlorophenyl-propionanilide.	Chem Rice, DPA, *Stam F-34*, Surcopur.
C. Arsonate Derivatives			
80. DSMA	....	Disodium methanearsonate.	*Ansar*, 8100, Ansar DSMA liquid, Arrhenal, Arsinyl, Crab-E-Rad, Dal-E-Rad 100, Ditac, DMA, DMA 100, Methar, Sedar,

Common Name		Chemical Name	Other Names
III. Weedicides (Contd.)			
81. MSMA	....	Monsodium methanearsonate.	Ansar 170, H.C., Bueno, *Ansar 529.*
D. Ether Derivative			
82. Nitrofen	....	2,4-Dichlorophenyl P-Nitrophenylether.	*Tok.*
E. Fatty Acid Derivatives			
83. Dalapon	....	2,2-Dichloropropionic Acid Sodium salt.	Ded-Weed, *Dowpon,* Gramevin, Radapon, Unipon.
84. TCA	....	Sodium Trichloroacetic acid.	Sodium TCA.
F. Pyridyl Derivatives			
85. Diquat	....	6,7-Dihydrodipyridol (1,2'-a : 1'·C) pyra-zidilium dibromide.	Aquacide, Dextrone, Reglone.
86. Paraquat (dichloride salt)	....	1, 1'-Dimethyl-4', 4-bipyridilium ion.	*Gramoxone,* Weedol.
87. Paraquat (dimethyl sulphate)	....	1, 1'-Dimethyl-4, 4'-bipyridilium ion.	*Dual paraquat.*
G. Thiocarbamate Derivatives			
88. Eptam	....	S Ethyl-n, n-dipropylthiocarbamate.	EPTC
89. Tillam	....	S-Propyl, butylethylthiocarbamate.	*Pebulate,* PEBC
90. Triallate	....	S-(-2,2,3-Trichloroallyl) diisopropylthiocar-bamate.	Avadex BW, Far-Go.

Common Name		Chemical Name	Other Names
III. Weedicides (Contd.)			
H. Triazines			
91. Atrazine	....	2-Chloro-4-(ethylamino)-6-(isopropylamino)-S-triazine.	Atrex, Gesaprim, P' matol A.
92. Lambast	....	2, 4-bis [(3-methoxypropyl) amino]-6-methylthio-S-triazine.	....
93. Prometon	....	2,4-bis (Isopropylamino)-6-methoxy-S-triazine.	Gesafram, Pramitol, Prometone.
94. Prometryne	....	2,4-bis (Isopropylamino)-6-methylthio-S-triazine.	Caparol, Gesagard, Primatol Q
95. Propazine	....	2-Chloro-4, 6-bis (isopropylamino)-s-triazine	Gesamil, Milogard, Primatol P.
96. Simazine	....	2-Chloro-4, 6-bis (ethylamino)-s-triazine	Gesatop, Princep, Primatol S.
I. Others			
97. Ammonium Sulphamate	....	H_3NSO_2O NH_4	Ammate, Amcide, AMS
98. Borax	....	Sodium tetraborate decahydrate ($Na_2B_4O_7$ $10H_2O$)	Borascu, Gerstley Borate, Neobor, Trona-bor.

Common Name		Chemical Name	Other Names
III. Weedicides (Contd.)			
99. Banvel D	...	3, 6-Dichloro-o-anisic acid	Benex, *Dicamba*, Mediben.
100. Banvel T	...	3, 5, 6-Trichloro-o-anisic acid	Metriben, *Tricamba*.
101. PCP	....	Pentachlorophenol.	Dowicide, 7, Penchlorol, Penta, **Sinituho**, Wedone.
102. Sinbar	...	3-tert-Butyl-5-chloro-6-methyluracil	*Terbacil*.
103. Sirmate	...	3, 4 and 2, 3-Dichlorobenzyl N-methylcar. bamate	*Rowmate*.
104. Trifluralin	...	a, a, a,-Trifluoro-2, 6-dinitro-N, N-dipro-pyl-p-toluidine.	*Treflan*, Elancolan.
105. Tenoranon	...	3-[p-(p-Chlorophenoxy)-phenyl]-1, 1-di-methyl urea.	*Chloroxuron*.
J. Sub-Ureas			
106. Diuron	...	3-(3, 4-Dichlorophenyl)-1, 1-dimethylurea.	DMU, Mermer.
107. Monuron	...	3-(p-Chlorophenyl)-1, 1-dimethylurea.	Chlorfenidim, Telvar.
IV. Nematicides			
108. D.B.C.P.	...	1, 2-Dibromo-3-chloropropane	Dibromochloropropane, Fumazone, Nema-fume, *Nemagon*.

Common Name		Chemical Name	Other Names
IV. Nematicides (Conted)			
109. D.D.	...	1, 2,-Dichloropropane ; and 1, 3-Dichloropropene	Nemafene.
110. Methan N Sodium	...	Sodium N-methyldithiocarbamate	Metham, Trimaton, VPM, *Vapam*.
111. Telone	...	1, 3-Dichloropropene	*Dichloropropene*.
112. Terracur	...	5-Carboxymethyl-3-methyl-2H-1, 3, 5-thiadiazine-2-thione.	Thiadiazinthion.
113. Terracur P	...	0,0-Diethyl 0[4-(methylsulfinyl)phenyl] phosphorothioate.	*Dasanit*, Fensulfothion.
V. Acaricides			
114. Azobenzene	...	Diphenyl diamide.	*Azobenzide*
115. Chlorobenzilate	...	Ethyl 4, 4'-dichlorobenzilate.	Akar, Acaraben, Folbex, Kop-Mite, *Benzilan*.
116. Ethion	...	0,0,0,0-Tetraethyl S, S-methylene bisphosphorodithioate.	Nialate.
117. Kelthane	...	1,1 bis(chlorophenyl)-2,2,2-trichloroethanol	Acarin, *Dicofol*.

Common Name		Chemical Name	Other Names
V. Acaricides (*Contd.*)			
118. Morestan	....	6-Methylquinoline-2, 3-diyl dithiolocarbonate	Chinomethionate, Forstan, Oxythioquinox, Quinomethionate.
119. Morocide	....	2-sec-Butyl-4, 6 dinitrophenyl-3-methyl-2-butenoate.	Acricid, Ambox, *Binapacryl*, Dinoseb methacrylate, Endosan.
120. Omite	....	2-(p-ter-Butylphenoxy) cyclohexyl-2-propynyl sulfite.	
121. Tedion	....	S-p-Chlorophenyl-2, 4,5-trichlorophenyl sulfone.	*Tetradifon.*
122. Trithion	....	S-[(p-Chlorophenylthio) methyl] 1, 0,0-diethyl-phosphorodithioate.	*Carbophenothion*, Dagadio, Garathion
VI. Antibiotics			
123. Aureomycin	....	Chlortetracycline.	Acronize.
124. Terramycin	....	Oxytetracycline	Biostat PA
Streptocycline			
125. Streptomycin	**....**	Streptomycin (sulphate or nitrate)	*Agrimycin*, Agri Strep.
Others			
126. RLA-S	....	Blasticidin-S benzylaminobenzen-sulfonate	*Blasticidin.*

Common Name		Chemical Name	Other Names
VI. Antibiotics (Contd)			
127. Kasumin	...	D.3-0-[2-amino-4-(1-carboxyimino-methyl) amino]-2, 3,4,6-tetradeoxy-l-D-arabinohexopyranosyl]-D-chiro-inositol.	*Kasugamycin.*
VII. Rodenticides			
128. Barium Carbonate	...	$BaCO_3$	....
129. Coumafuryl	...	3(a-acetonylfurfuryl)-4-hydroxycoumarin	*Fumarin*
130. Cynodust	...	Calcium cyanide, Sodium cyanide	
131. Strychnine hydrochloride	...		
132. Warfarin	...	3-(α-acetonylbenzyl)-4-hydroxycoumarin.	*Coumafene,* Kypfarin, Zoocoumarin, Ratox.
133. Zinc phosphide	...	Zn_3P_2	
VIII. Fumigants			
134. Aluminium Phosphide	...	AlP	Phostoxin, *Celphos.*
135. Ethylene Dibromide	...	1,2-Dibromoethane	Bromofume, *Celmide,* Dowfume, W-85, Kopfume, Nephis, Soilbrom 85.
136. Ethylene Dichloride	...	1,2-Dichloroethane	EDC

Common Name	Chemical Name	Other Names
VIII. Fumigants (Contd)		
137 E D/CT ...	Mixture of Ethylene Dichloride & Carbon Tetrachloride.	
IX. Plant growth regulants		
138. Alar ...	Succinic acid 2,2 dimethyl-hydrazide.	B-Nine
139. Cycocel ...	(2-Chloroethyl) trimethylammonium chloride.	Chlormequat chloride.
140. Ethephon ...	2-(Chloroethyl) phosphonic acid.	*Ethrel.*
141. Gibberellic Acid ...	2, 4a, 7, Trihydroxy-1-methyl-8-methyl-enegibb-3-ene-1, 10 Carboxylic acid-1......4 lactone.	Activol, Berelex, *Gibberellin*, Gibrel, Gibsol, Gib-Tabs, Grocel, Pro-Gibb.
142. Maleic hydrazide ...	1,2-Dihydropyridazine-3, 6-dione.	KMH, Maintain 3, Retard, Slo-gro; Sprout-stop, Sucker-Stuff, Super Sucker-Stuff, Vondalhyde.

Note :—In column 'other names' most common names are in italics.

Some Records of Felt Loss Due to Diseases in India

Crop	Disease	Loss	Reference
1	2	3	4
paddy	bacterial leaf blight (*Xanthomonas oryzae*)	22.7%	Anon. 1964. quoted from Srivastava, 1972 *Indian Phytopath.* 25 : 1-16
		50%	Rao and Kauffman 1971 ; same as above.
		6-60%	Srivastava et al., 1966 ; *Indian Fmg.* 16 : 15, Srivastava & Rao, 1968. *Proc. 1st Summer Inst. Plant Dis. Control,* ICAR, 1968, p. 180.
	Helminthosporiose (*Cochliobolus miyabeanus*)	15-19%	Padmanabhan et al., 1948, *Indian Phytopath.* 1 : 34-37.
	Udbatta (*Ephelis oryzae*)	10-11%	Mohanty, 1964. *ibid,* 17 : 308-316.
	bacterial streak (*Xanthomonas oryzicola*)	5-30% in MP	Chand et al., 1971. *Proc. Symp. Epidem. Forecast. Control Plant. Dis.* INSA Bull. 46 1973. p. 309-313.
wheat	rusts	10-15%	Pal, 1968. *Proc. 1st Summer Inst. Plant Dis. Control.* ICAR, p. 1-4.
	black rust (*Puccinia graminis tritici*)	2 million tonnes	Joshi et al., 1974. *Current Trends in Plant Path.* Ed. S. P. Raychaudhury & J. P. Varma, Dept. Bot. Univ. Luck., p. 150-159.

1	2	3	4
	brown rust (*Puccinia recondita*)	1 million tonnes	-Ditto-
	rust	Rs. 40 million annually	Barclay, 1890, quoted from Joshi *et. al.*, 1974, *ibid* p. 150-159.
		Rs. 60 million annually	Mehta, 1940, quoted from Joshi *et al.*, 1974, *ibid.*, p. 150-159.
		8-20%	Joshi *et al.*, 1974, *ibid*, p. 150-159.
	loose smut (*Ustilago nuda tritici*)	1 million tons	Suryanarayana, 1971. *2nd Intern. Symp. Plant Path.* (Abstr.) **IPS** IARI. p. 22-23.
	tundu (*Anguina tritici* and *Coryne bacterium tritici*)	2-5% 50-60% (1965-66)	Raychaudhuri, 1968. *Indian Phytopath.* 21 : 1-13.
maize	all diseases	370, 645 tonnes	Payak and Renfro 1974. *Current Trends, in Plant Path.* Ed. S. P. Raychaudhuri & J. P. Varma Dept. Bot., Univ. Luck. p. 165-170.
		10-12%	Payak and Renfro, 1968. *Proc. 1st Summer Inst. Plant Dis. Control.* ICAR p. 163.
		5-12%	Payak *et al.*, 1973. *Indian Fmg.* July, 1973.
	mosaic (virus)	32%	Seth *et al.*, 1968. *Proc. 1st Summer inst. Plant Dis. Control* ICAR p. 165-166.

1	2	3	4
jowar	grain smut (*Sphacelotheca sorghi*)	50%	Lucy Channamma and Delvi 1966. *Proc. Symp. Dis. Rice, Maize, Sorghum and Millets*, IPS Bull No. 3 p. 1-2.
		326,000 tons	Suryanarayana, 1971. *2nd Intern. Symp. Plant Path.* (Abstr.) IPS, IARI. p. 22-23.
bajra	ergot (*Claviceps microcephala*)	2-3%	Sundaram, 1968. Proc. *1st Summer Inst. Plant Dis. Control* ICAR p. 164-165.
	green ear (*Sclerospora graminicola*)	5-10%	Suryanarayana, 1962, *Indian Phytopath.* 15 : 247-249.
barley	stem and stripe rusts	600,000- 700,000 tons	Mehta, 1949, quoted from Ahmed et. al., 1972. *Indian Phytopath,* 25 : 434-488.
	covered smut (*Ustilago hordei*)	2-5% 20-40%	Bedi and Singh, 1972, *ibid.* 25 : 101-103.
cereals	smuts	2-5%	Pal, 1968. Proc. *1st Summer Inst. Plant Dis. Control,* ICAR p. 1-4.
sugar cane	all diseases	Rs. 300- 450 million p.a.	Singh, 1968. *ibid,* p. 150.
	grassy shoot	50%	Seth, 1968, *ibid,* p. 151.
potato	late blight (*Phytophthora infestans*)	15-25%	Pal, 1968, *ibid,* p. 1-4.

1	2	3	4
		31.5% (during 17 years in Simla Hills)	Dutt, 1968, *ibid*, p. 65.
		55.8% in plains	Dutt, 1968, *ibid*, p. 65.
		15-50%	Majid, 1950, quoted from Roy & Das, 1968. *Indian Phytopath.* 21(2) : 232-233.
		10-65%	Majid, 1952, quoted from Roy & Das, 1968 *ibid*.
Potato	brown rot (*Pseudomonas solanacearum*)	50%	Dutt et. al., 1968. Proc. *1st Summer Inst. Plant Dis. Control* ICAR p. 128.
Groundnut	tikka (*Cercospora personata, C. arachidicola*)	15-50%	Chohan and Singh, 1973, quoted from Chohan, 1974. *Current Trends Plant Path.* Ed. S. P. Raychaudhuri & J. P. Varma, Dept. Bot., Univ., Luck., p. 171-184.
	stem rot (*Sclerotium rolfsii*)	27%	Mathur, 1953, Singh and Mathur, 1953 and Kang, 1957, quoted from Chohan, 1974. *Current Trends Plant Path.* Ed. S. P. Raychaudhuri & J. P. Varma, Dept. Bot., Univ., Luck., p. 171-184.
	collar rot (*Aspergillus niger, A. pulverulentus*)	40%	Chohan, 1974, *ibid.* p. 171-184.

1	2	3	4
Grape	anthracnose (*Gleosporium ampelophagum* or *Sphaceloma ampelinum*)	15%	Bedi, et al., 1969. *Indian Phytopath.* 22(1) : 155-156.
Cotton	bacterial leaf blight *Xanthomonas malvacearum*)	20-30%	Singh and Verma ; 1973. *Pesticides.* 7 : 16-17.
Tea	blister blight (*Exobasidium vexans*)	180 million pounds in India for 6 years since 1946	Venkata . Ram, 1971. *Proc. Epidem Forecast Control Plant Dis.* INSA Bull. 46 1973. p. 377-383.

Some Records of Felt Loss[1] Due to Insect Pests in India

Crop or crop product	Pest	Loss %
1	2	3
sugarcane	all pests	10
sugarcane (as raw sugar)	pyrilla, borers, termites	16
foodgrains	all pests	10-15
wheat	wheat weevils	1.6
	pests and other causes	3
rice	pests and other causes	1.1
	swarming caterpillar, rice stem borer, caterpillar, stem borers and army hopper, rice hispa	10
other cereals including jowar, bajra, ragi, small millets, wheat and barley	grasshoppers, termites, cutworms, hairy caterpillar, stem borers and army worm	7
jowar	pests and other causes	5
bajra	-do-	5
maize	-do-	5
ragi	-do-	5
barley	-do-	2
small millets	-do-	2.5
pulses (including gram)	red hairy caterpillar, pod borer, gram caterpillar and cutworms	5
gram	pests and other causes	2
other pulses	pests and other causes	2.5
potatoes	-do-	17
potatoes	jassids, aphids, cutworms, white grubs	5
tobacco	tobacco caterpillar, stem borer, aphids, cutworms	5
groundnut (nut in shell)	hairy caterpillar, termites	5
other oilseeds including mustard, castor sesamum and linseed	mustard aphids, hairy, caterpillar, semi-lopper, capsule borer, leaf and pod caterpillar	5

[1] Pradhan, S. 1964. Assessment of Losses caused by Insect Pests of Crops and Estimation of Insect Population. Entomology in India. 1964, Silver Jubilee Number of the Indian Journal of Entomology.

Some Records of Felt Loss and Loss Estimation Due to
Nematodes in India

Felt loss

Crop	Loss	Reference
1	2	3
wheat	Rs. 40,000,000/- in Rajasthan	Seshadri, 1973. *Natn. Symp. Agric. Res. & Dev. since Independence*, March, 1973.
barley	Rs. 30,000,000/- in Rajasthan	-do-

Loss Estimation

Crop	Nematode	Loss	Reference
1	2	3	4
wheat	car cockle *Anguina-tritici*)	Rs. 75,000,000	Seshadri, 1973 *National Symp. Agric. Res. and Dev. since Independence*, March, 1973.
coffee	*Pratylenchus coffee*	Rs. 20,000,000	-do-
sugar-cane	Nematodes	22-78%	Singh, 1968. *Proc. 1st Summer Inst. Plant Dis. Control*, ICAR, p. 158.

Some Loss Estimation Due to Rats under Field Conditions
Carried out in India

Crop	Loss	Reference
1	2	3
Jowar	5.85%	Report of the Coordinated Scheme for Research on the Study of Field Rats, 1965-66. ICAR.
Groundnut	20.1 Kg/ac*	-Do-
Wheat	2.36%	-Do-
Paddy	1.14-30.5%	-Do-
Coconut	upto 17%	-Do-
Barley	147 Kg/ha	Srivastava, 1966. *Labdev J. Sci Tech.* 4 : 197-200
Sugar cane	1.52-16.7%	Srivastava, 1968. Rodent Control for Increased Food Production. Rotary Club (West) Kanpur, p. 152.
Paddy	0.63-7.14%	-Do-
Jowar	5.9%	Srivastava and Pandya, 1968. *Proc. Intern. Symp. Bionomics & Control of Rodents.* Ed. S. L. Perti, Y. C. Wal & C. P. Srivastava Sci. Tech. Soc. Kanpur, May, 1971. p. 32-34.
Arhar	6.5%	-Do-
Barley	4.9%	-Do-
Gram	4.0%	-Do-
Wheat	10.29 Kg/ac.	Bindra and Prem Sagar, 1968 ibid, pp. 28-31.
Groundnut	12-31 Kg/ac.	-Do-
Sugar cane	65-230 Kg/ac (*gur*)	-Do-

* for calculating loss in kg/ha multiply by 2.47

(Source—Report of National Commission on Agriculture Vol. X 1976 Ministry of Agriculture +Irrigation, Govt. of India)

PLANNING FOR PESTICIDE APPLICATION

For the control of a pest, both the choice of an appropriate pesticide and its application are equally important. When the number of pesticides were comparatively few, knowledge on usefulness of different methods of application was limited and machineries involved were simpler, problem was also comparatively less complex. At present a large array of pesticides including systemic ones are available. Considerable improvement has also taken place in the method of application.

Choice of an appropriate pesticide for control of a pest or a pest complex needs consideration of several factors—namely pest or pests concerned, relative efficacy in terms of immediate and persistent effect, comparative cost, easeness of application, toxicity hazards involved, easy availability, etc. Benefit/cost ratio has also to be taken care of. The decision has to be taken by the subject matter specialist. Very often a number of alternative pesticides has to be suggested because availability of pesticides and their cost may often prove to be limiting factors. In view of availability of a large number of pesticides in the market, it may not be difficult to select a few instead of a particular one.

For treatment of soil insects including termites gamma-BHC, chlordane, heptachlor, aldrin, carbofuran or dimethoate are often suggested. Gamma-BHC often imparts a taint to tuber and underground organs meant for consumption. Hence it needs to be recommended with caution. Gamma-BHC or DDT are commonly used insecticides and have a wide-spectrum activity. They are still popular in spite of the adverse effects reported. They are available both in the form of dust and wettable powder. Nevertheless other insecticides are also being advocated.

For the control of gall midges or agromyzid fly, apart from DDT, phorate, carbofuran or quinalphos or disulphoton is recommended.

Similarly for sucking, leaf-eating and borer insects, apart from gamma-BHC and DDT, methyl parathion, or quinalphos, or phosphomidon, carbaryl, endosulphan, or monocrotophos, malathion, and in some cases carbofuran is suggested. In specific cases monocrotophos, carbaryl, quinalphos, phosalone, dimethoate or demeton-s-methyl is recommended.

Against mites and arachnid pests, lime sulphur or endosul-plan or kelthane is advocated, besides specific nematicides like binapacryl, morestan, tedion, trithion which are also equally effective.

For control of locusts in the egglaying or gregarious phase gamma-BHC or dieldrin is used, besides a number of other chemicals have shown promise.

In the use of more potent, systemic insecticides with greater toxic hazards, care has to be used in recommending against fruits and vegetables which are consumed direct.

For treatment of seeds against seed-borne pathogenic fungi a number of different chemicals are available in the market. Normally various preparations or formulations of organomercurial compounds are used because of their broad spectrum of activity. Thiram also recommended in place of organomercurials has wide range of activity. Besides various other dithiocarbamates and captan are also advocated for use in specific cases. For specific diseases, e.g., bunt, hexachloro-benzene or sclerotia infestation quintozene, for smut diseases, benomyl, thiabendazol, or oxathin compounds and for powdery mildew chloroneb, or ethirimol are recommended for use.

In cases of specific infections of bacteria, streptocycline or agrimycin or aureofungin treatment of seeds is advised.

Normally dust formulations are used for treatment of seeds, but for the treatment of the bulky vegetable propagative organs, dips in liquid are used. In this context organomercurials are the best. Alternatively thiram may be used.

As spray fungicides, copper compounds, particularly Bordeaux Mixture, copper oxychloride compounds are in use for a long time. Alternatively at present thiocarbamate fungicides, zineb, maneb, ziram (in some cases), captan, difoltan are recommended. In recent times, systemic fungicides, e.g., bavistin, calixin, etc., are also finding their use, specifity being determined in individual cases. Copper and thiocarbamate fungicides and captan have a wider coverage.

For treatment against soil-borne pathogens, apart from use of formaldehyde solution for sterilisation of soil prior to planting drenching with copper or thiocarbamate fungicides is in use in many cases, copper being particularly useful against

Phycomycetous fungi, quintozene compounds which may be applied as dust are particularly useful against sclerotia-forming fungi, e.g., *Rhizoctonia* spp., *Sclerotium* spp.

For the control of powdery mildews, sulphur preparations, namely, dusting or wettable sulphur have been in use, recently dinocap (karathane) is finding its use. A few systemic fungicides namely ethirimol, methirimol, chloroneb have been found to be useful against powdery mildews, particularly of cereals.

In rust diseases, copper fungicides are of no value. Formerly sulphur used to be recommended. At present zineb is finding its place in the package of practices. Systemic fungicides, particularly benomyl compounds have been found to be of value in many cases.

Control of bacterial diseases may be sought in the use of copper fungicides in most cases, but many plants like high yielding varieties of rice are shy to copper, hence agrimycin, streptocycline or aureofungin may be applied in such cases.

In the use of pesticides-fungicides and insecticides, different formulations may be present. So the actual dosage in terms of active ingredient has to be worked out in individual cases.

In pesticides, the use of sublethal doses does not contribute materially to the control of pests, on the other hand it may create the problem of giving rise to resistant strains.

Methods of application are often determined by the formulation of pesticides concerned, and exigencies of the circumstances.

In the control of plant diseases, affecting aerial organs, spraying is normally taken recourse to excepting for control of powdery mildews for which dusting sulphur is used in many cases. Spraying is advocated mainly for two reasoss, namely, action is protective in nature, pesticide requires an uniform and thorough coverage besides it must have persistent effect for some time. At present low volume spraying is being advocated for greater efficacy. Knapsac or pneumatic compression sprayers fitted with low volume nozzles are used. If they cannot be made available, pneumatic compression sprayers may be used. Power sprayers are useful for taller crops like sugar cane or in orchards. At present mist blowers which may serve purpose of both dusting and spraying are being advocated instead of power sprayers. In power sprayers of conventional types, hydraulic or compression types are preferred. Low volume preparations cannot be used in copper or sulphur preparations because of coarseness or larger size of particles. In those cases conventional hand or power sprayers with coarse nozzles will have to be used.

So far insecticides are concerned, previously dusting was largely used. Nowadays with the development of persistent and systemic insecticides, spraying is being advocated. At present a larger number of formulations are in emulsified concentrate form. In case of insecticides, fine sprays are always advocated hence low volume nozzles are to be used. Mist blowing is generally advocated at present instead of conventional power dusters or sprayers. Where mist blowers may not be available pneumatic compression sprayers with low volume nozzles should be used. Hand equipments may be of value in field crops, but in tall crops like jute, sugar cane or in orchards power operated ones are needed.

In case of small allotments use of rocking or bucket sprayers may be advocated, but good or uniform results may not be expected from them. Dusting is normally done by bellow or rotary dusters no doubt, but mist blowers should be used. In small allotments rotary dusters should be used. Other forms of dusters are of little value.

In the case of herbicides, hand sprayers with special nozzles should be used. In case special nozzles are not available low volume nozzles should be in operation, for minimising drift hazards. Mist application may be taken recourse to.

When any pesticide-fungicide, insecticide or herbicide has to be incorporated into the soil, normally it is done in the form of dust. Such dusting should be done, with the help of hand duster in rows or furrows. Blowers or power dusters will not be of value in such cases. In the case of mechanised cultivation, special hoppers containing pesticides may be fitted. Many fungicides are incorporated in the soil as drenches. Normally such drenching should be done with the help of hand sprayers which may be fitted with coarse nozzles so that an uniform coverage and economy of fluid can be effected. Somewhat similar effect may be obtained with watering cans or rosaries, provided the volume of application can be controlled by an experienced worker.

Efficacy of pesticides

Complaints regarding not having the desired result in control of pests by application of pesticides are often received from the users. The reasons for not achieving the expected level of success in reducing pest population may be manifold, a few of which are enliated below.

1. *Selection of the pesticide*

Very often right type of pesticide is not used. It is not uncommon to observe that a fungicide is used for control of insect pest or insecticide for control of fungal pathogen.

2. *Application of pesticide*

(a) *Proper dosage* may not be used.

(b) *Spray volume* used for coverage is not adequate as a result desired quantity of pesticide is not deposited.

(c) *Compatibility*—Many farmers have the propensity to mix more than one pesticides and spray the mixture for control of a pest. Compatibility of the pesticides used are never taken into consideration. Consequently poor performance is obtained.

(d) *Incorrect timing* of application.

(e) *Lack of moisture* in application of granular pesticides. In case of granular pesticides, sufficient moisture must be present in soil to allow them to be dissolved and taken up by roots of plants.

3. *Weather conditions*

(a) If weather is too humid, intermittant rains occur the spray will get washed, hence repeated applications may be necessary.

(b) Spraying should be done in the morning or afternoon, not in hot sun, because spraying in hot sun is injurious to plants.

(c) Specific weather conditions favour specific diseases or pests—e.g.

(i) potato late blight is favoured by cool humid rainy or dizzing weather.

(ii) Dry weather will favour rice hispa—(*Dicladispa armigera*).

(iii) Rice blast is favoured by moderately cool temperature with high humidity.

4. *Quality of water*

(a) With hard or brackish water, the suspensibility or emulsion stability may be affected with consequent effect on distribution of the pesticide on the sprayed surface. In such cases constant agitation may be necessary.

(b) If the water used is alkaline, then the efficacy of the pesticide may be adversely affected, as most pesticides are degraded under alkaline conditions.

Besides there may be problem of substandard pesticide. Quite often inefficacy of pesticides is ascribed to resistance of pests to pesticides, but in India pesticide resistance has not posed any problem as yet.

DISSIPATION OF PESTICIDES IN SOILS

Pesticide	Location	Soil Type	Dosage used (kg a.i/ha)	Formulation used	Method & no. of application	Crop	Dissipation rate (ppm) after days	Halflife days	Reference	Method of analysis	Remarks
1	2	3	4	5	6	7	8	9	10	11	12
Lindane	Bangalore	Sandy loam	100 ppm			Puddling	93.2(7),94.3(14),91.3(21), 88.5(28),86.8(35)84.2(42)		Mithyantha Pot (1973)	"	
						Field Capacity	98.0(7),98.0(14),96.4(21), 92.8(28),90.7(35),39.6(42)		"	"	
						1 bar tension	100(7),98.0(14),98.0(21), 96.6(28),94.0(35),94.0(42)		"	"	
	Mangalore	Laterite sandy clay loam	100 ppm			Puddling	98.6(7),94.8(14),92.6(21), 89.6(28),86.3(35),85.0(42)		"	"	
						Field Capacity	100(7),98.6(14),96.4(21), 94.8(28),92.6(35),92.6(42)		"	"	
						1 bar tension	100.0(7),100.0(14),98.2(21), 95.4(28),94.8(35),94.8(42)		"	"	
	Nargund	Calcareous saline sodic black clay (karl)	100 ppm			Puddling	98.2(7), 96.4(14),93.8(21), 90.2(28),97.6(35),86.7(42)		"	"	
						Field Capacity	100(7),99.0(14),96.6(21), 95.0(28),94.6(35),93.8(42)		"	"	
						1 bar tension	100.0(7)100.0(14),100.0(21), 98.0(28),98.0(35),97.4(42)		"	"	

Pesticide	Loca-tion	Soil Type	Dosage used (kg a.i/ha)	For-mula-tion used	Method & no. of applica-tion	Crop	Dissipation rate (ppm after days)	Halflife days	Reference	Method of analysis	Remarks
1	2	3	4	5	6	7	8	9	10	11	12
	Dharwar	Black clay	100 ppm			Puddling	100.0(7)98.2(14),96.4(21), 89.6(28),89.3(35),85.6(42)		Mithyantha (1973)	Pot	
						Field Capacity	100.0(7),99.2(14),96.8(21), 94.7(28),94.0(35),93.7(42)		"	"	
						I bar tension	100.0(7),100.0(14),98.2(21), 97.6(28),96.0(35),95.0(42)		"	"	
BHC (Lindane)	Delhi		10.0	5% dust	Furrow applica-tion	Cowpea & greengram	16.0(0),13.2(20),8.2(30), 3.5(40),2.8(5),1.8(70), 0.4(100)		Agnihothri et al. (1974e)	GLC	
Aldrin	Delhi		10.0	"	"	"	12.0(0),8.0(20),6.0(30) 2.5(55),1.5(70),0.95(100)			"	as Aldrin
							0.0(0),2.6(20),2.5(30), 1.6(55),1.0(70)		"		as Dieldrin
BHC	Udaipur	Clay loam	5	5% dust	Soil in	Maize	3.17-4.38(0),1.5-2.8(15) 0.56-2.19(30),0.52-1.34(60) 0.42-0.44(90)	35.0-53.75	Srivastava & Yadav (1977)	C & B	Results of 3 years
			15	"	"		9.3-12.1(0).4.3-7.0(15), 1.4-5.6(60),1.1-3.0(60), 1,4-3.4(90)	35.0-36.71			

Pesticide	Location	Soil Type	Dosage used (kg a.i/ha)	Formulation used	Method & no. of application	Crop	Dissipation rate (ppm after days)	Halflife days	Reference	Method of analysis	Remarks
1	2	3	4	5	6	7	8	9	10	11	12
			30	5% dust	Soil in		17.5-24.4(0);7.8-13.0(15), 2.2-10.3(30);1.6-5.5(60), 1.4-3.4(90)	32.0-36.71			
BCH	Delhi	Sandy loam	3.75	10% dust	Whorl applica-tion	Sorghum	0 24-0.54(1),0 to 0.22(82)		Kathpal et al. (1968a)	C	
Carbo-furan	Banga-lore	Red sandy-loam	100 ppm AR	Mixing grade with soil		Puddled condition	88.8(7),78.8(14),70.0(21), 62.1(28),55.2(35),49.0(42)	40.25	Mithyantba (1973)	C	Laboratory study
						field capacity	92.7(7),86.0(14),79.7(21), 73.9(28),68.5(35),63.6(42)	63.35			
						l bar	92.9(7),86.3(14),73.9(28), 74.5(21),68.2(35),64.3(42)				
	Manga-lore	Laterite	,,	,,	,,	Puddled condition	88.3(7)78.0(14),68.8(21), 60.7(28),53.6(35),47.4(42)	38.85			
						Field capacity	92.9(7),86.3(14),80.4(21), 74.5(28),69.3(35),64.3(42)	64.05			
						l bar	93.1(7),89.7(14),80.4(21), 75.2(28),70.0(35),65.1(42)	67.20			

Pesticide	Location	Soil Type	Dosage used (kg a.i/ha)	Formulation used	Method & no. of application	Crop	Dissipation rate (ppm after days)	Halflife days	Reference	Method of analysis	Remarks
1	2	3	4	5	6	7	8	9	10	11	12
	Dharwar	Black clay	100 ppm AR grade		Mixing with soil	Puddled condition	89.0(7),79.3(14),70.5(21), 62.9(28),.55.9(35),50.0(4)	42.00			
						Field capacity	90.6(7),82.0(14),74.2(21), 67.2(28),60.4(35),55.0(42)	48.48			
						I bar	91.7(7),84.2(14),77.2(21), 70.8(28),64.4(35),59.7(42)	56.35			
	Nargund	Karl clay	"	"	"	Puddled condition	88.2(7),78.0(14),68.8(21), 60.8(28),53.8(35),47.4(42)	38.50			
						Field capacity	90.3(7),83.1(14),75.0(21), 68.2(28),61.8(35),56.2(42)	49.50			
						I bar	92.1(7),84.9(14),78.2(21), 72.1(28),66.4(35),61.2(42)	58.45			
Phorate	Bangalore	Red sandy loam	"	"	"	Puddling	72.7(7),52.7(14),38.4(21), 28.1(28),20.4(35),13.1(42)	15.05	Mithyantha & Perur (1974)	C	
						Field capacity	81.8(7),66.9(14),54.8(21), 44.8(28),36.6(35),27.6(42)	24.80	Mithyantha & Perur (1974)	C	
						Mustard	83.4(7),69.6(14),58.1(21), 48.6(28),40.4(35),31.2(42)	26.60			

Pesticide	Location	Soil Type	Dosage used (kg a.i/ha)	Formulation used	Method & no. of application	Crop	Dissipation rate (ppm after days)	Halflife days	Reference	Method of analysis	Remarks
1	2	3	4	5	6	7	8	9	10	11	12
	Manga-lore	Laterite	100 ppm	AR grade	Mixing with soil	Puddling	80.7(7),65.2(14),52 6(21),42.6(28),34.2(35),21.2(42)	22.23			
						Field capacity	80.2(7), 64.2(14),51.6(21),41.4(28), 33.2(35),24.2(42)	21.70			
						I bar	83.5(7),.69.7(14),58.1(21),48.6(28),.33.2(35),24.2(42)	25.55			
	Dhar-war	Black clay	100 ppm	AR	Mixing grade with soil	Puddling	80.3(7),64.2(14),51.6(21),41.3(28),33.1(35),24.1(42)	21.88			
						Field capacity	84.4(7),71.2(14),60.0(21),50.6(28),42.7(35),30.3(42)	28.35			
						I bar	85.4(7),73.0(14),62.3(21),53.2(28),45.5(35),36.5(42)	30.03			
	Nargund	Karl clay	100 ppm	AR	Mixing grade with soil	Puddling	80.0(7),65.2(14),52.6(21),42.5(28),34.3(35),25.2(42)	22.23			
						Field capacity	85.2(5), 72.9(14),62.2(21),53.1(28),45.4(35),36.2(42)	30.80			
Phorate	Delhi	Sandy loam	2.30	10% Gr.	Furrow applica-tion	Mustard	5.02(25), 4.08(30),1.38(50),0.14(60) ND(70)		Dixit et al. (1974)	C	
									"		

Pesticide	Location	Soil Type	Dosage used (kg a.i/ha)	Formulation used	Method & no. of application	Crop	Dissipation rate (ppm after days)	Halflife days	Reference	Method of analysis	Remarks
1	2	3	4	5	6	7	8	9	10	11	12
Phorate	Coimbatore	Black clay	5.00	10% Gr.	Furrow application	Mastard	9.34(25),8.42(30),4.37(50),0.27(60),0.07(70),ND(80)		Dixit et al. (1974)		
			1.25	"	Soil incorporation	Paddy	3.02(1),2.90(3),1.36(5),0.94(10),0.20(20),ND(30)		Rajukka-nnu et al. (1977a)	C	1st application 1 DAT
			2.50	"			3.46(3),3.16(33),2.32(35),1.25(40),0.15(50),ND(30)			"	application 2nd application 30 DAT
Phorate	Coimbatore	Black clay	0.25	"	Soil incorporation	Paddy	2.90(3),136(5),0.94(10),0.25(20),0.07(30)		Kandasamy et al. (1975)	C	
			1.25	"	"	"	3.16(3),1.25(10),0.15(20),ND(30)		"	"	
Aldicarb	Delhi	Sandy loam	1.50	"	Furrow application	Pea	13.70(0),8.50(15),4.87(30),2.77(45),1.19(60),ND(80)	17.80	Dixit et al (1976)	C	
Aldicarb	Coimbatore	Black clay	0.50	10% Gr.	Furrow application	Bhindi	1.35(1),0.26(15),0.19(30),0.09(45),0.05(60),ND(75)		Rajukka-nnu et al. (1975)	C	
			1.00	"	"	"	2.12(1),0.73(15),0.60(30),0.42(45),0.11(60),0.09(75)		"	"	
Carbofuran	"	"	0.625	3% Gr.	Soil incorporation	Rice	7.33(3),5.20(5),2.86(10),1.93(20),1.55(30)		Kandaswamy et al (1977)	C	1st application
			1.250	"	"	"	8.44(3),3.72(10),2.66(20),		"	"	2nd application

1	2	3	4	5	6	7	8	9	10	11	12
Pesticide	Location	Soil Type	Dosage used (kg a.i/ha)	Formulation used	Method & no.of application	Crop	Dissipation rate (ppm after days)	Halflife days	Reference	Method of analysis	Remarks
Carbofuran	Cuttack	Alluvial	50 ppm	Technical	Soil incorporation	Rice	36.8(0),20.5(20),7.38(40)	17.2	Venkatesarlu et al (1977)	TLC-C	
		Laterite	,,	,,	,,	,,	32.13,(0)24.6(20),10.1(40)	23.9	,,	,,	
		Pokkali	,,	,,	,,	,,	28.3(0),21.0(20),10.0(40)	26.6	,,	,,	
		Kari	,,	,,	,,	,,	28.5(0),27.0(20),22.0(40)	107.1	,,	,,	
PCNB (Quinto- zene)	Banga- lore	Red sandy loam	100 ppm AR	AK grade	Soil incorporation	Puddling	73.5(7),54.0(14),39.7(21), 29.2(28),21.1(35),12.5(42)	16.1	Mithyantha (1973)	C	
						Field capacity	93.0(7),86.9(14),80.2(21), 74.8(28),69.5(35) 62.5(42)	62.3	,,	,,	
						1 bar	96.4(7),92.8(14),89.5(21), 86.4(28),83.9(35),78.8(42)	120.4	,,	,,	
	Manga- lore	Laterite	100 ppm AR	AK grade	Soil incorpora- tion	Puddling	72.8(7),53.1(14),38.7(21), 28.2(28),20.5(35),14.9(42)	15.4	Mithyantha (1973)	C	
						Field capacity	92.6(7),85.7(14),79.4(21), 73.4(28),68.8(35),69.0(42)	63.4	,,	,,	
						1 bar	96.3(7),92.7(14),89.3(21), 86.0(28),82.8(35),79.7(42)	125.3	,,	,,	
	Dhar- war	Black clay	,,	,,	,,	Puddling	79.7(7),63.6(14),50.6(21), 40.2(28),32.2(35),25.7(42)	21.7	,,	,,	

Pesticide	Location	Soil Type	Dosage used (kg a.i/ha)	For-mula-tion used	Method & no. of applica-tion	Crop	Dissipation rate (ppm after days)	Halflife days	Reference	Method of analysis	Remarks
1	2	3	4	5	6	7	8	9	10	11	12
						Field capacity	94.2(7),88.2(14),83.5(21), 78.6(28),74.0(35),69.0(42)	79.1	Mithyantha (1973)	C	
						I bar	97.1(7),94.3(14),91.6(21), 88.9(28),86.3(35),83.8(42)	152.6	,,	,,	
	Nar-gund	Karl clay	100 ppm AR		Soil in-grade corpora-tion	Puddling	78.4(7),61.4(14),48.2(21), 37.8(28),129.6(35),24.3(42)	20.0	,,	,,	
						Field capacity	94.4(7),89.2(14),84.4(21), 79.5(28),75.0(35),70.9(42)	85.4	,,	,,	
						I bar	97.2(7),94.4(14),91.7(21), 89.2(28),86.6(35),84.2(42)	153.0	,,	,,	

APPENDIX

Effect of Processing on Residues

Crop	Compound	Part of Plant	Residue in the raw Produce (ppm)	Process involved	Residue after Processing	Percentage reduction	Tolerance limit (ppm)	Reference
1	2	3	4	5	6	7	8	9
Spinach	BHC	Leaves	17.45 8.30	Washing "	7.60 2.04		3.0 3.0	Jaglan & Chopra (1971?) "
	DDT	"	26.56 16.32	" "	14.12 8.22		7.0 7.0	" "
Tomato	Carbaryl	Fruit	17.32 11.50 8.65 6.29	" " Cooking after washing	8.65 6.29 5.26 3.60	50.0 45.0 64.0 69.0	10.0 " " "	Rajukkannu et al. (1976) " " "
Potato	Aldrin	Tubers	0.19 0.15	Washing, Cooking, Peeling	0.03 0.03	84.2 80.0	0.10	Misra et al. (1977) "
	Dieldrin	Tubers	0.25 0.21	" "	0.05 0.08	80.0 80.5		Misra et al. (1977) "
	Heptachlor	"	0.18 0.23	" "	0.06 0.07	66.7 69.6	0.05	" "
Cauliflower	Endosulfan	Leaves	22.00 33.00 23.10 6.30 2.05	Washing " " " "	6.80 14.50 10.40 2.40 1.35	59.9 56.1 50.0 61.9 34.10	0.5 " " " "	Awastbi et al. (1974) " " " "

Crop	Compound	Part of Plant	Residue in the raw Produce (ppm)	Process involved	Residue after Processing	Percentage reduction	Tolerance limit (ppm)	Reference
1	2	3	4	5	6	7	8	9
		Curds	11.25	,,	6.90	38.7	0.5	Awasthi et al. (1977)
			18.60	,,	10.50	43.5	0.5	,,
			1.95	,,	1.35	30.7	0.5	,,
Cauliflower	Endosulfan	Curds	0.22		N.D.	100.0	0.5	,,
			0.15	,,	N.D.	100.0	0.5	,,
			0.60	,,	N.D.	100.0	0.5	,,
			11.25	Washing, Cooking	5.5	50.7	0.5	,,
			5.80	,,	3.70	36.2	,,	,,
			1.20	,,	N.D.	100.0	,,	,,
Bhindi	Endosulfan	Fruit	6.89	Washing	4.56	34.67	2.0	Nath et al. (1974)
			6.89	Cooking	5.15	26.22	,,	,,
			6.89	Dehydration	2.96	57.59	,,	,,
Greengram	Endosulfan	Leaves	11.58	Washing	3.41	70.60	2.0	Verma & Pant (1976a)
			6.26	,,	2.62	58.20	,,	,,
			1.47	,,	0.68	53.90	,,	,,
		Pod	5.08	,,	2.28	55.20	,,	,,
			4.44	,,	2.45	44.70	,,	,,
			2.80	,,	1.43	45.50	,,	,,
Greengram	Endosulfan	Pod	0.85	,,	0.66	39.20	,,	,,
		Pod	7.76	,,	3.90	49.70	,,	,,
			6.74	,,	3.42	49.30	,,	,,
			1.90	,,	1.07	43.80	,,	,,

Crop	Compound	Part of Plant	Residue in the raw Produce (ppm)	Process involved	Residue after Processing	Percentage reduction	Tolerance limit (ppm)	Reference
1	2	3	4	5	6	7	8	9
Chillies	Dimethoate	Greenfruit	7.89	Washing	0.63		2.0	Anon (1977a)
Cauliflower	Fenitrothion	Curds	4.69	Washing, Cooking	1.85	60.4	0.3	Attri (1977)
			1.96	"	1.07	45.4	0.3	"
			0.83	"	0.53	36.0	0.3	"
Okra	Malathion	Fruits	2.58	Washing	0.28	89.2	3.0	Jat & Srivastava (1973)
			2.58	Open Cooking	0.34	86.8	3.0	"
			2.58	Steam Cooking	0.62	76.0	3.0	"
			2.58	Dehydration	0.21	91.9	3.0	"
Mustard	Malathion	Leaves	3.61	Washing	0.85	76.4	8.0	Dixit et al. (1974)
			2.46	"	0.47	80.8	8.0	"
		Pods	4.15	"	0.85	81.9	8.0	"
Cowpea	Monocrotophos	Pods	3.73	"	0.50		0.2	Awasthi et al. (1977b)
			3.73	Boiling	0.73		0.2	"
			5.83	Washing	0.69		0.2	"
			5.53	Boiling	0.91		0.2	"
	Methyl parathion	Curds	5.01	Washing	2.79	44.3	0.75	Attri & Rattan Lal (1974c)
			1.09	"	0.77	29.4		
			0.63	"	0.58	7.9		
			5.01	Washing, Cooking	2.09	58.3		
			2.92	"	1.24	57.5		
			1.09	"	0.56	48.6		

Crop	Compound	Part of Plant	Residue in the raw Produce (ppm)	Process involved	Residue after Processing	Percentage reduction	Tolerance limit (ppm)	Reference
1	2	3	4	5	6	7	8	9
Redgram	Endosulfan	Leaves	18.93	Washing	5.49	71.00	2.0	Verma & Pant (1976a)
			4.95	"	1.63	66.80	"	"
		Pod	5.75	"	3.19	44.50	"	"
			1.59	"	1.18	25.90	"	"
Cauliflower	Endosulfan	Curds	15.84	"	7.92	50.0	"	Verma & Rattan Lal(1976)
			15.84	Cooking	4.99	68.5	"	"
Ber	Endosulfan	Fruits	2.14	Washing	N.D.	100.0	"	Kathpal et al. (1977)
			5.14	"	N.D.	100 0	"	"
Cabbage	Dimethoate	Heads	5.07	"	2.22	56.21	"	Krishnaiah & Rattan Lal (1973a)
			3.17	"	2.06	35.01	"	"
			1.28	"	1.01	21.09	"	"
			6.99	"	3.37	51.79	"	"
Cauliflower	Dimethoate	Curds	4.46	Washing	2.30	48.43	"	"
			2.79	"	2.05	26.52	"	"
			1.78	"	1.45	18.54	"	"
Frenchbeans	Dimethoate	Green pods	1.27	Cooking	0.24		"	Anon (1977a)
French beans	Dimethoate	Green pods	1.27	Cooking	0.20		2.0	Anon (1977a)
			2.35	"	0.11		2.0	"
			2.24	"	0.16		2.0	"
			2.67	"	0.12		2.0	"
Tomato	Dimethoate	Fruit	1.78	Washing	0.48		2.0	"

Crop	Compound	Part of Plant	Residue in the raw Produce (ppm)	Process involved	Residue after Processing	Percentage reduction	Tolerance limit (ppm)	Reference
1	2	3	4	5	6	7	8	9
	Methyl parathion	Pods	1.85	Washing	1.06	42.7	0.75	Attri & Rattan Lal(1974c)
			0.88	Washing, "	0.53	40.0	0.75	
			1.85	Washing, Cooking	0.56	69.7	0.75	
			0.88	"	0.36	59.1	0.75	
Mustard	Phorate	Plant (leaves and stem)	8.77	Washing	4.82	45.0	0.50	Dixit et al. (1974)
			13.45	"	6.31	53.0	0.5	"
			15.83	"	1.50	90.5	0.5	"
			3.96	"	N.D.	100.0	0.5	"
			7.18	"	N.D.	100.0	0.5	"
			13.45	Washing, Cooking	3.03	77.4	0.5	"
			10.12	"	0.40	96.0	0.5	"
			11.14	"	N.D.	100.0	0.5	"
Potato	Phorate	Tubers	0.40	Peeling	0.07		0.5	Awasthi et al. (1977c)
			1.50	"	1.20		0.5	"
			2.15	"	1.65		0.5	"
			1.50	Washing	N.D.		0.5	"
Potato	Phorate	Tubers	1.95	Washing, Boiling, Peeling	N.D.		0.5	"
Okra	Carbaryl	Fruits	11.07	Washing	3.17		10.0	Krishnaiah et al. (1975)
			7.63	"	2.38		10.0	"
Tomato	Carbaryl	Fruits	17.32	"	8.65	50.0	10.0	Rajukkannu et al. (1976c)

Crop	Compound	Part of Plant	Residue in the raw Produce (ppm)	Process involved	Residue after Processing	Percentage reduction	Tolerance limit (ppm)	Reference
1	2	3	4	5	6	7	8	9
Sponge gourd	Carbaryl	Fruit	13.818	Washing	0.509	99.93	10.0	Singh et al. (1976)
Cowpea	Carbaryl	Pods	24.16	Boiling	2.15	91.10	5.0	Awasthi et al. (1977b)
			24.16	"	3.75	84.40	5.0	"
Bringal	Carbofuran	Fruit	0.021	Cooking	N.D.	100.0	0.2	Mithyantha et al. (1977a)
			0.074	"	N.D.	100.0	0.2	"
Potato	Carbofuran	Tubers	0.191	"	0.010		0.2	Mithyantha et al. (1977b)
			0.295	"	0.133		0.2	"
Cabbage	Carbofuran	Head	0.078	"	0.051		0.2	Anon (1977a)
Lady's finger	Carbofuran	Fruit	0.175	"	0.061		0.2	"
			0.071	"	0.035		0.2	"
Tomato	Captafol	Fruit	1.231	Washing	0.214		2.0	Anon (1977a)
Chillies	Captafol	Greenfruit	6.745	"	0.237		2.0	"

COMMON PESTICIDES, THEIR PROPERTIES AND TOXICITY TO NON TARGET SPECIES

Insecticides and Nematicides

Common Name	Trade names	Group	Action	Water solubility (ppm)	Acute toxicity		Fish LC$_{50}$ (ppm)	Bee (ug/bee)	Bird (mg/kg)
					Mammalian (LD$_{50}$ mg/kg)				
					Oral	Dermal			
1	2	3	4	5	6	7	8	9	10
Acephate	Orthene	Organo Phosphate	Systemic & contact insecticide	6.5x10^5	361-2025	>2000	1000-9550	1.2	140-852
Aldicarb	Temik	Carbamate	Systemic & contact nematicide and insecticide	6.0x10^3	0.6-09	3.0-5.0	0.45-1.60		4.4
Aldrin	Aldrin Aldrex	Cyclodiene	Contact insecticide	0.027	67		0.005-0.019	0.35	37-520
Aluminium phosphide	Celphos Phosfume Phostoxin	Inorganic	Fumigant	2.6x10^5	5.0				
Azinphos-methyl	Guthion	Organo Phosphate	Contact Insecticide & Acaricide	33	16.4-80.0		0.014-0.235	0.42	136-488
Bendiocarb	Garvox Ficam	Carbamate	Contact Insecticide	40	40-179	566 >1000	0.5-2.2	0.1	80-1466

Common Name	Trade names	Group	Action	Water solubility (ppm)	Mammalian (LD$_{50}$ mg/kg) Oral	Mammalian (LD$_{50}$ mg/kg) Dermal	Fish LC$_{50}$ (ppm)	Bee (μg/bee)	Bird (mg/kg)
1	2	3	4	5	6	7	8	9	10
B.H.C.	Hexidole Gammax-ene Hexamar etc.	Chlorinated Hydro-carbon	Contact Insecticide	BHC is a mixture of five isomers and their properties differ widely. The insecticidally active isomer is gamma-BHC (see lindane)					
Carbaryl	Sevin Killex carbaryl Caravet Hexavin	Carbamate	Contact Insecticide	40	280-850	>2000	1.0	1.34	56-197
Carbofuran	Furadan Hexafuran	Carbamate	Contact & Systemic Insecticide & Nematicide	700	8-19	>10,200	48-55	0.16	7-37
Carbopheno-thion	Trithion	Organo-Phosphate	Contact Acaricide	<40	32-91	1270	—	1.4	57
Chlordane	Chlordane Termex	Cyclodiene	Contact Insecticide	Insolu-ble	457-590	700	0.42-69.0	1.4-8.8	220-1200
Chlorfen-vinphos	Birlane	Organo Phosphate	Contact Insecticide	145	117-200	417-12000	0.36	4.1	29

Common Name	Trade names	Group	Action	Water solubility (ppm)	Mammalian (LD$_{50}$ mg/kg) Oral	Mammalian (LD$_{50}$ mg/kg) Dermal	Acute toxicity Fish LC$_{50}$ (ppm)	Bee (ug/bee)	Bird (mg/kg)
1	2	3	4	5	6	7	8	9	10
Chlorpyriphos	Durzban Coroban	Organo Phosphate	Contact Insticide	2	135-163	2000	0.003-0.007	0.114	32-80
Cypermethrin	Ripcord Cymbush	Synthetic Pyrethroid	Contact Insecticide	0.02	242-303	3000	0.0028		
D.D.T.	Tafidex Tafarol	Chlorinated Hydrocarbon	Contact Insecticide	0.007	113-118	2150	0.0016-0.019	5.36	611-2240
Deltamethrin	Decis	Synthetic pyrethroid	Contact Insecticide	Insoluble	128-138	>2000	0.008-0.10	0.51	
Diazinon	Ditaf Basudin	Organophosphate	Contact insecticide & Acaricide	40	300 850	>2150	0.90	0.22-0.37	52.2
Dichlorvos	Nuvan Vapona	Organophosphate	Contact insecticide & fumigant	10,000	56-80	75-107	1.00	0.52	298
Dicrotophos	Bidrin	Organophosphate	Systemic Insecticide & acaricide	Highly soluble	16.5-22	168-224	2.8-13.0	0.30	32
Diflubenzuron	Dimilin		Contact Insecticide	0.20	>2150	>2000	135	>30	>4640

Common Name	Trade names	Group	Action	Water solubility (ppm)	Acute toxicity				
					Mammalian (LD$_{50}$ mg/kg)		Fish LC$_{50}$ (ppm)	Bee (ug/bee)	Bird (mg/kg)
					Oral	Dermal			
1	2	3	4	5	6	7	8	9	10
Dimethoate	Rogor Hexagon Cygon Devigor Vijaygor Corothate	Organo-phosphate	Systemic & Contact Insecticide & acaricide	2.5x10^4	320-600	650-1500	3.14-40.0	0.12-0.19	37
Disulfoton	Disyston Solvirex	Organo-phosphate	Systemic Insecticide, Acaricide & Nematicide	25	2.6-8.6	20	0.04	4.3-5.14	6.5
Endosulfan	Thiodan Hildan Hexasulfan Endocel	Cyclodiene	Contact Insecticide	Insoluble	80-110	167-35	0.008	7.1-7.8	200-750
Ethion	Tafethion	Organo-phosphate	Contact Acaricide and Insecticide	Slight	208	915	7.9	20.6	331>10.00)
Fenamiphos	Nemacur	Organo-phosphate	Contact and Systemic Nematicide	700	15-20	500			12
Fenitrothion	Sumithion Folithion Accothion	Organo-phosphate	Contact insecticide	Slight	250-1062	>3250	6.0-14.9	0.18-0.38	27-2550

Common Name	Trade names	Group	Action	Water solubility (ppm)	Mammalian (LD$_{50}$ mg/kg) Oral	Dermal	Acute toxicity Fish LC$_{50}$ (ppm)	Bee (µg/bee)	Bird (mg/kg)
1	2	3	4	5	6	7	8	9	10
Fenthion	Lebaycid Baytex	Organo-phosphate	Contact Insecticide	54-56	190-615	150-650	0.93-2.44	0.308	1.8-40.0
Fensulfothion	Dasanit	Organo-phosphate	Contact & systemic Insecticide & Nematicide	1540	4.6-10.5	3.5-30.0	0.055	0.35	0.7-35.0
Fenvalerate	Sumicidin Belmark Pydrin	Synethetic pyrethroid	Contact Insecticide	<1.0	200-451	>1000	0.003-0.20	0.077	>5000
Formothion	Anthio	Organo-phosphate	Contact & systemic insecticide & acaricide	500	375-535	>1000			
Heptachlor	Heptaf	Cyclodiene	Contact insecticide	0.056	100.162	195-250	0.008-0.056	0.53	92-2000
Iodofenphos	Nuvanol	Organo-phosphate	Contact insecticide	<2	2100-3000	>2000	0.25-0.75		
Isofenphos	Oftanol	Organo-phosphate	Contact insecticide	20	38-150	188-1000	1-4		5-21

Common Name	Trade names	Group	Action	Water solubility (ppm)	Acute toxicity				
					Mammalian (LD$_{50}$ mg/kg)		Fish LC$_{50}$ (ppm)	Bee (µg/bee)	Bird (mg/kg)
					Oral	Dermal			
1	2	3	4	5	6	7	8	9	10
Lindane	Lintaf	Chlorinated hydrocarbon	Contact insecticide	10	88-91	900-1000	0.031-0.46	0.2-0.76	>2000
Malathion	Cythion Malathion	Organophosphate	Contact insecticide	145	857-2800	4100	0.12-7.60	0.70	>850
Mephosfolan	Cytrolane	Organophosphate	Contact & systemic insecticide	Soluble	8.9-11 3	9.7			
Methomyl	Lannate	Carbamate	Contact & systemic insecticide & nematicide	5.8 × 10^4	17-24	>5000	0.87-3.4		1890-3680
Methamidophos	Monitor Metataf Tamaron	Organophosphate	Contact & systemic insecticide & acaricide	Soluble	30-50	50-110	46-118	1.37	2558
Methyl parathion	Metacid Pairataf Folidol-M	Organophosphate	Contact insecticide	55	14-42	67-300	2750-7580	0.29	7.0-10.0

Common Name	Trade names	Group	Action	Water solubility (ppm)	Acute toxicity				
					Mammalian (LD_{50} mg/kg)		Fish LC_{50} (ppm)	Bee (μg/bee)	Bird (mg/kg)
					Oral	Dermal			
1	2	3	4	5	6	7	8	9	10
Methyl bromide			Fumigant insecticide	1.3×10^4	50	8000	70-350		
Monocrotophos	Nuvacron Azodrin Monocil	Organophosphate	Systemic & contact insecticide and acaricide	Soluble	8-23	112-354	12-23	0.35	3.4
Naled	Dibrom	Organophosphate	Contact insecticide & acaricide	Insoluble	430	1100	2-6		
Nicotine	Nicotine sulphate	Plant Product	Contact insecticide	Soluble	50-60	50	1.0		75
Oxamyl	Vydate	Carbamate	Systemic insecticide & nematicide	2.8×10^5	5.4	2960	4.2-5.6	0.31	4.6
Oxydemeton methyl	Metasystox Hexasytox	Organophosphate	Systemic & contact insecticide & acaricide:	3300	65-75	250	0.65-18.24	0.54	

Common Name	Trade names	Group	Action	Water solubility (ppm)	Mammalian (LD$_{50}$ mg/kg) Oral	Dermal	Fish LC$_{50}$ (ppm)	Bee (µg/bee)	Bird (mg/kg)
1	2	3	4	5	6	7	8	9	10
Permethrin	Ambush Pounce Lee Permasect	Synthetic pyrethroid	Contact insecticide	Insoluble	>4000	>4000	0.006-0.10	0.11	>10,000
Phenthoate	Elsan Cidial Phendal	Organo-phosphate	Contact insecticide	Insoluble	300-400	>4800	0.3-4.5	0.12	218-300
Phosphamidon	Dimecron	Organo-phosphate	Systemic & contact insecticide and acaricide	Highly soluble	20	374	8-163	1.46	3-24
Phorate	Thimet Phorate Phoratox	Organo-phosphate	Systemic & contact insecticide and nematicide	50	1.6-3.7	2.5-6-2	0.009	10.1	0.6
Phosalone	Zolone	Organo-phosphate	Contact insecticide and acaricide	Insoluble	120-170	1000-1500	0.084		290

Common Name	Trade names	Group	Action	Water solubility (ppm)	Acute toxicity				
					Mammalian (LD$_{50}$ mg/kg)		Fish LC$_{50}$ (ppm)	Bee (µg/bee)	Bird (mg/kg)
					Oral	Dermal			
1	2	3	4	5	6	7	8	9	10
Pirimiphos-methyl	Actellic	Organo-phosphate	Contact acaricide and insecticide	5	1180-2050	>2000	1.4-1.6	0.39	
Propoxur	Baygon	Carbamate	Contact insecticide	2000	90-128	800 1000	206	1.35	2.6
Pyrethrum		Plant product	Contact insecticide	(Mixture of 6-7 compounds)	584-900	>1500	0.027-0.093	0.3-0.29	>10,000
Quinalphos	Ekalux	Organo-phosphate	Contact insecticide & acaricide	22	62-137	1250-1400	1.55-3.72	1.6mg/kg	
Thanite		Organic thiocyanate	Contact insecticide		200-630	6000			
Tetrachlor-vinphos	Gardona	Organo-phosphate	Contact insecticide	11	2500-5000	>2500			>2000
Thiometon (Morphothion)	Ekatin	Organo-phosphate	Systemic insecticide and acaricide	200	120-130.	700-900	3.82	0.55	

1	2	3	4	5	6	7	8	9	10
Common Name	Trade names	Group	Action	Water solubility (ppm)	Mammalian (LD$_{50}$ mg/kg) Oral	Mammalian (LD$_{50}$ mg/kg) Dermal	Acute toxicity Fish LC$_{50}$ (ppm)	Bee (ug/bee)	Bird (mg/kg)
Toxaphene	Anatox	Chlorinated terpene	Contact insecticide	3	80-90	1075	0.011-0.018	50.4	31
Trichlorphon	Dipterex	Organo-phosphate	Contact insecticide	1.54×10^5	560-630	>2000	0.26-1.4	59.8	500
Vamidothion	Kilval	Organo-phosphate	Systemic insecticide & acaricide	Highly soluble	34-105	1160	>10		35
Allethrin		Pyrethrin	Contact insecticide	Very low	480-920		0.019	3.4	>2000
Ethylene dibromide	EDB		Fumigant insecticide	4300	146	300	10.0		79
Menazon	Sayfos	Organo-phosphate	Systemic aphicide	240	1950	>800	220	4.3	87

II ACARICIDES

Chlorobenzitale	Akar	Chlorinated hydrocarbon	Contact acaricide	Insoluble	700-3100				

Common Name	Trade names	Group	Action	Water solubility (ppm)	Acute toxicity				
					Mammalian (LD$_{50}$ mg/kg)		Fish LC$_{50}$ (ppm)	Bee (ug/bee)	Bird (mg/kg)
					Oral	Dermal			
1	2	3	4	5	6	7	8	9	10
Carbophenothion—See under "Insecticides & Nematicides"									
Dicofol	Kelthane	Chlorinated hydrocarbon	Contact acaricide	Insoluble	668-842	1870			
Ethion	Tafethion	See under "Insecticide and Nematicides"							
Quionome-thionate	Morestan		Contact acaricide with fungicidal properties	Insoluble	2500-3000	>500		>80	980
Pirimiphos methyl	See under "Insecticides and Nematicides"								
Tetradifon	Tedion	Chlorinated	Contact acaricide	200	>14.700	>10.000			
III FUNGICIDES									
Benomyl	Benalate	Benzami dazole	Systemic	3.8	>10.000				>5000

Common Name	Trade names	Group	Action	Water solubility (ppm)	Mammalian (LD$_{50}$ mg/kg) Oral	Mammalian (LD$_{50}$ mg/kg) Dermal	Acute toxicity Fish LC$_{50}$ (ppm)	Bee (μg/bee)	Bird (mg/kg)
1	2	3	4	5	6	7	8	9	10
Captafol	Foltaf Difolatan	Nitrogen heterocyclic	Contact	1.4	5000-6200	15,400	0.5-3.0	>50	>10,000
Captan	Captaf	Nitrogen heterocyclic	Contact	3.3	9000	>15,000	1.387	>50	6581->10,000
Carbendazim	Bavistin	Benzamidazole	Systemic	5.8	>15000	>15,000	>1000		>10,000
Carboxin	Vitavax	Oxathiin derivative	Systemic	170	3850	>8000			
Copper oxychloride	Blitox Fycol Fytolan etc	Copper fungicide	Contact	Insoluble	1440				
Cuprous oxide	Perenox Copper-Sandoz	Copper fungioide	Contact	Insoluble	470				

Common Name	Trade names	Group	Action	Water solubility (ppm)	Mammalian (LD$_{50}$ mg/kg) Oral	Dermal	Fish LC$_{50}$ (ppm)	Bee (ug/bee)	Bird (mg/kg)
1	2	3	4	5	6	7	8	9	10
Dinocap	Karathane	Dinitrophenol	Contact fungicide with acaricidal properties	Insoluble	980-1190				
Edifenphos	Hinosan	Organophosphate	Contact	Insoluble	150-340				
Ferbam	Fermate	Dithiocarbamate	Contact	130	>4000	1000-3000			
Kitazin	Kitazin	Organophosphate	Systemic and contact	1000	600	400	5.1		
Mancozeb	Dithane M45	Dithiocarbamate	Contact	Insoluble	>8000		4.0		
Metalaxyl	Ridomil		Systemic and contact	7100	30-50				
Methoxy ethyl mercury chloride	Agallol Aretan Ceresan	Organomercurial	Contact	5x10^{-4}	30-47				

Common Name	Trade names	Group	Action	Water solubility (ppm)	Mammalian (LD$_{50}$ mg/kg) Oral	Mammalian (LD$_{50}$ mg/kg) Dermal	Fish LC$_{50}$ (ppm)	Bee (ug/bee)	Bird (mg/kg)
1	2	3	4	5	6	7	8	9	10
Oxycarboxin	Plantavax	Oxathiin derivative	Systemic	1000	2000	>16,000			
Phenyl mercury acetate	Agrosan	Organo-mercurial	Contact	4370			0.0086-0.025		
Quintozene	Brassicol Terrachlor	Chlorinated nitrobenzene	Contact	Insoluble	>12.000				
Thiram	Thiride Arasan	Dithio-carbamate	Contact	30	375-865		0.007		
Triadimefon	Bayleton		Systemic	260	363-1071	>1000	10-50		1750-2500
Tridemorph	Calixin	Oxazine	Systemic	Highly soluble	562-825	1350			
Sulphur	Sultaf Sulfex	Sulphur	Contact	Insoluble	500		1000		
Zineb	Dithane Z-78 Hexathaae	Dithio-carbamate	Contact	10	>5,200		250		

Common Name	Trade names	Group	Action	Water solubility (ppm)	Acute toxicity				
					Mammalian (LD$_{50}$ mg/kg)		Fish LC$_{50}$ (ppm)	Bee (µg/bee)	Bird (mg/kg)
					Oral	Dermal			
1	2	3	4	5	6	7	8	9	10
Ziram	Ziride Cuman	Dithiocarbamate	Contact	65	1400				
Maneb	Dithane M-22	Dithiocarbamate	Contact	Sight	6750				
Nickel chloride									
IV HERBICIDES.									
Alachlor	Lasso	Acetanilide	Pre-emergence	240	1200	3500			
Atrazine	Atrataf Gesaprim	s-triazine	Selective Pre and Post emergence	30	1859-3080	7500	12.6-26	>96	>2000
Benthiocarb (Thiobencarb)	Saturn Bolero	Carbamate	Selective Pre and Post emergence	30	560-1300	2900	3.6		
Butachlor	Machete	Acetanilide	Pre-emergence	20	3300	4000			
Diuron	Karmex	Substituted Urea	Pre-emergency	42	3400		3-60	>145	>2000

Common Name	Trade names	Group	Action	Water solubility (ppm)	Mammalian (LD$_{50}$ mg/kg) Oral	Mammalian (LD$_{50}$ mg/kg) Dermal	Fish LC$_{50}$ (ppm)	Bee (µg/bee)	Bird (mg/kg)
1	2	3	4	5	6	7	8	9	10
2,4-D	Taficide Fernoxone Weedol etc.	Phenoxy alkanoic acid/salt/ ester	Pre-emergence	620	375-805	700	0.39-2.1	>18	1000
Dalapon	Dowpon	Phenoxy alkanoic acid/salt/ ester	Selective Pre-emergence	Highly soluble	7570-9330		>50	>24	5600
Dinoterb	Tolkan	Dinitro-phenol	Selective Pre-emergence	Insolu-ble	25	150			
Fluchloralin	Basalin	Dinitro-aniline	Pre-emergence		750-1550				
Glyphosate	Round up	Organo-phosphate	Post-emergence	12000	4320	>7940			
Isoproturon	Arelon Graminon Phytosami-taire	Substitu-ted Urea	Selective Pre-emergence	55	>4640	>3170			

Common Name	Trade names	Group	Action	Water solubility (ppm)	Acute toxicity				
					Mammalian (LD$_{50}$ mg/kg)		Fish LC$_{50}$ (ppm)	Bee (μg/bee)	Bird (mg/kg)
					Oral	Dermal			
1	2	3	4	5	6	7	8	9	10
Linuron	Afalon Lorox	Substituted Urea	Selective Pre and Post-emergence	75	500-4000				
MCPA	Agrozone Agritox	Phenoxy acid	Systemic hormone like, non-selective	825	550-700		10.0		
Methabenzthiazuron	Tribunil	Substituted Urea	Selective Pre and Post-emergence	59	>2500	>500			
Metoxuron	Dosanex Purivel	Substituted Urea	Selective Pre and Post-emergence	678	3200	>2000			>1250
Monuron	Telvar	Substituted Urea	Pre-emergence Non-selective	230	3600		16.3-76	110	
MSMA	Ansar	Arsenical	Selective Pre-emergence	Highly soluble	900-1800		1000		
Nitrofen	Toke-25	Chlorophenoxy benzene	Selective Pre-emergence	1.2	630-755				

Common Name	Trade names	Group	Action	Water solubility (ppm)	Acute toxicity				
					Mammalian (LD$_{50}$ mg/kg)		Fish LC$_{50}$ (ppm)	Bee (ug/bee)	Bird (mg/kg)
					Oral	Dermal			
1	2	3	4	5	6	7	8	9	10
Paraquat	Gramaxone	Bipyridinium	Non-selective	Highly soluble	104-150	236			
PCP	Dowicide Santobrite	Chlorophenol	Defoliant	20	210		0.17		
Propanil	Stam F-34 Rogue	Amide	Post-emergence	225	1285-1483	7080	0.35-0.55		
Sirmate		Carbamate	Selective, Post emergence		1879				
Simazine	Tafazine Gesatop	S-triazine	Selective pre & Post emergence	3.5	>5000		25-100	>96	>5000
TCA		Chloroacetate	Selective	Highly soluble	3200-5000		>100	>48	4280
Triallate	Avadex Fargo	Carbamate	Selective Pre & Post emergence	4	1675-2165	2225-4050			

SUBJECT INDEX

ORGANISATIONAL CHART

DIRECTORATE OF PLANT PROTECTION QUARANTINE & STORAGE
MINISTRY OF AGRICULTURE
(DEPARTMENT OF AGRICULTURE & COOPERATION)

PLANT PROTECTION ADVISOR TO THE GOVERNMENT OF INDIA AND DIRECTOR, LOCUST CONTROL

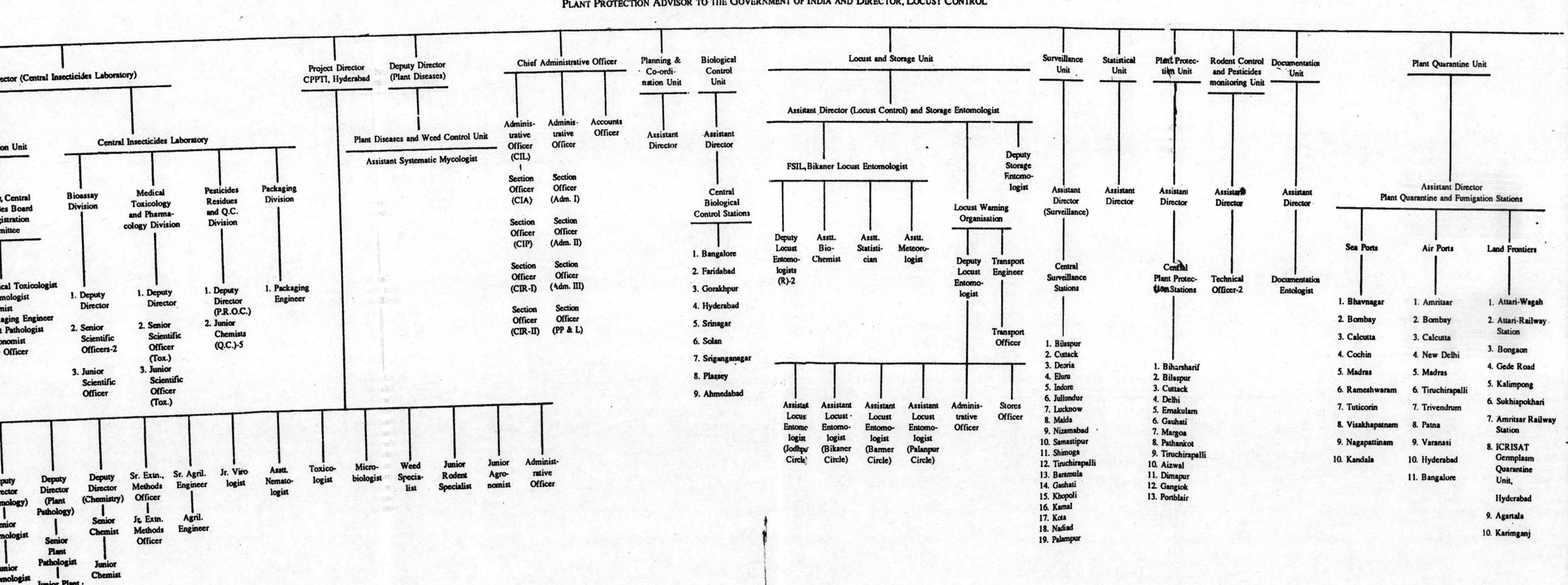